A handbook of numerical and statistical techniques
with examples mainly from the life sciences

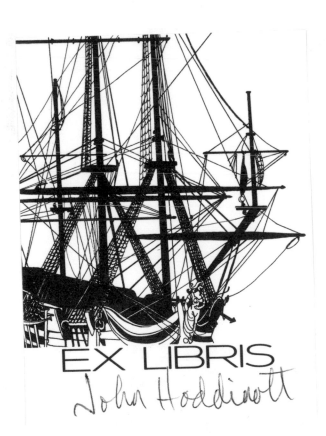

A HANDBOOK OF

Numerical and statistical techniques

WITH EXAMPLES MAINLY FROM
THE LIFE SCIENCES

J.H.POLLARD

CAMBRIDGE UNIVERSITY PRESS

CAMBRIDGE
LONDON - NEW YORK - MELBOURNE

Published by the Syndics of the Cambridge University Press
The Pitt Building, Trumpington Street, Cambridge CB2 1RP
Bentley House, 200 Euston Road, London NW1 2DB
32 East 57th Street, New York, NY 10022, USA
296 Beaconsfield Parade, Middle Park, Melbourne 3206, Australia

First published 1977
First paperback edition 1979

Printed in Great Britain at the
University Press, Cambridge

Library of Congress Cataloguing in Publication Data

Pollard, J. H.
A handbook of numerical and statistical techniques with examples
mainly from the life sciences.

Bibliography: p.
Includes index.
1. Biomathematics. 2. Medicine–Mathematics.
1. Title. II. Title: Numerical and statistical techniques.
QH323.5.P64   574'.01'51   76–27908
ISBN 0 521 21440 8 hard covers
ISBN 0 521 29750 8 paperback

TO CARYS

Contents

STATISTICAL TABLES

MATHEMATICAL TABLES

Preface

Although scientists nowadays have ready access to large-scale computer systems which incorporate sophisticated package programs for solving statistical and numerical problems, many of their day-to-day problems can be solved (and are possibly better solved) on the small programmable desk calculators and mini-computers now finding their way into every laboratory. These machines, with capacities not unlike those of computers two decades ago, are proving popular because of their ready availability, their cheapness and the fact that the research worker can 'play around' with his data as long as he likes. His final analysis is likely to be better than that produced by batch processing on a large computer. Package programs are available for these desk machines, but the research worker usually needs to tailor these programs to meet his own particular needs. This handbook is designed to aid him in this task.

Package programs on large-scale computer systems solve standard problems and they usually produce copious output. The scientist needs to understand something of the program methodology and he should be familiar with the various items output by the program. This is particularly true if he wants to modify or combine packages to solve non-standard problems. Many of the basic numerical and statistical techniques are described in this handbook.

Interactive computer terminals are now becoming more widely available. They have the advantage of the full-scale computer system and the convenience of a desk-top programmable calculator. The comments made in respect of programmable calculators and full-scale computors also apply to the interactive terminal.

A word of caution should be given. It is dangerous to accept package programs in blind faith. Write-ups can be misleading and programs are sometimes found to be faulty. The user is advised to test with known data any program he has not used previously.

This handbook is designed for ready reference by the scientist who has some knowledge of statistical method. It is not a textbook, although some of the techniques (for example, least squares) are explained in considerable detail, and exercises are given at the end of each chapter to amplify certain points. Each technique is demonstrated by at least one example. When he wants to solve a problem, the practical scientist will try a reasonable method. If it does not work, he will try another. One of them will usually work. With more difficult numerical and statistical problems he is advised to seek professional assistance.

The book is divided into three parts. Part I deals with numerical techniques and Part II statistical techniques. Part III is devoted to the method of least squares, which can be regarded as both a numerical technique and a statistical method.

Most of the writing of this book was done while on study leave in Europe during 1975-76, and I would like to thank Professors Gustav Feichtinger, Jan Hoem and Peter Whittle for their hospitality in Vienna, Copenhagen and Cambridge respectively. I would also like to thank Dr Odd Aalen, Dr Joe Gani, Mr Cameron Kirton, Dr Søren Johannsen and Professor Mark Williamson for their many helpful comments, and Miss Betty Thorn for drawing the diagrams. The book is dedicated to my wife who showed great patience during the many evenings when I was busy writing.

Macquarie University, Sydney   J.H.P.
December 1976

Part I
Basic numerical techniques

I
Introduction

Summary In this chapter we summarise some important mathematical results which are frequently required to solve numerical and statistical problems arising in scientific work. First, we discuss some Taylor series expansions for functions of one variable; we continue with the notion of functions of two or more variables and of partial differentiation. Lastly, we outline the concept of a matrix and the basic matrix operations: addition, subtraction, multiplication and the derivation of an inverse.

1.1 The level of mathematics required

The level of mathematics necessary for understanding the techniques described in this book is quite modest. Yet, using these techniques we are able to solve many difficult numerical and statistical problems. We are able to solve non-linear equations in one or more unknowns, integrate and differentiate a given function (including empirical functions), smooth crude data, fit curves and interpolate. Some of the techniques we develop can be used to solve other complex problems. For example, the method of finite differences (chapter 6) is used extensively for solving differential equations.

Readers will be familiar with the mathematical concepts of differentiation (obtaining the slope of a curve) and integration (finding the area under a curve). There are certain other basic concepts and formulae which bear repeating and these are summarised in the remainder of this chapter.

1.2 The Taylor series expansion

This formula uses the value of a function $f(x)$ and its derivatives at a particular point x to produce the value of that function at a neighbouring point $x + h$.

■
$$f(x + h) = f(x) + \frac{h}{1!} f'(x) + \frac{h^2}{2!} f''(x) + \dots . \qquad (1.2.1)$$

Further reading: Conte [17] 16; Grossman and Turner [38] 407–27; Nielsen [73] 9; Sokolnikoff and Sokolnikoff [92] 30–42.

1.3 The exponential series

The following series for e^x is valid for all values of x:

■
$$e^x = 1 + \frac{x}{1!} + \frac{x^2}{2!} + \frac{x^3}{3!} + \dots . \qquad (1.3.1)$$

This formula can be derived using the Taylor series expansion (1.2.1) and recalling that the derivative of e^x is e^x.

Further reading: Conte [17] 16; Grossman and Turner [38] 428–34, 457–60.

1.4 The logarithmic series

The following series for the natural logarithm of $1+x$ is valid whenever $|x|<1$:

■ $$\ln(1+x) = x - \frac{x^2}{2} + \frac{x^3}{3} - \dots . \qquad (1.4.1)$$

This formula can be derived using the Taylor series expansion (1.2.1) and recalling that the derivative of $\ln(1+x)$ is $(1+x)^{-1}$.

Further reading: Conte [17] 16; Grossman and Turner [38] 428–34.

1.5 The binomial expansion

The following series for $(1+x)^n$ is valid for all n when $|x|<1$:

■ $$(1+x)^n = 1 + \binom{n}{1}x + \binom{n}{2}x^2 + \dots , \qquad (1.5.1)$$

where

■
■ $$\left.\begin{array}{l} \binom{n}{1} = \frac{n}{1}, \quad \binom{n}{2} = \frac{n(n-1)}{1\times 2}, \\[2mm] \binom{n}{3} = \frac{n(n-1)(n-2)}{1\times 2\times 3}, \text{ etc.} \end{array}\right\} \qquad (1.5.2)$$

The formula may be derived using the Taylor series expansion (1.2.1).

When n is a positive integer, the series is valid for all x and consists of a sum of $n+1$ terms. Furthermore

■ $$\binom{n}{r} = \frac{n!}{r!\,(n-r)!}. \qquad (1.5.3)$$

Numerical values of the binomial coefficient (1.5.3) are often displayed in a *Pascal triangle* (table 1.5.1).

Table 1.5.1. *Binomial coefficients* $\binom{n}{r}$ *for small integer values of n.*

Each entry is equal to the sum of the entry immediately above and the one above to the left (for example, $70 = 35 + 35$)

	r								
n	0	1	2	3	4	5	6	7	8
0	1								
1	1	1							
2	1	2	1						
3	1	3	3	1					
4	1	4	6	4	1				
5	1	5	10	10	5	1			
6	1	6	15	20	15	6	1		
7	1	7	21	35	35	21	7	1	
8	1	8	28	56	70	56	28	8	1

Example 1.5.1. Use (1.5.1) to find the square root of 1.03 correct to six decimal places.

$$(1.03)^{\frac{1}{2}} = 1 + \frac{\frac{1}{2}}{1}(0.03) + \frac{\frac{1}{2}(-\frac{1}{2})}{1 \times 2}(0.03)^2 + \frac{\frac{1}{2}(-\frac{1}{2})(-\frac{3}{2})}{1 \times 2 \times 3}(0.03)^3 + \ldots$$

$$= 1 + 0.015 - 0.000\,1125 + 0.000\,0016875 - \ldots$$

$$= 1.014\,889 \text{ (correct to six decimal places)}.$$

Further reading: Grossman and Turner [38] 32–42; Nielsen [73] 8.

1.6 Partial differentiation

The slope of a function $f(x)$ in one dimension is the rate of increase of that function as x increases. A function of two variables x and y can be likened to a hill. The x-value may be compared with latitude, the y-value with longitude and the height at a particular point with the function value $f(x,y)$.

If we were to start walking at the bottom of the hill and head straight towards the summit, we might find the climbing rather difficult, because the slope is rather steep. Alternatively, we might zig-zag up the hill along paths which are less steep. Clearly, the slope at a given point (x,y) depends upon the direction in which we are heading.

In a mathematical context, two directions are important: the direction of constant y and increasing x (constant longitude and increasing latitude) and the direction of constant x and increasing y (constant latitude and increasing longitude). These two directions are at right angles.

The slope in the direction of constant y and increasing x is referred to as the partial derivative of $f(x,y)$ with respect to x and is denoted by $\partial f/\partial x$. It is obtained by treating y as a constant and differentiating $f(x,y)$ with respect to x in the usual way. Likewise, the slope in the direction of constant x and increasing y is referred to as the partial derivative of $f(x,y)$ with respect to y and is denoted by $\partial f/\partial y$. It is obtained by treating x as a constant and differentiating $f(x,y)$ with respect to y.

At the top of the hill, the slope in any direction is zero, including the direction (y constant, x increasing) and the direction (x constant, y increasing). At a maximum point of the function $f(x,y)$ (or minimum point, for that matter) $\partial f/\partial x$ and $\partial f/\partial y$ are both zero.

Example 1.6.1. Find the minimum value of

$$f(x,y) = x^2 + y^2 - x + y + xy - 3.$$

We equate to zero the partial derivatives with respect to x and y:

$$2x - 1 + y = 0,$$
$$2y + 1 + x = 0.$$

We solve these two simultaneous equations in two unknowns and find that the minimum is -4 at the point $(1, -1)$. This stationary point is a minimum point because the solution is unique and $f(x,y) \to \infty$ as both x and y tend to $\pm\infty$.

Further reading: Grossman and Turner [38] 435–8; Sokolnikoff and Sokolnikoff [92] 123–6, 160–6.

1.7 The two-dimensional Taylor series

The two-dimensional Taylor series takes the form

■
$$f(x+h, y+k) = f(x,y) + \left(h\frac{\partial}{\partial x}f(x,y) + k\frac{\partial}{\partial y}f(x,y)\right) + \frac{1}{2!}\left(h^2\frac{\partial^2}{\partial x^2}f(x,y)\right.$$
$$\left. + 2hk\frac{\partial^2}{\partial x \partial y}f(x,y) + k^2\frac{\partial^2}{\partial y^2}f(x,y)\right) + \dots,$$

$$(1.7.1)$$

and it may be generalised to three or more variables.

Further reading: Sokolnikoff and Sokolnikoff [92] 155–7.

1.8 The concept of a matrix

A matrix is a rectangular array of elements. These elements are usually numbers. For example,

$$\begin{pmatrix} -2 \\ 3 \end{pmatrix}, \quad \begin{pmatrix} 1.3 & 1.2 \\ 1.7 & 0.6 \\ 0.2 & 1.3 \end{pmatrix}, \quad \begin{pmatrix} 4.3 & 0.0 \\ 1.9 & 0.3 \end{pmatrix}.$$

By the *dimension* of a matrix we mean the number of rows and columns in that matrix. Thus, a matrix of m rows and n columns is called $m \times n$ matrix. The above matrices are of dimension 2×1, 3×2 and 2×2 respectively.

For two matrices to be *equal* they must have the same number of elements arranged in exactly the same pattern and have the same elements in the same places. Clearly

$$\begin{pmatrix} 1 & 2 \\ 3 & 1 \end{pmatrix} = \begin{pmatrix} 1 & 2 \\ 3 & 1 \end{pmatrix} \quad \text{and} \quad \begin{pmatrix} 1 & 2 \\ 3 & 1 \end{pmatrix} \neq \begin{pmatrix} 1 & 2 & 0 \\ 3 & 1 & 4 \end{pmatrix}.$$

If two matrices **A** and **B** have the same dimension, the *sum* of the two matrices is a matrix **C**, obtained by adding the corresponding elements of **A** and **B**. Thus, if

$$\mathbf{A} = \begin{pmatrix} 1 & 2 \\ -3 & 4 \end{pmatrix} \quad \text{and} \quad \mathbf{B} = \begin{pmatrix} -3 & 2 \\ 4 & 1 \end{pmatrix},$$

$$\mathbf{C} = \mathbf{A} + \mathbf{B} = \begin{pmatrix} -2 & 4 \\ 1 & 5 \end{pmatrix}.$$

In a similar manner,

$$\mathbf{D} = \mathbf{A} - \mathbf{B} = \begin{pmatrix} 4 & 0 \\ -7 & 3 \end{pmatrix}.$$

The sum and difference are not defined when **A** and **B** are not of the same dimension.

A matrix with all its elements zero is called a *zero matrix* and is usually denoted by **0**.

When a matrix is multiplied by a constant, the result is a matrix each of whose elements is equal to the corresponding element of the original matrix multiplied by the constant. Thus,

$$-3\begin{pmatrix} 1 & 2 \\ 3 & 4 \end{pmatrix} = \begin{pmatrix} -3 & -6 \\ -9 & -12 \end{pmatrix} = \begin{pmatrix} -1 & -2 \\ -3 & -4 \end{pmatrix} 3.$$

The *product of two matrices* is denoted by writing two matrices alongside each other *without* a multiplication sign. For example, **AB**. The product **AB** only exists if the number of columns in **A** is equal to the number of rows in **B**. The resulting product matrix **AB** has the same number of rows as **A** and the same number of columns as **B**. The element in the rth row and the sth column of **AB** is obtained by multiplying each element in the rth row of **A** by the corresponding element in the sth column of **B** and adding the products. Thus, the element in the second row and first column of the product

$$\text{second row} \rightarrow \begin{pmatrix} 2 & 3 & 5 \\ 1 & 9 & 2 \\ 4 & 7 & 6 \\ 14 & 12 & 13 \end{pmatrix} \begin{pmatrix} -1 & -5 \\ 12 & 11 \\ 8 & 0 \end{pmatrix}$$
$$\uparrow \text{first column}$$

is $1 \times (-1) + 9 \times 12 + 2 \times 8 = 123$.

In this case the first matrix is of dimension 4×3 and the second of 3×2. The product matrix is of dimension 4×2, and it takes the form

$$\begin{pmatrix} 74 & 23 \\ 123 & 94 \\ 128 & 57 \\ 234 & 62 \end{pmatrix}.$$

This method of defining a matrix product may seem complicated, but it is extremely useful, because we often need sums of the products of pairs of numbers. Examples are given at the end of this section. Although the product **AB** may be defined, **BA** may not; the above example shows this. Furthermore, even when **AB** and **BA** are both defined, the two products are not in general equal.

A square matrix ($n \times n$ say) with ones down the *principal diagonal* and zeros elsewhere is called the *unit matrix* and is usually denoted by **I**. For example,

$$\mathbf{I} = \begin{pmatrix} 1 & 0 & 0 \\ 0 & 1 & 0 \\ 0 & 0 & 1 \end{pmatrix}.$$

It is easy to verify that whenever the product is defined

$$\mathbf{AI} = \mathbf{IA} = \mathbf{A}.$$

A square matrix $\mathbf{A}$ is said to have an *inverse* $\mathbf{A}^{-1}$ if there exists $\mathbf{A}^{-1}$ such that

$$\mathbf{A}^{-1}\mathbf{A} = \mathbf{A}\mathbf{A}^{-1} = \mathbf{I}.$$

A square matrix without an inverse is said to be *singular*. Methods for determining the inverse of a matrix are given in section 1.10.

A matrix of dimension $1 \times n$ is often referred to as a *row vector* or *vector*. A matrix of dimension $n \times 1$ may be referred to as a *column vector* or *vector*. The above rules of addition, subtraction and multiplication still apply.

A square matrix with zeros everywhere except on the principal diagonal is called a *diagonal matrix*. If all the elements on the diagonal are non-zero, the matrix has an inverse, and the inverse is also diagonal. The inverse is obtained by taking the reciprocals of the diagonal elements. For example,

$$\begin{pmatrix} -2 & 0 & 0 \\ 0 & 3 & 0 \\ 0 & 0 & 4 \end{pmatrix}^{-1} = \begin{pmatrix} -\frac{1}{2} & 0 & 0 \\ 0 & \frac{1}{3} & 0 \\ 0 & 0 & \frac{1}{4} \end{pmatrix}.$$

The *transpose matrix* $\mathbf{A}'$ of a matrix $\mathbf{A}$ is obtained by interchanging rows and columns. Thus,

$$\mathbf{A} = \begin{pmatrix} 1 & 5 & 9 \\ 2 & 6 & 10 \\ 3 & 7 & 11 \\ 4 & 8 & 12 \end{pmatrix} \text{ has transpose } \mathbf{A}' = \begin{pmatrix} 1 & 2 & 3 & 4 \\ 5 & 6 & 7 & 8 \\ 9 & 10 & 11 & 12 \end{pmatrix}.$$

A square matrix $\mathbf{A}$ is *symmetric* if $\mathbf{A}' = \mathbf{A}$. For example,

$$\begin{pmatrix} 2 & 5 & 6 \\ 5 & -3 & -7 \\ 6 & -7 & 4 \end{pmatrix}.$$

Example 1.8.1. One important use of matrix algebra is in the study of systems of linear equations. To see why, let us consider the matrix product

$$\begin{pmatrix} 1 & 2 & 3 \\ 4 & 0 & 2 \\ 3 & -1 & -2 \end{pmatrix} \begin{pmatrix} x \\ y \\ z \end{pmatrix}.$$

The result is a 3×1 matrix or column vector

$$\begin{pmatrix} x + 2y + 3z \\ 4x + 2z \\ 3x - y - 2z \end{pmatrix}.$$

If we equate this vector to the column vector

$$\begin{pmatrix} 5 \\ 8 \\ 0 \end{pmatrix},$$

we are in effect expressing in matrix notation the three simultaneous equations

$$\left.\begin{array}{l} x + 2y + 3z = 5, \\ 4x \quad\;\; + 2z = 8, \\ 3x - \; y - 2z = 0. \end{array}\right\}$$

Let us write

$$\mathbf{A} = \begin{pmatrix} 1 & 2 & 3 \\ 4 & 0 & 2 \\ 3 & -1 & -2 \end{pmatrix}, \quad \mathbf{X} = \begin{pmatrix} x \\ y \\ z \end{pmatrix}, \quad \mathbf{C} = \begin{pmatrix} 5 \\ 8 \\ 0 \end{pmatrix}.$$

A and **C** are known and **X** unknown. To solve the three simultaneous equations, we need to find **X** such that

$$\mathbf{AX} = \mathbf{C}.$$

If **A** has an inverse, we can premultiply both sides by $\mathbf{A}^{-1}$ and obtain

$$\mathbf{X} = \mathbf{A}^{-1}\mathbf{C}.$$

We now have the solution to the three simultaneous linear equations. The reader should note that inverting a matrix is usually an inefficient way of obtaining the solution to a system of equations, but the short-hand matrix notation is very convenient for mathematical purposes.

Example 1.8.2. The cost of a new house depends upon the type of structure and the price of land, labour and materials. House *A* requires 10 000 units of land, 1000 units of labour and 2000 units of materials. House *B*, on the other hand, requires 7500 units of land, 1500 units of labour and 1500 units of materials. These components may be summarised in a 3 × 2 matrix.

$$\begin{array}{cc} \text{House } A & \text{House } B \\ \begin{pmatrix} 10\,000 & 7\,500 \\ 1\,000 & 1\,500 \\ 2\,000 & 1\,500 \end{pmatrix} & \begin{array}{l} \text{Land} \\ \text{Labour} \\ \text{Materials} \end{array} \end{array}$$

The prices of the components depend upon the locality, and the unit prices (in dollars) in each of four localities are summarised in the following 4 × 3 matrix:

$$\begin{array}{ccc} \text{Land} & \text{Labour} & \text{Materials} \\ \begin{pmatrix} 3 & 4 & 2 \\ 2 & 4 & 3 \\ 2 & 3 & 4 \\ 1 & 3 & 5 \end{pmatrix} & & \begin{array}{l} \text{Locality 1} \\ \text{Locality 2} \\ \text{Locality 3} \\ \text{Locality 4} \end{array} \end{array}$$

The family man does not want to know all this detail. He merely wants to know the total cost of each type of dwelling in each of the four localities. The total cost of House *A* in locality 2 may be calculated as follows:

$$2 \times 10\,000 + 4 \times 1\,000 + 3 \times 2\,000 = 30\,000.$$

Total costs for both types of house in all four localities may be summarised in a 4×2 matrix, and it is soon apparent that this matrix is the product of the above two matrices. Thus,

$$
\begin{array}{cc}
\text{House } A & \text{House } B \\
\end{array}
$$
$$
\begin{array}{c}
\text{Locality 1} \\
\text{Locality 2} \\
\text{Locality 3} \\
\text{Locality 4}
\end{array}
\begin{pmatrix}
38\,000 & 31\,500 \\
30\,000 & 25\,500 \\
31\,000 & 25\,500 \\
23\,000 & 19\,500
\end{pmatrix}
=
\begin{pmatrix}
3 & 4 & 2 \\
2 & 4 & 3 \\
2 & 3 & 4 \\
1 & 3 & 5
\end{pmatrix}
\begin{pmatrix}
10\,000 & 7\,500 \\
1\,000 & 1\,500 \\
2\,000 & 1\,500
\end{pmatrix}.
$$

The first matrix on the right-hand side gives the unit cost of materials, etc. When it is multiplied by the second matrix (the number of units required), we obtain the total cost matrix.

Further reading: Conte [17] 144–6, 148–50, 152–4; Grossman and Turner [38] 105–35; Hartree [41] 152; Sokolnikoff and Sokolnikoff [92] 114–20.

1.9   Determinants and cofactors

Every square matrix **A** has a number associated with it known as its *determinant*. This may be written as det **A** or $|\mathbf{A}|$. In the case of a 1×1 matrix, the determinant is merely the numerical value of the single element. For a 2×2 matrix, the determinant is equal to the product of the two diagonal elements minus the product of the two off-diagonal elements. Thus, for example,

$$
\begin{vmatrix}
2 & 1 \\
3 & 4
\end{vmatrix}
= 2 \times 4 - 3 \times 1 = 5.
$$

For matrices of higher dimension, we choose a row or column. (Any row or column will do.) We then multiply each element in that row or column by a number known as its cofactor and add the products. The sum is the determinant of the matrix.

To find the *cofactor* of the element in the ith row and jth column of an $n \times n$ matrix **A**, we delete the entire ith row and the entire jth column from that matrix to produce an $(n-1) \times (n-1)$ matrix and evaluate the determinant of this smaller matrix. This is the cofactor we require when $i+j$ is even. When $i+j$ is odd, the cofactor is obtained by changing the sign of this determinant.

Sometimes the complete *matrix of cofactors* of a matrix **A** is required. We then replace each element of **A** by its cofactor.

Example 1.9.1. The matrix of cofactors of the 2×2 matrix

$$
\mathbf{A} = \begin{pmatrix} 2 & 1 \\ 3 & 4 \end{pmatrix} \text{ is } \begin{pmatrix} 4 & -3 \\ -1 & 2 \end{pmatrix}.
$$

A rule for calculating the determinant of a 2×2 matrix was given above. This rule is really only a special case of the more general rule for an $n \times n$ matrix.

Let us now calculate the determinant of **A** using the general rule.

By the first row: det $\mathbf{A} = 2 \times 4 + 1 \times (-3) = 5$;
by the second row: det $\mathbf{A} = 3 \times (-1) + 4 \times 2 = 5$;
by the first column: det $\mathbf{A} = 2 \times 4 + 3 \times (-1) = 5$;
by the second column: det $\mathbf{A} = 1 \times (-3) + 4 \times 2 = 5$.

The same answer is obtained whichever row or column is used.

Example 1.9.2. The matrix of cofactors of the 3×3 matrix

$$\mathbf{A} = \begin{pmatrix} -2 & 5 & 3 \\ 4 & 7 & 1 \\ 6 & 9 & 8 \end{pmatrix} \text{ is}$$

$$\begin{pmatrix} \begin{vmatrix} 7 & 1 \\ 9 & 8 \end{vmatrix} & -\begin{vmatrix} 4 & 1 \\ 6 & 8 \end{vmatrix} & \begin{vmatrix} 4 & 7 \\ 6 & 9 \end{vmatrix} \\ -\begin{vmatrix} 5 & 3 \\ 9 & 8 \end{vmatrix} & \begin{vmatrix} -2 & 3 \\ 6 & 8 \end{vmatrix} & -\begin{vmatrix} -2 & 5 \\ 6 & 9 \end{vmatrix} \\ \begin{vmatrix} 5 & 3 \\ 7 & 1 \end{vmatrix} & -\begin{vmatrix} -2 & 3 \\ 4 & 1 \end{vmatrix} & \begin{vmatrix} -2 & 5 \\ 4 & 7 \end{vmatrix} \end{pmatrix} = \begin{pmatrix} 47 & -26 & -6 \\ -13 & -34 & 48 \\ -16 & 14 & -34 \end{pmatrix}.$$

The determinant of **A** may be calculated using any row or column. For example,

by the second column: det $\mathbf{A} = 5 \times (-26) + 7 \times (-34) + 9 \times 14 = -242$;
by the third row: det $\mathbf{A} = 6 \times (-16) + 9 \times 14 + 8 \times (-34) = -242$.

The reader should verify that the same answer is obtained whichever row or column is used.

Note that it is not necessary to compute the complete matrix of cofactors to find the determinant of a matrix.

Further reading: Conte [17] 146–8; Grossman and Turner [38] 145–52; Scheid [89] 347–50; Sokolnikoff and Sokolnikoff [92] 102–9.

1.10 Matrix inversion

Sophisticated programs for matrix inversion are available on all computer systems and programs are also available on many desk machines. The reader is advised to make use of these programs. There are occasions, however, when it is convenient to invert a small matrix by hand. We now outline two methods by which the inverse of a non-singular square matrix may be obtained. Most general matrix inversion programs are similar to the first method we describe. The second method is useful for hand computation with small matrices, but it is inefficient (in terms of the number of operations required) for large matrices.

The matrix representation of a system of linear equations was described in example 1.8.1. Let us now look at a simple elimination procedure for solving a pair of simultaneous linear equations and deduce a method for inverting a

2×2 matrix. The two simultaneous equations are numbered (1) and (2) below, and the steps in the elimination procedure are given alongside the relevant equation. A concise tableau[1] arrangement of the work is also given.

Linear equation	Equation number and elimination step	Tableau arrangement of work		
$2x + y = 4$	(1)	2	1	4
$3x + 4y = 11$	(2)	3	4	11
$x + \frac{1}{2}y = 2$	$(3) = (1) \div 2$	1	$\frac{1}{2}$	2
$x + \frac{4}{3}y = \frac{11}{3}$	$(4) = (2) \div 3$	1	$\frac{4}{3}$	$\frac{11}{3}$
$x + \frac{1}{2}y = 2$	$(5) = (3)$	1	$\frac{1}{2}$	2
$0 + \frac{5}{6}y = \frac{5}{3}$	$(6) = (4) - (3)$	0	$\frac{5}{6}$	$\frac{5}{3}$
$x + \frac{1}{2}y = 2$	$(7) = (5)$	1	$\frac{1}{2}$	2
$0 + y = 2$	$(8) = (6) \div \frac{5}{6}$	0	1	2
$x + 0 = 1$	$(9) = (7) - \frac{1}{2}(8)$	1	0	1
$0 + y = 2$	$(10) = (8)$	0	1	2

The solution is $x = 1, y = 2$.

The same sequence of operations can be used to invert the matrix

$$\mathbf{A} = \begin{pmatrix} 2 & 1 \\ 3 & 4 \end{pmatrix}.$$

We use a tableau arrangement, and insert the 2×2 identity matrix instead of the above right-hand sides:

2	1	1	0	(1)
3	4	0	1	(2)
1	$\frac{1}{2}$	$\frac{1}{2}$	0	$(3) = (1) \div 2$
1	$\frac{4}{3}$	0	$\frac{1}{3}$	$(4) = (2) \div 3$
1	$\frac{1}{2}$	$\frac{1}{2}$	0	$(5) = (3)$
0	$\frac{5}{6}$	$-\frac{1}{2}$	$\frac{1}{3}$	$(6) = (4) - (3)$
1	$\frac{1}{2}$	$\frac{1}{2}$	0	$(7) = (5)$
0	1	$-\frac{3}{5}$	$\frac{2}{5}$	$(8) = (6) \div \frac{5}{6}$
1	0	$\frac{4}{5}$	$-\frac{1}{5}$	$(9) = (7) - \frac{1}{2}(8)$
0	1	$-\frac{3}{5}$	$\frac{2}{5}$	$(10) = (8)$

The inverse matrix is

$$\mathbf{A}^{-1} = \begin{pmatrix} \frac{4}{5} & -\frac{1}{5} \\ -\frac{3}{5} & \frac{2}{5} \end{pmatrix}.$$

[1] This type of arrangement is recommended when solving simultaneous linear equations by hand.

It is easy to verify that $AA^{-1} = A^{-1}A = I$.

The *method of cofactors* is often useful with small matrices, but it is very inefficient with larger matrices:

1. Write down the matrix: $\begin{pmatrix} 2 & 1 \\ 3 & 4 \end{pmatrix}$.

2. Calculate the matrix of cofactors (section 1.9): $\begin{pmatrix} 4 & -3 \\ -1 & 2 \end{pmatrix}$.

3. Compute the determinant of the original matrix (multiply the elements of one row or column by the corresponding elements of the matrix of cofactors):

determinant $= 1 \times (-3) + 4 \times 2 = 5$.

4. Divide the elements of the cofactor matrix by the determinant:[2]

$\begin{pmatrix} \frac{4}{5} & -\frac{3}{5} \\ -\frac{1}{5} & \frac{2}{5} \end{pmatrix}$.

5. Transpose this matrix. We then have the inverse: $\begin{pmatrix} \frac{4}{5} & -\frac{1}{5} \\ -\frac{3}{5} & \frac{2}{5} \end{pmatrix}$.

All general methods for inverting matrices involve a large number of arithmetic operations, and rounding errors (section 2.3) are often a problem, particularly when the determinant is small compared with some of the elements. Double-precision arithmetic may help (section 2.3).

Certain matrices have very simple inverses (for example, diagonal matrices), and it may be very inefficient to use a general program to invert such matrices. Special methods are also available for symmetric matrices, which are commonly encountered in statistical work.

Further reading: Conte [17] 151, 169–76; Grossman and Turner [38] 136–44; Hartree [41] 162–3; Scheid [89] 335–6, 343–7.

1.11 Exercises

1. Without recourse to logarithms, find $(35)^{\frac{1}{5}}$ correct to five decimal places. *Hint*: $(35)^{\frac{1}{5}} = 2(1 + \frac{3}{32})^{\frac{1}{5}}$.

2. Use (1.5.1) to obtain the expansion of $(1 + x)^{-\frac{3}{2}}$ up to and including the term involving x^3.

3. Write down the partial derivatives with respect to x and y of $f(x,y) = x/y$.

4. Find the matrix $Z = X(Y + W)$, where

$$X = \begin{pmatrix} 1 & 2 & 3 \\ 6 & 5 & 4 \end{pmatrix}, \quad Y = \begin{pmatrix} 3 & 2 \\ 0 & 0 \\ 1 & 4 \end{pmatrix}, \quad W = \begin{pmatrix} -2 & 3 \\ 1 & 5 \\ 0 & -2 \end{pmatrix}.$$

5. Find the matrix of cofactors, the determinant, and the inverse of the matrix A in example 1.8.1.

6. Use the result of question 5 to solve the three simultaneous equations in example 1.8.1.

[2] A square matrix with zero determinant is *singular*. It has no inverse.

2

Errors, mistakes and the arrangement of work

Summary The result of a numerical calculation may differ from the exact answer because of truncation errors, round-off errors and mistakes. In this chapter, we describe some simple techniques for detecting and reducing truncation and round-off errors. We also suggest ways of avoiding mistakes.

2.1   Introduction

The result of numerical calculation may differ from the exact answer to the mathematical problem for one or more of three basic reasons:

1. The calculation formula may be derived by cutting off an infinite series after a finite number of terms; the errors introduced in this manner are called *truncation errors.*

2. A calculating device is only able to retain a certain number of decimal digits and the less significant digits are dropped; errors introduced in this manner are called *round-off errors.*

3. *Mistakes* may be made by man or machine in performing the calculation or recording the result. The word 'mistake' is used to distinguish this source of discrepancy due to human or mechanical fallibility from the largely unavoidable 'errors' caused by the necessity to truncate an infinite series or the finite capacity of the calculating device.

The research worker must ensure that the final results of a calculation are not rendered useless by errors or mistakes. Intermediate checks are advisable in a long calculation.

> *Further reading:* Conte [17] 6–14; Hartree [41] 1–8; Hildebrand [43] 1–8;
> Lyon [62] 266–93; Nielsen [73] 2–4; Ralston [84] 1–8.

2.2   Truncation errors

It is often possible to estimate the magnitude of truncation errors and so ensure that they are kept within reasonable bounds. Numerical methods of integration and differentiation, for example, usually make use of polynomial approximations to the function involved, and a formula with a finite number of terms results. The effectiveness of the polynomial approximation can be evaluated by varying the distance between ordinates, and the truncation error reduced by using a smaller interval.

> *Further reading:* Hartree [41] 5–8.

14

2.3 Round-off errors

Let us consider two numbers X and Y stored in a calculating device as x and y respectively. Because of round-off, we know that

$$\left.\begin{array}{l} x-a < X < x+a, \\ y-b < Y < y+b, \end{array}\right\} \tag{2.3.1}$$

where a and b are both positive. Clearly

$$x+y-(a+b) < X+Y < x+y+(a+b),$$
$$x-y-(a+b) < X-Y < x-y+(a+b),$$

and we deduce the following rule of error analysis.

When two numbers are added (subtracted), the maximum possible error in the sum (difference) is equal to the sum of the maximum possible errors. This rule tells us that we must avoid, if possible, any calculation which involves the difference between two large, nearly equal numbers.

To see what happens with two rounded numbers under multiplication and division, we note that the inequalities (2.3.1) may be written:

$$x(1-a/x) < X < x(1+a/x),$$
$$y(1-b/y) < Y < y(1+b/y).$$

The maximum proportionate errors a/x and b/y should be small, and the product $(a/x)(b/y)$ can therefore be ignored. It follows that the product XY lies in the range

$$xy\left\{1\pm\left(\frac{a}{x}+\frac{b}{y}\right)\right\}$$

and the quotient X/Y lies in the range

$$\frac{x}{y}\left\{1\pm\left(\frac{a}{x}+\frac{b}{y}\right)\right\}.$$

We have thus deduced the following rule of error analysis:

When two numbers are multiplied (divided), the maximum percentage error in the product (quotient) is equal to the sum of the maximum percentage errors.

It should be noted that the two rules we quote apply when the answer itself is not rounded. Further analysis is needed when the answer is rounded and an example is given below.

In a long calculation involving many numbers, the actual error in the answer might be expected to be much smaller than the upper bound provided by the above rules because some of the errors will be positive and some negative. A probabilistic approach to round-off is given, for example, by Ralston [84] 8–11.

Most modern desk calculators provide twelve or more decimal digits. Modern computers provide about nine decimal digits and they usually have facilities for double-precision arithmetic as well (eighteen decimal digits). With this accuracy, round-off tends to be less of a problem. Nevertheless, it does crop up from time to time, particularly with large near-singular matrices. In doubtful situations a comparison of results using single-precision and double-precision arithmetic will usually indicate whether a round-off problem exists.

Example 2.3.1. A floating point calculator has a five-decimal-digit capacity. After any arithmetic operation, the result is rounded to the nearest five-digit figure. The value of $(1.05)^{32}$ is to be evaluated by repeatedly squaring the number in the accumulator. Determine the answer the machine will give and the bounds within which the true value must lie.

The working may be set out as follows:

$$(1.05)^2 = 1.1025 \quad \text{(exact)}.$$
$$(1.05)^4 = (1.1025)^2$$
$$= 1.2155 \pm 0.00005 \qquad (\pm 0.004\%).$$
$$(1.05)^8 = (1.2155 \pm 0.004\%)^2$$
$$= (1.2155)^2 \pm 0.008\%$$
$$= (1.4774 \pm 0.00005) \pm 0.008\%$$
$$= 1.4774 \pm 0.00017 \qquad (\pm 0.012\%).$$
$$(1.05)^{16} = (1.4774 \pm 0.012\%)^2$$
$$= (1.4774)^2 \pm 0.024\%$$
$$= (2.1827 \pm 0.00005) \pm 0.024\%$$
$$= 2.1827 \pm 0.00057 \qquad (\pm 0.026\%).$$
$$(1.05)^{32} = (2.1827 \pm 0.026\%)^2$$
$$= (2.1827)^2 \pm 0.052\%$$
$$= (4.7642 \pm 0.00005) \pm 0.052\%$$
$$= 4.7642 \pm 0.00253 \qquad (\pm 0.053\%).$$

The answer the machine will give is 4.7642 and the true value must lie in the range 4.7642 ± 0.0025. Note that $(1.05)^{32} = 4.7649$ (correct to four decimal places).

Further reading: Hartree [41] 5–8; Hildebrand [43] 14–17; Ralston [84] 8–10, 12–17.

2.4   Mistakes and the arrangement of work

The scientist will be most meticulous to record the results of his experiments in an orderly manner. He must continue to be orderly when he performs his numerical work. Calculations should not be done on odd scraps of paper, but laid out systematically and in such a way as to show how the intermediate and final results were obtained. This orderly approach will help to avoid mistakes and to locate and correct any that do happen to be made.

Mistakes tend to be made when figures are copied from one page to another. Loose sheets are preferable to a book because they can be brought close together for copying purposes. It is worth noting that columns of figures can be copied quickly and inexpensively with a photocopying machine and then pasted to the next working sheet.

Two types of mistake are very common in transcribing numbers. The first is to interchange adjacent digits; for example, 358 764 may be written instead of 385 764. The second is to duplicate the wrong digit in a number with two digits the same; for example, 48 335 may be written as 48 835. A method for detecting and correcting errors in a table of numbers is given in section 6.3.

Checks on calculations made by the same individual using the same method tend to be almost worthless. The individual is likely to repeat his mistakes. Whenever possible, a different method should be used for checking. If the same method must be used, another person should perform the check calculations.

Further reading: Hartree [41] 8–9.

2.5   Exercises

1. The following computation is to be performed on the five-digit calculator of example 2.3.1:

   $$\frac{100.23}{300.59} - \frac{200.47}{601.19}.$$

   Determine the answer the machine will give and obtain bounds within which the correct answer must lie. Assume that the above figures are exact.

2. Find the answer to the numerical problem in question 1 correct to four significant figures.

3. Suggest a method, suitable for use with the five-digit calculator, which will solve the numerical problem in question 1 and give an answer correct to at least three significant digits.

4. The above-mentioned five-digit calculator is to be used to calculate $e^{-0.5}$ by formula (1.3.1). Determine the answer the machine will give and obtain bounds within which the correct answer must lie. Compare these results with the true answer.

3

The real roots of
non-linear equations

Summary   In this chapter we describe several different methods for finding the real roots of non-linear equations. Each has its own advantages and disadvantages. The method of false position (section 3.2) is the only one which always converges. The Newton–Raphson method of section 3.6 for solving two simultaneous non-linear equations can be generalised to three or more variables.

3.1   Introduction

Non-linear equations often arise in scientific work. Graphical methods may be used to solve these equations, but the order of accuracy is usually such that the answer obtained can only be regarded as a first approximation. Trial-and-error methods can also be used. *Iterative methods* are usually most convenient. These are procedures whereby x_n, the nth approximation to the root is obtained by evaluating a function of the earlier approximation x_{n-1}.

The iterative procedures described in this chapter are usually quite satisfactory, but the reader is warned that they may fail in certain circumstances; for example, with multiple roots and close roots. The practical research worker will try a reasonable method. If this fails, he will try another method. If this also fails, he may be advised to seek the assistance of a numerical analyst.

Further reading: Conte [17] 19–70; Fröberg [28] 17–26; Hartree [41] 190–7; Hildebrand [43] 443; Nielsen [73] 169–223; Sokolnikoff and Sokolnikoff [92] 95–6.

3.2   The method of false position

Let us imagine that we wish to solve the non-linear equation $f(x) = 0$ and we have found that $f(a)$ is positive whilst $f(b)$ is negative. The relevant root lies somewhere in-between. We therefore try a value between a and b, but which value? If $f(b)$ is closer to 0 than $f(a)$, it would be logical to try a value closer to b than to a and conversly. One approach is to choose a new point c using linear interpolation:

$$c = \frac{f(a)b - f(b)a}{f(a) - f(b)}. \tag{3.2.1}$$

We now repeat the iterative process using c as one of our points and as the other point whichever of a and b lies on the opposite side of the solution to c.

18

Unlike the other methods outlined in this chapter, this method always converges, but it usually requires a larger number of iterations. Round-off can be a problem near the final solution because of the need to subtract two nearly-equal quantities.

Example 3.2.1.[1] Solve the quadratic equation

$$f(x) = x^2 - 2 = 0.$$

We begin by trying

$$a = 1.5, \quad f(a) = 0.25;$$
$$b = 1.4, \quad f(b) = -0.04.$$

The new point is

$$c = \{0.25 \times 1.4 - (-0.04) \times 1.5\}/\{0.25 - (-0.04)\}$$
$$= 1.4138.$$

We now repeat the process with

$$a = 1.5, \qquad f(a) = 0.25;$$
$$b = 1.4138, \quad f(b) = -0.0012.$$

The new point is

$$c = \{0.25 \times 1.4138 - (-0.0012) \times 1.5\}/\{0.25 - (-0.0012)\}$$
$$= 1.4142.$$

When the process is repeated, the same answer (to four decimal places) is obtained. We have arrived at the solution.

Further reading: Hildebrand [43] 446; Whittaker and Robinson [100] 92.

3.3 The Newton–Raphson method

Let us imagine that we need to solve the non-linear equation $f(x) = 0$ with unknown solution x_∞. We start with an initial value x_0, which may be found by trial and error. If we can find an adjustment h_0 such that $x_0 + h_0 = x_\infty$, we have solved the non-linear equation.

From the Taylor series expansion (1.2.1), we know that

$$0 = f(x_\infty) = f(x_0 + h_0) \doteqdot f(x_0) + h_0 f'(x_0),$$

so that

$$h_0 \doteqdot -f(x_0)/f'(x_0),$$

---

[1] This example has been chosen because of its simplicity. Modern calculators and computers incorporate the automatic square root facility.

and a better approximation to the root is given by

$$x_1 = x_0 - f(x_0)/f'(x_0).$$

We now repeat the process using x_1 instead of x_0 to obtain an even better approximation x_2. The process is continued until the root of the equation is obtained to the required degree of accuracy.

The iterative process may be represented by the equation

∎ $\quad x_{n+1} = x_n - f(x_n)/f'(x_n).$ (3.3.1)

Example 3.3.1.[2] Solve the quadratic equation

$$f(x) = x^2 - 2 = 0.$$

We note that $f'(x) = 2x$. Let us try $x_0 = 1.5$ as our initial value. Then

$$x_1 = 1.5 - f(1.5)/f'(1.5)$$
$$= 1.5 - (0.25)/(3.0)$$
$$= 1.4167 \text{ (to four decimal places)}.$$

The second approximation is

$$x_2 = 1.14167 - f(1.4167)/f'(1.4167)$$
$$= 1.4167 - (0.00704)/(2.8334)$$
$$= 1.4142 \text{ (to four decimal places)}.$$

The same answer is obtained when we repeat the process, so the root of the equation (correct to four decimal places) is 1.4142.

Example 3.3.2. The size of a population can be estimated using capture/recapture data. Let us denote the number of individuals caught n times by f_n ($n = 1, 2, ...$). According to Craig [18], an estimate of the population size is given by $\hat{P}$, where

$$\ln (\hat{P}) - \ln (\hat{P} - \sum_n f_n) = (\sum_n n f_n)/\hat{P}. \tag{3.3.2}$$

The following data, collected by Macquarie University ecology students at Smith Lake in May 1974, relate to a population of lizards:

n	1	2	3	$\geqslant 4$
f_n	91	29	1	0

Let us use the Newton-Raphson method to find $\hat{P}$. Equation (3.3.2) may be written as

$$f(\hat{P}) = 0,$$

[2] See footnote 1.

where

$$f(\hat{P}) = \ln(\hat{P}) - \ln(\hat{P} - \Sigma f_n) - (\Sigma n f_n)/\hat{P},$$

and

$$f'(\hat{P}) = \frac{1}{\hat{P}} - \frac{1}{\hat{P} - \Sigma f_n} + (\Sigma n f_n)/\hat{P}^2.$$

We note that

$$\Sigma f_n = 91 + 29 + 1 = 121,$$

and

$$\Sigma n f_n = 1 \times 91 + 2 \times 29 + 3 \times 1 = 152.$$

As our initial value, let us try $\hat{P}_0 = 300$. An improved estimate is given by

$$\hat{P}_1 = 300 - (5.703\,78 - 5.187\,39 - 0.506\,67)$$
$$\div (0.003\,33 - 0.005\,59 + 0.001\,69) = 317.$$

A second approximation is given by

$$\hat{P}_2 = 317 - (5.758\,90 - 5.278\,11 - 0.479\,50)$$
$$\div (0.003\,15 - 0.005\,10 + 0.001\,51) = 320.$$

A third approximation is given by

$$\hat{P}_3 = 320 - (5.768\,32 - 5.293\,30 - 0.475\,00)$$
$$\div (0.003\,13 - 0.005\,03 + 0.001\,48) = 320,$$

which is our estimate of the size of the lizard population.

Further reading: Conte [17] 30–9; Fröberg [28] 19–26; Hartree [41] 193; Henrici [42] 77–90; Hildebrand [43] 447; Nielsen [73] 171–6; Ralston [84] 332; Sokolnikoff and Sokolnikoff [92] 97–102.

3.4   The secant method

The Newton–Raphson method requires the evaluation of the derivative for each successive approximation. This is sometimes difficult and often tedious. With the secant method, the derivative is replaced by the ratio

$$(f(x_n) - f(x_{n-1}))/(x_n - x_{n-1}),$$

and the following iterative equation is used:

■ $$x_{n+1} = x_n - (x_n - x_{n-1}) f(x_n)/(f(x_n) - f(x_{n-1})). \qquad (3.4.1)$$

This equation is identical with (3.2.1) when we write a, b and c instead of x_n, x_{n-1} and x_{n+1} respectively. But the two methods are not identical. The method of false position requires that $f(a)$ and $f(b)$ have opposite signs; with the secant method, $f(x_n)$ and $f(x_{n-1})$ sometimes have the same sign.

Example 3.4.1.[3] Solve the quadratic equation

$$f(x) = x^2 - 2 = 0.$$

By trial and error, we know that $x_0 = 1.5$ and $x_1 = 1.4$ lie in the neighbourhood of the root. Using (3.4.1),

$$x_2 = 1.4 - (1.4 - 1.5)(-0.04)/\{(-0.04) - 0.25\} = 1.4138,$$

$$x_3 = 1.4138 - (1.4138 - 1.4)(-0.0012)$$
$$\div \{(-0.0012) - (-0.04)\} = 1.4142.$$

The same answer is obtained when we repeat the process; so the root of the equation (correct to four decimal places) is 1.4142.

Further reading: Conte [17] 39–40; Ralston [84] 323–8.

3.5 Simple iterative methods

Sometimes a simple rearrangement of the non-linear equation provides a suitable iterative equation. The iterative equations in sections 14.9–14.12 for estimating the parameters of censored and truncated normal and Poisson populations were obtained in this manner. The following example arises in the context of population mathematics and appears originally in a paper by A. J. Lotka in 1931.

Example 3.5.1. Find the root between zero and one of the polynomial equation

$$x = 0.4982 + 0.2103x + 0.1270x^2 + 0.0730x^3$$
$$+ 0.0418x^4 + 0.02411x^5 + 0.0132x^6 + 0.0069x^7$$
$$+ 0.0035x^8 + 0.0015x^9 + 0.0005x^{10}. \tag{3.5.1}$$

The coefficients on the right-hand side of (3.5.1) sum to one; so $x = 1$ is a root of the equation. If we substitute $1 - y$ for x in (3.5.1), we obtain a polynomial equation for y with no constant term. The unwanted root $y = 0$ ($x = 1$) can be eliminated by dividing by y. The y-equation can then be rearranged as follows:

$$y = 0.107\,00 + 1.009\,96y^2 - 0.841\,80y^3 + 0.560\,21y^4$$
$$- 0.288\,15y^5 + 0.109\,87y^6 - 0.02915y^7$$
$$+ 0.004\,80y^8 - 0.000\,37y^9. \tag{3.5.2}$$

It is known that the y-root we seek is much closer to zero than to one. A first approximation to the root may be found by ignoring powers of y above the second and solving the quadratic

$$1.009\,96y^2 - y + 0.107\,00 = 0. \tag{3.5.3}$$

---

[3] See footnote 1.

The first approximation is found to be $y = 0.122$. When this is substituted into the right-hand side of equation (3.5.2), we obtain the second approximation $y = 0.120\,61$. Third and fourth approximations are also obtained via (3.5.2). They are $y = 0.120\,34$ and $y = 0.120\,27$ respectively. We conclude that the required root of equation (3.5.1) is $x = 0.8797$ (correct to four decimal places).

Further reading: Lotka [60], [61]; Nielsen [73] 177–8; Pollard [83] 102–3.

3.6 The two-dimensional Newton–Raphson method

Let us imagine that we need to solve the two simultaneous non-linear equations

$$f_1(x,y) = 0 \text{ and } f_2(x,y) = 0 \qquad (3.6.1)$$

with unknown solution (x_∞, y_∞). We start with an initial point (x_0, y_0), which may be found by trial and error. If we can find adjustments h_0 and k_0 such that

$$x_0 + h_0 = x_\infty \text{ and } y_0 + k_0 = y_\infty,$$

we have solved the non-linear system of equations. More generally, at the nth iteration, we have a point (x_n, y_n) and we seek adjustments h_n and k_n such that

$$x_n + h_n = x_\infty \text{ and } y_n + k_n = y_\infty. \qquad (3.6.2)$$

We make use of the two-dimensional Taylor series (1.7.1) and equate the expansions of f_1 and f_2 at $(x_n + h_n, y_n + k_n)$ to zero. Then

$$\left.\begin{array}{l}
\left(\dfrac{\partial}{\partial x} f_1(x_n, y_n)\right) h_n + \left(\dfrac{\partial}{\partial y} f_1(x_n, y_n)\right) k_n \doteqdot -f_1(x_n, y_n), \\[2ex]
\left(\dfrac{\partial}{\partial x} f_2(x_n, y_n)\right) h_n + \left(\dfrac{\partial}{\partial y} f_2(x_n, y_n)\right) k_n \doteqdot -f_2(x_n, y_n).
\end{array}\right\} \qquad (3.6.3)$$

These two simultaneous *linear* equations in h_n and k_n are then solved and we obtain an improved solution to the non-linear system (3.6.1):

$$\left.\begin{array}{l}
x_{n+1} = x_n + h_n, \\[1ex]
y_{n+1} = y_n + k_n.
\end{array}\right\} \qquad (3.6.4)$$

The process is repeated until the desired degree of accuracy is obtained.

The generalisation to three or more variables is straightforward.

Example 3.6.1. Solve the two simultaneous non-linear equations

$$x^2 + (y-4)^2 - 9 = 0,$$

$$(x-4)^2 + y^2 - 9 = 0,$$

using the initial values $x_0 = 2.4$ and $y_0 = 2.6$.

We note that

$$\frac{\partial f_1}{\partial x} = 2x, \quad \frac{\partial f_2}{\partial x} = 2(x-4);$$

$$\frac{\partial f_1}{\partial y} = 2(y-4), \quad \frac{\partial f_2}{\partial y} = 2y.$$

At the first iteration therefore, we need to solve the simultaneous linear equations

$$4.8\,h_0 - 2.8\,k_0 = 1.28,$$

$$-3.2\,h_0 + 5.2\,k_0 = -0.32.$$

We find that $h_0 = 0.36$ and $k_0 = 0.16$ so that a better approximation is given by $x_1 = 2.76$, $y = 2.76$.

When the operation is repeated, the following results are obtained:

n	x_n	y_n
2	2.7089	2.7089
3	2.7071	2.7071
4	2.7071	2.7071

A solution is therefore $x = y = 2.7071$ (a result which can be verified geometrically).

Further reading: Conte [17] 45–6; Henrici [42] 105–8; Hildebrand [43] 451; Nielsen [73] 200–10; Whittaker and Robinson [100] 90–2.

3.7   Exercises

1. Use the method of false position to find a solution (correct to four decimal places) to the equation

   $$e^x + \ln x = 3.4.$$

   A solution lies somewhere between $x = 1.1$ and $x = 1.2$.

2. Find a solution to the equation in question 1 using the Newton–Raphson method.

3. Use the two-dimensional Newton–Raphson method to solve the simultaneous non-linear equations

$$\tfrac{1}{2}x^2 + \ln y = 1.3,$$

$$\tfrac{1}{2}y^2 + \ln x = 0.825.$$

There is a solution in the vicinity of the point $(2, 0.6)$.

4
Simple methods for smoothing crude data

Summary In this chapter, we describe two simple methods for smoothing crude data: running averages and spline functions.

4.1 Introduction

The number of different methods designed for smoothing crude data and curve fitting is immense. The graphical method is perhaps the most popular, particularly when the curve is a straight line, and many non-linear problems can be reduced to this form by a suitable transformation or the use of logarithmic paper. Some methods are very specialised. In this chapter we describe two simple techniques. The most important method by far is that of least squares, and we devote Part III of this book to that method.

4.2 Running-average smoothing formulae

It is convenient in this section to regard any experimental value y_i as consisting of two parts, the true or universal value $f(x_i)$ and a superimposed statistical error e_i. Thus

$$y_i = f(x_i) + e_i.$$

Our aim in smoothing a set of data is to reduce the errors $\{e_i\}$ as much as possible.

Statisticians, economists, biologists and others make use of *moving averages* or *running averages*. Let us apply this technique to the observed data in table 4.2.1 and use the running-average formula

$$Y_i = \tfrac{1}{5}(y_{i-2} + y_{i-1} + y_i + y_{i+1} + y_{i+2}).$$

The symbol Y_i denotes the smoothed value at the ith point. The results are shown graphically in fig. 4.2.1, and the improvement in smoothness is immediately apparent.

There is a difficulty, however: as well as smoothing a set of data, running averages distort values that are already smooth. To see this, consider the quadratic curve $y = 1100 + 2x - 5x^2$. Numerical values of this function are given in table 4.2.2. The smoothed values in the table were obtained using a

Table 4.2.1. *The smoothing effect of a running average*

x	Observed value y	Smoothed value Y	x	Observed value y	Smoothed value Y
35	95	–	53	1274	1425
36	638	–	54	1579	1500
37	191	324	55	1873	1642
38	419	381	56	1482	1789
39	278	331	57	2002	2039
40	381	425	58	2011	2009
41	384	437	59	2827	2301
42	665	584	60	1722	2575
43	477	727	61	2942	2720
44	1015	822	62	3374	2701
45	1093	845	63	2735	3221
46	860	922	64	2731	3618
47	779	909	65	4323	3766
48	862	864	66	4926	4039
49	951	913	67	4114	4659
50	866	1016	68	4099	5015
51	1109	1098	69	5835	–
52	1291	1224	70	6103	–

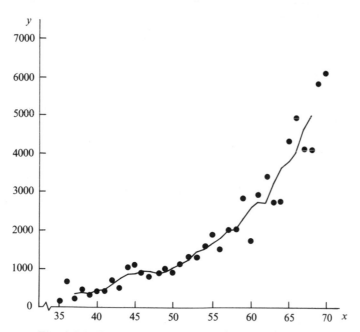

Fig. 4.2.1. The smoothing effect of a running average.

Table 4.2.2. *The distortion caused by a simple running average*

x	Quadratic value y	Smoothed value Y	x	Quadratic value y	Smoothed value Y
0	1100		5	985	975
1	1097		6	932	922
2	1084	1074	7	869	859
3	1061	1051	8	796	
4	1028	1018	9	713	

running average of five terms, and the distortion is obvious: each numerical value has been reduced by 10.

We therefore note the following three points concerning running averages:

1. They reduce irregularities and fluctuations.
2. They distort values already smooth.[1]
3. They do not provide values at the beginning and end of a table.

The smoothed values in fig. 4.2.1 can be made even smoother by repeating the running-average process, but further values will be lost from both ends of the table and the danger of distortion becomes greater.

It is not necessary to restrict our smoothing formulae to strict averages; one can also use *weighted* averages. Instead of the above five-term strict average, one might use, for example, the following five-term weighted-average formula:

$$Y_i = \tfrac{1}{9}(y_{i-2} + 2y_{i-1} + 3y_i + 2y_{i+1} + y_{i+2}).$$

Nor is it necessary to insist that all the weights be positive. Formulae can be designed which do not distort polynomials of degree three or less, and these formulae contain negative terms. Most smooth functions can be represented reasonably well by a series of low-order polynomial arcs, and the danger of distortion is slight when these 'non-distorting' formulae are used. Non-symmetric formulae with the same properties are also available for completing the ends of the table.

Let us write y_0 instead of y_i, y_1 instead of y_{i+1}, y_2 instead of y_{i+2}, etc. Hildebrand [43] quotes the following non-distorting five-term formulae:

■ $\qquad Y_0 = \tfrac{1}{35}(-3y_{-2} + 12y_{-1} + 17y_0 + 12y_1 - 3y_2),$ $\qquad$ (4.2.1)

■ $\qquad Y_1 = \tfrac{1}{35}(2y_{-2} - 8y_{-1} + 12y_0 + 27y_1 + 2y_2),$ $\qquad$ (4.2.2)

■ $\qquad Y_2 = \tfrac{1}{70}(-y_{-2} + 4y_{-1} - 6y_0 + 4y_1 + 69y_2).$ $\qquad$ (4.2.3)

[1] A method for reducing distortion is described in example 7.3.1.

Formula (4.2.1) is used for the main body of the table, while (4.2.2) and (4.2.3) are used to complete the final two entries. The formula for Y_{-1} is obtained from (4.2.2) by writing y_2 instead of y_{-2}, y_1 instead of y_{-1}, y_{-1} instead of y_1, and y_{-2} instead of y_2. The formula for Y_{-2} is obtained by changing the signs of the subscripts in (4.2.3).

Longer formulae tend to have greater smoothing power, and Hildebrand gives the following seven-term formulae:

$$\left.\begin{aligned}
Y_0 &= \tfrac{1}{21}(-2y_{-3} + 3y_{-2} + 6y_{-1} + 7y_0 + 6y_1 + 3y_2 - 2y_3), \\
Y_1 &= \tfrac{1}{42}(y_{-3} - 4y_{-2} + 2y_{-1} + 12y_0 + 19y_1 + 16y_2 - 4y_3), \\
Y_2 &= \tfrac{1}{42}(4y_{-3} - 7y_{-2} - 4y_{-1} + 6y_0 + 16y_1 + 19y_2 + 8y_3), \\
Y_3 &= \tfrac{1}{42}(-2y_{-3} + 4y_{-2} + y_{-1} - 4y_0 - 4y_1 + 8y_2 + 39y_3).
\end{aligned}\right\} \qquad (4.2.4)$$

Non-distorting formulae have also been developed with optimal smoothing properties, and the reader may be advised to use these formulae in preference to the above for smoothing the main body of a table (Miller [67] 68–72). The weights for the optimal smoothing formulae of range 5, 7, 9, 11, 13, 15, 17, 19, 21 and 23 are given in table 4.2.3. The formulae are symmetric, and K_0 refers to the central weight, while K_1 refers to the weight immediately on either side of the central term.

Optimal smoothing formulae for completing the ends of a table have been given by Greville [35], [36]. The above non-symmetric five- and seven-term formulae are usually adequate for this purpose, however, and additional smoothness can always be achieved by repeating the smoothing process.

> **Example 4.2.1.** Apply the optimal nine-term smoothing formula in table 4.2.3 to the crude data in table 4.2.1.

This task is very simple on a moderate-sized programmable calculator, but a little tedious by hand. The smoothed value at $x = 39$ is calculated in the following manner:

$$\begin{aligned}
Y_{39} = &-0.040\,724 \times 95 \\
&-0.009\,873 \times 638 \\
&+0.118\,470 \times 191 \\
&+0.266\,557 \times 419 \\
&+0.331\,140 \times 278 \\
&+0.266\,557 \times 381 \\
&+0.118\,470 \times 384 \\
&-0.009\,873 \times 665 \\
&-0.040\,724 \times 477 \\
= &\ 337.
\end{aligned}$$

Table 4.2.3. *Coefficients of optimal-smoothing running-average formulae*

	Number of terms in formula									
	5	7	9	11	13	15	17	19	21	23
K_0	0.559 441	0.412 587	0.331 140	0.277 945	0.240 057	0.211 541	0.189 231	0.171 266	0.156 469	0.144 060
K_1	0.293 706	0.293 706	0.266 557	0.238 693	0.214 337	0.193 742	0.176 390	0.161 691	0.149 136	0.138 318
K_2	−0.073 427	0.058 741	0.118 470	0.141 267	0.147 356	0.145 904	0.141 112	0.134 965	0.128 423	0.121 949
K_3		−0.058 741	−0.009 873	0.035 723	0.065 492	0.082 918	0.092 293	0.096 658	0.097 956	0.097 395
K_4			−0.040 724	−0.026 792	0.000 000	0.024 027	0.042 093	0.054 685	0.063 038	0.068 303
K_5				−0.027 864	−0.027 864	−0.014 134	0.002 467	0.017 475	0.029 628	0.038 933
K_6					−0.019 350	−0.024 499	−0.018 640	−0.008 155	0.003 119	0.013 430
K_7						−0.013 730	−0.020 370	−0.018 972	−0.012 896	−0.004 948
K_8							−0.009 960	−0.016 601	−0.017 614	−0.014 527
K_9								−0.007 378	−0.013 455	−0.015 687
K_{10}									−0.005 570	−0.010 918
K_{11}										−0.004 278

Note: The formula for K_r in the $(2m-3)$-term running average is

$$K_r = \frac{315\left\{(m-1)^2-r^2\right\}\left\{m^2-r^2\right\}\left\{(m+1)^2-r^2\right\}\left\{(3m^2-16)-11r^2\right\}}{8m(m^2-1)(4m^2-1)(4m^2-9)(4m^2-25)}.$$

Table 4.2.4. *The smoothing effect of a nine-term formula*

x	Observed value y	Smoothed value Y	x	Observed value y	Smoothed value Y
35	95	176	53	1274	1397
36	638	386	54	1579	1542
37	191	426	55	1873	1631
38	419	370	56	1482	1786
39	278	337	57	2002	1947
40	381	357	58	2011	2083
41	384	429	59	2827	2281
42	665	549	60	1722	2528
43	477	725	61	2942	2687
44	1015	973	62	3374	2805
45	1093	941	63	2735	3045
46	860	930	64	2731	3462
47	779	876	65	4323	3853
48	862	835	66	4926	4219
49	951	882	67	4114	4426
50	866	971	68	4099	4559
51	1109	1074	69	5835	5022
52	1291	1229	70	6103	6395

The same procedure is used to obtain smoothed values at $x = 40, 41, ..., 66$. The optimal seven-term formula can be used to obtain smoothed values at $x = 38$ and $x = 67$. Smoothed values at the remaining points can be obtained using formulae (4.2.4). The results are given in table 4.2.4 and the improved smoothness is evident in fig. 4.2.2. The process can be repeated to obtain greater smoothness.

Further reading: Greville [35], [36]; Hartree [41] 247–51; Hildebrand [43] 295–302; Miller [67] 25–33, 68–72; Nielsen [73] 287–93; Tetley [96] 176–80.

4.3　Spline functions
It is always possible to fit a polynomial through a set of n points (provided, of course, no point is vertically above another point), but the polynomial will generally be of order $n - 1$ and if n is large, the polynomial will usually display a marked undulatory behaviour. The appearance of such a curve is hardly 'smooth'.

This problem can often[2] be overcome by the use of a *spline function*. (The

[2] But not always. See example 4.3.2.

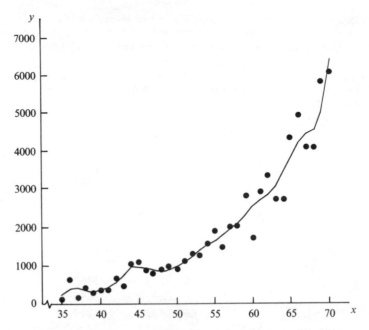

Fig. 4.2.2. The smoothing effect of a nine-term formula.

word 'spline' refers to a device used by draughtsmen to draw a smooth curve, consisting of a strip of some flexible material, to which weights can be attached at certain points in order to constrain the curve to pass through or close to certain given data points.)

We give the name *spline function* to a function defined by piecewise polynomial arcs in such a way that derivatives up to and including the order one less than the degree of the polynomials used are continuous everywhere. For purposes of interpolation the use of such functions offers substantial advantages. By employing polynomials of relatively low degree one can often avoid the marked undulatory behaviour that commonly arises from fitting a single polynomial exactly to a large number of empirical observations. On the other hand, much greater smoothness is obtained than with the traditional piecewise interpolation procedures, which give rise to discontinuities in the first derivative of the interpolating function. A spline function obviously provides continuity of the greatest possible number of derivatives of the interpolating function consistent with the use of polynomials of lower degree than would be required to fit all data points exactly by a single polynomial.

Consider the set of data points $(a, f(a))$, $(b, f(b))$, $(c, f(c))$, $(d, f(d))$ and imagine that we have already fitted a polynomial of degree three or less through $(a, f(a))$ and $(b, f(b))$. We now wish to fit a third-degree polynomial

$$y = A + B(x-b) + C(x-b)^2 + D(x-b)^3 \qquad (4.3.1)$$

through $(b, f(b))$ and $(c, f(c))$. We need to determine four constants A, B, C and D, and the $f(b)$ and $f(c)$ points impose two constraints. We can therefore arrange that the slope and second derivative of our polynomial be the same as the slope and second derivative of the previous polynomial at the point $x = b$. We now have four equations for the four unknowns A, B, C and D, and it is easy to see that

$$\begin{aligned}
f(b) &= A, \\
f(c) &= A + B(c-b) + C(c-b)^2 + D(c-b)^3, \\
f'(b) &= B, \\
f''(b) &= 2C,
\end{aligned}$$

so that

- $A = f(b),$
- $B = f'(b),$
- $C = \frac{1}{2}f''(b),$ (4.3.2)
- $D = \{f(c) - A - B(c-b) - C(c-b)^2\}/(c-b)^3.$

We can now interpolate between b and c. When this is complete, we move on to the next section of the curve through $f(c)$ and $f(d)$. To do this, we need the slope and second derivative of our polynomial at $x = c$, namely

- $f'(c) = B + 2C(c-b) + 3D(c-b)^2,$
- $f''(c) = 2C + 6D(c-b).$ (4.3.3)

Example 4.3.1. Let us demonstrate this process using the data in table 4.2.1. First we need to select the pivotal points or 'knots'. The graph in fig. 4.2.1 suggests that a smooth curve might be obtained if we use the running-average values at $x = 37$, $x = 42$ and $x = 55$ and the original data point at $x = 69$. Let us fit a spline curve through these points, namely

x	y
37	324
42	584
55	1642
69	5835

It would seem that a straight line might be used between $x = 37$ and $x = 42$, in which case the slope at $x = 42$ is $(584-324)/5 = 52$ and the second derivative is zero. We now fit a polynomial of degree three between $x = 42$ and $x = 55$. From (4.3.2),

$$\begin{aligned}
A &= 584, \\
B &= 52, \\
C &= 0,
\end{aligned}$$

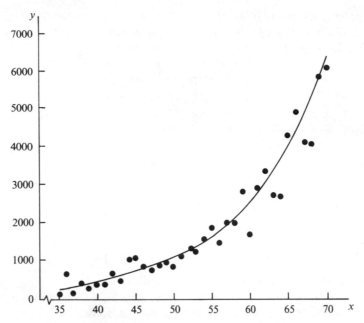

Fig. 4.3.1. Spline curve with knots at $x = 37$, $x = 42$, $x = 55$ and $x = 69$.

$$D = (1642 - 584 - 52 \times 13 - 0 \times 13^2)/13^3$$
$$= 0.173\,8734.$$

We can now interpolate between $x = 42$ and $x = 55$ using (4.3.1).

To determine the next section of the curve, we need to calculate the slope and second derivative at $x = 55$. According to (4.3.3)

$$f'(55) = 52 + 2 \times 0 \times 13 + 3 \times 0.173\,8734 \times 13^2 = 140.153\,81,$$
$$f''(55) = 2 \times 0 + 6 \times 0.173\,8734 \times 13 = 13.562\,125.$$

For the next section of the curve therefore:

$$A = 1642,$$
$$B = 140.153\,81,$$
$$C = \tfrac{1}{2}(13.562\,125) = 6.781\,063,$$
$$D = (5835 - 1642 - 140.153\,81 \times 14 - 6.781\,063 \times 14^2)/14^3$$
$$= 0.328\,6291.$$

Values can now be calculated for x between 55 and 69. The complete curve is given in fig. 4.3.1. The value at $x = 70$ has been obtained by extrapolating the final section of the curve, while the values at $x = 35$ and $x = 36$ were obtained by extrapolating the original straight line. Note the smoothness of the final curve. The actual (rounded) values obtained from the spline function are given in table 4.3.1.

Table 4.3.1. *Curve obtained using spline functions*

x	f(x)	x	f(x)	x	f(x)
35	220	47	866	59	2332
36	272	48	934	60	2553
37	324	49	1008	61	2798
38	376	50	1089	62	3068
39	428	51	1179	63	3365
40	480	52	1278	64	3692
41	532	53	1387	65	4050
42	584	54	1508	66	4442
43	636	55	1642	67	4868
44	689	56	1789	68	5332
45	745	57	1952	69	5835
46	803	58	2132	70	6379

The reader should note the rather arbitrary choice of pivotal values in this example. Some workers vary the values and positions of their knots in order to optimise certain criteria. A computer program is necessary for this purpose.

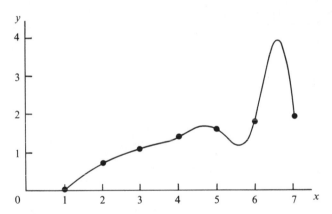

Fig. 4.3.2. A spline curve through the values of ln x at the integer points $x = 1, 2, \ldots, 7$.

Example 4.3.2. This example is given as a warning. Values of ln x are shown in fig. 4.3.2 for integer values of x. A quadratic was fitted through the first three points and the spline method was used to continue the curve through the remaining points. The resulting curve can hardly be described as 'smooth'.

Further reading: Greville [37]; Tetley [96] 265–7.

4.4 Exercises

1. Apply formulae (4.2.4) to a polynomial of degree three, and verify the fact that the formulae do not distort such a polynomial.

2. Verify the fact that the optimal five-term smoothing formula in table 4.2.3 does not distort a polynomial of degree three.

3. Use the method of example 4.2.1 to improve the smoothness of the smoothed values in table 4.2.4.

4. Example 4.3.1 makes use of four pivotal values. Fit a quadratic through the first three points by the method of section 6.10, and then use the spline method to complete the curve.

5
The area under a curve

Summary   Numerical methods are necessary to obtain the integral of a function when the function is defined by a series of empirical values and the integral of a mathematical function when the integral does not have an explicit mathematical form. Suitable formulae are given in this chapter.

5.1   Introduction

The integral of a function gives the area between the curve representing the function and the horizontal axis. If the function is always positive, the area will be positive; if the function is always negative, the area will be negative. If the function has an explicit mathematical form, the area *may* have an explicit mathematical form; in many instances the area does not (for example, the normal curve of error). In these circumstances, some form of numerical integration is necessary; likewise, when the function is given in the form of empirical values rather than a mathematical formula, numerical integration is necessary. Suitable formulae are given in this chapter.

It should be noted that when we use a numerical integration formula to find the area under a curve, we actually evaluate the total area under a series of approximating polynomial arcs and these arcs oscillate above and below the actual curve. In spite of this, the area estimate is usually quite accurate, because integration (like summation) is a smoothing process.

5.2   The trapezoidal rule

If the function $f(x)$ is a straight line in the interval $(0, 1)$

$$\int_0^1 f(x)\,dx = \tfrac{1}{2}\{f(0) + f(1)\}.$$

It is a trivial matter to prove this result. The formula is unlikely to provide an accurate value for the integral when the function deviates substantially from a straight line. For an interval of width h, the formula becomes

■ $\int_0^h f(x)\,dx = \tfrac{1}{2}h\{f(0) + f(h)\}.$ (5.2.1)

Over an interval of width nh, $n + 1$ ordinates can be used in the extended trapezoidal rule:

■ $\int_0^{nh} f(x)\,dx = \tfrac{1}{2}h\{f(0) + 2f(h) + 2f(2h) + 2f(3h)\dots$
$$+ 2f((n-1)\,h) + f(nh)\}.$$ (5.2.2)

37

When the curve is positive and concave upwards, the trapezoidal rule will overestimate the area and the converse is true when it is convex upwards. For accurate results, h must be small.

> **Example 5.2.1.** Use the trapezoidal rule to obtain the area under the curve $y = \exp(1/x)$ between $x = 1$ and $x = 2$, correct to four significant digits.

Let us begin by selecting $h = 1$ and use two ordinates. The integral is approximated by

$$\tfrac{1}{2}(e^{1/1} + e^{1/2}) = 2.183\,50.$$

When we repeat the calculation using $h = \tfrac{1}{2}$ and three ordinates, we obtain the improved approximation

$$\frac{0.5}{2}(e^{1/1} + 2e^{1/1.5} + e^{1/2}) = 2.065\,62.$$

Repeated halving of the interval h yields the following results:

$$h = \tfrac{1}{4} \text{ (5 ordinates)} \quad \text{area} \doteqdot 2.031\,89;$$
$$h = \tfrac{1}{8} \text{ (9 ordinates)} \quad \text{area} \doteqdot 2.023\,05;$$
$$h = \tfrac{1}{16} \text{ (17 ordinates)} \quad \text{area} \doteqdot 2.020\,81;$$
$$h = \tfrac{1}{32} \text{ (33 ordinates)} \quad \text{area} \doteqdot 2.020\,25;$$
$$h = \tfrac{1}{64} \text{ (65 ordinates)} \quad \text{area} \doteqdot 2.020\,11.$$

The last two approximations are equal to four significant digits; it would seem therefore that the area under the curve is 2.020 correct to four significant digits. Note that the curve is convex upwards and successive approximations decrease in value. A large number of ordinates was needed to calculate the area with the required precision.

> *Further reading*: Conte [17] 119–26; Hildebrand [43] 73; Whittaker and Robinson [100] 156–8.

5.3 Simpson's rule

The trapezoidal rule will provide an accurate value for an integral if the ordinates are chosen very close together. The amount of arithmetic involved is then considerable. We are therefore led to study alternative integration rules.

Let us imagine that $f(x)$ is a third-degree polynomial. It is not difficult to prove that

$$\blacksquare \qquad \int_{-h}^{h} f(x)\,dx = \tfrac{1}{3}h\{f(-h) + 4f(0) + f(h)\}. \tag{5.3.1}$$

This is Simpson's formula, and it is much more useful than the trapezoidal rule. It will produce accurate results whenever the function being integrated can be accurately approximated over the range of integration by a polynomial of degree three. The extended form of Simpson's rule is

■ $\int_0^{2nh} f(x)\, dx = \frac{1}{3}h\{f(0) + 4f(h) + 2f(2h) + 4f(3h) + \ldots$
$$+ 4f((2n-1)h) + f(2nh)\}. \qquad (5.3.2)$$

Example 5.3.1. Let us use Simpson's rule to obtain the area under the curve $y = \exp(1/x)$ between $x = 1$ and $x = 2$, correct to four significant digits. We start with $h = 0.5$ and three ordinates.

$h = 0.5$				
t	$\exp\{1/(1+t)\}$	Simpson factor	Product	
0.0	2.718 2818	1	2.718 2818	
0.5	1.947 7340	4	7.790 9360	
1.0	1.648 7213	1	1.648 7213	
		Total	12.157 9391	

The area is approximately $\frac{1}{3}(0.5)(12.1579391) = 2.026\,32$. We now repeat the computation with $h = 0.25$ and five ordinates.

$h = 0.25$				
t	$\exp\{1/(1+t)\}$	Simpson factor	Product	
0.00	2.718 2818	1	2.718 2818	
0.25	2.225 5409	4	8.902 1636	
0.50	1.947 7340	2	3.895 4680	
0.75	1.770 7950	4	7.083 1800	
1.00	1.648 7213	1	1.648 7213	
		Total	24.247 8147	

The area is approximately $\frac{1}{3}(0.25)(24.247\,8147) = 2.020\,65$. Repeated halving of the interval h yields the following results:

$h = 0.125$ (9 ordinates):   area $\doteq 2.020\,10$,
$h = 0.0625$ (17 ordinates): area $\doteq 2.020\,06$.

The last two approximations are equal to four significant digits and it would seem therefore that the area under the curve is 2.020 correct to four significant figures. Note the smaller amount of labour involved compared with the trapezoidal rule.

Further reading: Conte [17] 134–6; Freeman [27] 174–80; Hildebrand [43] 73–5, 141, 145–6; Nielsen [73] 118–20; Whittaker and Robinson [100] 156–8.

5.4   The three-eighths rule

Simpson's rule can be used for numerical integration whenever the number of equally-spaced ordinates is an odd number greater than or equal to three. We then have an even number of intervals. In some empirical situations however, the number of ordinates is even and Simpson's rule cannot be used. The three-eighths rule makes use of four ordinates and, using it in conjunction with Simpson's rule, we can cover all contingencies.

Let us again assume that $f(x)$ is a third-degree polynomial. It is not difficult to prove that

■ $$\int_0^{3h} f(x) \, dx = \tfrac{3}{8}h\{(f(0) + 3f(h) + 3f(2h) + f(3h)\}. \tag{5.4.1}$$

The extended form of this formula is obvious.

Example 5.4.1. A smooth curve $y = f(x)$ passes through the following points: (1, 3.724), (2, 4.492), (3, 5.359), (4, 6.332), (5, 7.415) and (6, 8.615). Find the area under this curve between $x = 1$ and $x = 6$.

It is not possible to use Simpson's rule to integrate with six ordinates. Nor can we use the three-eighths rule. The trapezoidal rule could be used, but it would not give an accurate answer. All is not lost, however, because it is possible to integrate between $x = 1$ and $x = 3$ using Simpson's rule, and between $x = 3$ and $x = 6$ using the three-eighths rule. The working may be set out as follows:

$h = 1$

x	$f(x)$	Simpson factor	Product
1	3.724	1	3.724
2	4.492	4	17.968
3	5.359	1	5.359
		Total	27.051

The area under the curve and between $x = 1$ and $x = 3$ is approximately $\tfrac{1}{3}(27.051) = 9.017$.

$h = 1$

x	$f(x)$	Three-eighths factor	Product
3	5.359	1	5.359
4	6.332	3	18.996
5	7.415	3	22.245
6	8.615	1	8.615
		Total	55.215

The area under the curve and between $x = 3$ and $x = 6$ is approximately $\frac{3}{8}(1)(55.215) = 20.706$. The total area between $x = 1$ and $x = 6$ is therefore 29.723.

Further reading: Freeman [27] 181; Hildebrand [43] 73; Whittaker and Robinson [100] 156–8.

5.5   Other integration formulae – Weddle's rule

Simpson's rule and the three-eighths rule are adequate for most purposes. Numerous other formulae are available, however, and mention might be made of Weddle's rule:

■ $$\int_0^{6h} f(x)\,dx = \tfrac{3}{10}h\{f(0) + 5f(h) + f(2h) + 6f(3h)$$
$$+ f(4h) + 5f(5h) + f(6h)\}. \tag{5.5.1}$$

This formula is exact if $f(x)$ is a polynomial of degree five or less throughout the range of integration. The extended form of the formula is straightforward.

Example 5.5.1. A smooth curve $y = f(x)$ passes through the following points: $(0, 3.049)$, $(1, 3.724)$, $(2, 4.492)$, $(3, 5.359)$, $(4, 6.332)$, $(5, 7.415)$ and $(6, 8.615)$. Use Weddle's rule to find the area under the curve between $x = 0$ and $x = 6$.

The working may be set out as follows:

$h = 1$			
x	$f(x)$	Weddle factor	Product
0	3.049	1	3.049
1	3.724	5	18.620
2	4.492	1	4.492
3	5.359	6	32.154
4	6.332	1	6.332
5	7.415	5	37.075
6	8.615	1	8.615
		Total	110.337

The area is approximately $\frac{3}{10}(1)(110.337) = 33.101$.

Further reading: Conte [17] 137; Freeman [27] 182–5; Hildebrand [43] 73 160–1; Nielsen [73] 118–20; Sokolnikoff and Sokolnikoff [92] 554–8.

5.6   Unequally-spaced ordinates

In experimental situations, problems often arise because the ordinates are not equally spaced. The trapezoidal rule can always be used, but the final result is unlikely to be very accurate. Simpson's rule and the three-eighths rule can sometimes be used. For example, if the ordinates corresponding to $x = 1, 1.5, 2, 2.5, 3, 4, 5$ and 6 are given, the extended Simpson's rule can be used over the range $x = 1$ to $x = 3$ and the three-eighths rule over the range $x = 3$ to $x = 6$.

More often than not, however, some other approach is necessary. Equally-spaced ordinates can be estimated using interpolation methods (chapter 6) or the spline function method (section 4.3), and Simpson's rule and the three-eighths rule can then be applied. If the spline-function method is used, we actually determine a series of polynomial arcs and the area is probably better obtained by the direct integration of each polynomial. Care should be exercised because the fitted curve can sometimes display an unsatisfactory undulatory behaviour, and the reader is advised to sketch some of the interpolated values. Example 4.3.2 serves as a warning.

Example 5.6.1. A smooth curve passes through the following points: $(0, 1.0000)$, $(0.5, 0.7788)$, $(2, 0.3679)$, $(3, 0.2231)$ and $(5, 0.0821)$. Find the area under the curve between $x = 0$ and $x = 5$.

Using the method of section 6.10, we find that the following quadratic curve passes through the first three points:

$$y = 1.0000 - 0.484\,51\dot{6}x + 0.084\,2\dot{3}x^2.$$

The area under this parabola between $x = 0$ and $x = 2$ is

$$[x - 0.242\,258\dot{3}x^2 + 0.028\,0\dot{7}x^3]_0^2 = 1.2556.$$

The spline-function method of section 4.3 can be used to fit the following cubic curve through the points $(2, 0.3679)$ and $(3, 0.2231)$:

$$y = 0.3679 - 0.147\,583\,(x - 2) + 0.084\,2\dot{3}\,(x - 2)^2$$
$$- 0.081\,45\,(x - 2)^3.$$

Writing $X = x - 2$, we see that the area under this cubic between $x = 2$ and $x = 3$ is

$$[0.3679X - 0.147\,583\,(\tfrac{1}{2}X^2) + 0.084\,2\dot{3}\,(\tfrac{1}{3}X^3) - 0.081\,45\,(\tfrac{1}{4}X^4)]_0^1 = 0.3018.$$

The spline function through the final two points is

$$y = 0.2231 - 0.223\,4\dot{6}\,(x - 3) + 0.160\,11\dot{6}\,(x - 3)^2 - 0.041\,81\dot{6}(x - 3)^3,$$

and the appropriate area under this curve turns out to be 0.2590.

We conclude that the total area under the unknown curve, between $x = 0$ and $x = 5$ is approximately equal to $1.2556 + 0.3018 + 0.2590$ or 1.82 to three significant digits. The points in this particular example actually lie on the curve $y = \exp{(-\tfrac{1}{2}x)}$ and the area under this curve is known to be 1.84 to three significant digits. Our answer has an error of 1 per cent.

The reader should note the large amount of work involved in determining the area under a curve when the usual formulae are not immediately applicable.

5.7   Exercises

1. Use the trapezoidal rule to find the area under the curve in example 5.6.1.

2. Use Simpson's rule to find the area under the curve in example 5.5.1.

3. Use the three-eighths rule to find the area under the curve in example 5.5.1.

4. Use two quadratic curves, one through the first three points and the other through the last three points to estimate the area under the curve in example 5.6.1.

5. A hill is to be levelled to form a rectangular playing field, and the height of the hill above the final level is $f(x, y)$ at the point (x, y). If the playing field is to be of dimension $2h \times 2k$, prove that the net volume of soil to be removed from the site is approximately

$$\tfrac{1}{9}hk\left\{ (f_{0,0} + f_{0,2} + f_{2,0} + f_{2,2}) + 4(f_{0,1} + f_{1,0} + f_{1,2} + f_{2,1}) + 16f_{1,1}\right\},$$

where $f_{r,s} = f(rh, sk)$. *Hint*: use Simpson's rule.

6

Finite differences, interpolation and numerical differentiation

Summary   In this chapter we study the properties of finite differences and demonstrate the use of the finite-difference calculus in numerical differentiation and interpolation. Ordinary finite-difference methods fail when ordinates are unequally spaced, and divided differences are introduced in section 6.7 to overcome this problem. Simple methods for determining the straight line through two points and the quadratic through three points are given in sections 6.9 and 6.10.

6.1   Finite-difference processes

The slope of the curve $y = f(x)$ at the point x is given by

$$\frac{dy}{dx} = f'(x) = \lim_{h \to 0} \frac{f(x+h) - f(x)}{h}.$$

This slope (or *derivative* of y with respect to x) is usually readily calculable when the functional form of $f(x)$ is known. In many experimental situations, however, a smooth curve is obtained, but the curve does not have a known functional form. An approximation to the slope at the point x may be obtained using $\{f(x+h) - f(x)\}/h$ rather than the limit of this expression as $h \to 0$ (fig. 6.1.1). The difference $f(x+h) - f(x)$ is called a *finite difference.*

Even in situations where the functional form of $f(x)$ is known, a finite-difference method of differentiation may be preferable because of the complicated form of the derivative (question 2 of section 6.11).

Finite-difference methods are also used to obtain numerical solutions for differential equations, the derivatives in these equations being replaced by approximately equivalent finite differences.

Another area of application is *interpolation.* When forming a table of a continuous function, we are faced with the problem of making the table as compact as possible, while at the same time ensuring that values of the function can be obtained from the table with the required degree of accuracy. Finite-difference methods allow us to 'read between the lines of the table' and interpolate.

6.2   The difference table

Finite-difference processes always produce exact results when applied to polynomials. Let us therefore begin by considering a difference table for the

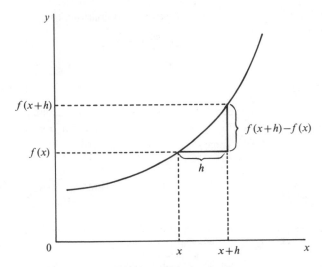

Fig. 6.1.1. A finite-difference approximation to the slope of a curve.

Table 6.2.1. *A difference table*

x	$f(x)$	$\Delta f(x)$	$\Delta^2 f(x)$	$\Delta^3 f(x)$	$\Delta^4 f(x)$
0.0	1				
		0			
0.5	1		6		
		6		6	
1.0	7		12		0
		18		6	
1.5	25		18		0
		36		6	
2.0	61		24		
		60			
2.5	121				

third-degree polynomial $f(x) = 8x^3 - 2x + 1$. Values of this function for $x = 0$, 0.5, 1, 1.5, 2 and 2.5 are listed in the second column of the difference table 6.2.1, and it is a simple matter to calculate the first-order differences

$$\Delta f(0) = f(0.5) - f(0),$$
$$\Delta f(0.5) = f(1) - f(0.5),$$
$$\Delta f(1) = f(1.5) - f(1),$$

etc.

These values appear in the third column of table 6.2.1, and if divided by the *differencing interval* 0.5, they would provide rough approximations to the slope of the curve at the various points.

In the general situation when the differencing interval is h, the first-order difference of $f(x)$ at the point $x = a$ is defined by

■ $\qquad \Delta f(a) = f(a+h) - f(a).$ $\qquad\qquad\qquad\qquad$ (6.2.1)

In the differential calculus, the second derivative (or slope of the slope curve) is often needed. Likewise, in the difference calculus, we make use of second differences (the differences of the first-order differences). Thus, in our polynomial example

$$\Delta^2 f(0) = \Delta(f(0.5) - f(0)) = (f(1) - f(0.5)) - (f(0.5) - f(0)),$$
$$\Delta^2 f(0.5) = \Delta(f(1) - f(0.5)) = (f(1.5) - f(1)) - (f(1) - f(0.5)).$$

Numerical values are given in the fourth column of table 6.2.1. In the general case with differencing interval h,

■ $\quad \Delta^2 f(a) = \Delta(f(a+h) - f(a)) = (f(a+2h) - f(a+h)) - (f(a+h) - f(a)),$

$$(6.2.2)$$

and the generalisation to higher-order differences is obvious.

Note how, in the difference table, differences are placed at a level midway between the two values whose difference is taken. The value $\Delta^n f(a)$ falls in the column of nth differences n half-lines below $f(a)$. Thus, for example, $\Delta^2 f(1.5)$ falls in the column of second differences one full line below $f(1.5)$. The value shown in table 6.2.1 is 24.

In table 6.2.1, we observe that the fourth (and higher) differences are zero, and third differences are constant. The generalisation is obvious, namely that for a polynomial of degree j, the jth differences are constant and higher-order differences are zero. For a function which is not a polynomial, successive orders of differences normally decrease in absolute value and then begin to increase again. The increase usually begins after about the third order of differences. The higher-order differences are not used in numerical work, and this is equivalent to approximating the function by a low-order polynomial.

Further reading: Conte [17] 82–7; Freeman [27] 1–6; Hartree [41] 36–48; Hildebrand [43] 91–4; Nielsen [73] 27–30; Ralston [84] 46–52; Sokolnikoff and Sokolnikoff [92] 527–8.

6.3 Checking numbers in a table

The difference table can be a useful tool for checking numbers in a table. If we know that the tabulated function is a polynomial, but do not know its degree, we may examine its differences. Let us imagine that the table is extensive and that all the fourth differences, except five, are zero. Then we may conclude that the tabulated function is a polynomial of degree three and

Table 6.3.1. *How mistakes fan out in a difference table*
– a polynomial example

x	x^3	$\Delta f(x)$	$\Delta^2 f(x)$	$\Delta^3 f(x)$	$\Delta^4 f(x)$
0	0				
		1			
1	1		6		
		7		6	
2	8		12		0
		19		6	
3	27		18		$0+\epsilon$
		37		$6+\epsilon$	
4	64		$24+\epsilon$		$0-4\epsilon$
		$61+\epsilon$		$6-3\epsilon$	
5	$125+\epsilon$		$30-2\epsilon$		$0+6\epsilon$
		$91-\epsilon$		$6+3\epsilon$	
6	216		$36+\epsilon$		$0-4\epsilon$
		127		$6-\epsilon$	
7	343		42		$0+\epsilon$
		169		6	
8	512		48		0
		217		6	
9	729		54		
		271			
10	1000				

that one of the tabulated values is wrong. As a result of the mistake, the first-difference column contains two wrong entries, the second-difference column contains three wrong entries, the third-difference column contains four wrong entries and the fourth-difference column contains five wrong entries. The errors in the difference table fan out from the original mistake with values proportional to the coefficients in the binomial expansion[1] $(1-x)^n$, where n is the order of the difference (table 6.3.1). It is easy to find the mistake and correct it.

When the underlying function is not a polynomial, higher-order differences will not be zero, but they should be small and they should progress in a regular manner. Mistakes are still readily detected and corrected.

Example 6.3.1. Certain of the following twenty consecutive values, corresponding to equally-spaced arguments, are incorrect because of typical copying mistakes (section 2.4). Locate the mistakes and correct them.

[1] See table 1.5.1.

1.7278	4.8818	7.9779	11.2630
2.3424	5.4440	8.6249	11.9398
2.9585	6.0723	9.2752	12.6246
3.5764	6.7041	9.9318	13.3180
4.1964	7.3398	10.5937	14.0206

We begin by forming the difference table 6.3.2. (Note how the table has been simplified by omitting decimal points.) This table suggests that three entries are wrong. If suitable adjustments are made to these entries, the column of third differences should progress in an orderly manner from 3 to 6.

The first third-difference inside the fan emanating from 4.8818 is given as 633, and this value is too large by about 630. The entry 4.8818 must therefore exceed the correct value by about 630×10^{-4}. We deduce that the correct entry is 4.8188, and the four third-differences then become 3, 4, 3 and 4 respectively.

In a similar manner, we deduce that the entry 7.9779 should read 7.9799 and the entry 9.9318 should read 9.9313 (digit 3 has been mistaken for an 8). When these two adjustments are made, the other incorrect third-differences become 5, 5, 4, 5, 5, 6 and 6 respectively.

Further reading: Hartree [41] 44–7; Hildebrand [43] 110–12; Nielsen [73] 38–40; Ralston [84] 50–2.

6.4 Newton's advancing-difference interpolation formula

A polynomial $f(x)$ is evaluated at the points a, $a + h$, $a + 2h$, ..., $a + nh$, and a difference table is formed using these values. Newton's advancing-difference interpolation formula tells us that the value of the polynomial at the point $a + rh$ is

■ $$f(a + rh) = f(a) + \binom{r}{1} \Delta f(a) + \binom{r}{2} \Delta^2 f(a) + ...$$ (6.4.1)

This formula is the finite-difference equivalent of the Taylor series expansion (1.2.1), and it is exact for all r when $f(x)$ is a polynomial. It also produces exact results for other functions when r is a positive integer less than or equal to n. The formula would be of little interest if its use were limited to these cases. Fortunately, we can apply the formula in other situations. (The second example demonstrates this.) The results are then approximate, but they can be very accurate.

It will be noted that the difference operators and the binomial coefficients[2] in (6.4.1) are really the terms in the expansion of $(1 + \Delta)^r$. Newton's formula can therefore be written in the symbolic form

■ $$f(a + rh) = (1 + \Delta)^r f(a).$$ (6.4.2)

[2] Section 1.5.

Table 6.3.2. *Using a difference table to detect and correct mistakes*

$f(x)$	$\Delta f(x)$	$\Delta^2 f(x)$	$\Delta^3 f(x)$
1.7278			
	6146		
2.3424		15	
	6161		3
2.9585		18	
	6179		3
3.5764		21	
	6200		633
4.1964		654	
	6854		−1886
4.8818		−1232	
	5622		1893
5.4440		661	
	6283		−626
6.0723		35	
	6318		4
6.7041		39	
	6357		−15
7.3398		24	
	6381		65
7.9779		89	
	6470		−56
8.6249		33	
	6503		30
9.2752		63	
	6566		−10
9.9318		53	
	6619		21
10.5937		74	
	6693		1
11.2630		75	
	6768		5
11.9398		80	
	6848		6
12.6246		86	
	6934		6
13.3180		92	
	7026		
14.0206			

Example 6.4.1. Use the difference table 6.2.1 to find the value of the polynomial $f(x) = 8x^3 - 2x + 1$ at the point $x = 0.25$.

In (6.4.1), we set $a = 0, h = 0.5$ and $r = 0.5$. According to the formula, then

$$f(0.25) = f(0) + 0.5\Delta f(0) - 0.125\Delta^2 f(0) + 0.0625\Delta^3 f(0)$$
$$= 1 + 0.5 \times 0 - 0.125 \times 6 + 0.0625 \times 6$$
$$= 0.625.$$

Alternatively, we could have used $a = 1, h = 0.5$ and $r = -1.5$. Formula (6.4.1) then gives us

$$f(0.25) = f(1) - 1.5\Delta f(1) + 1.875\Delta^2 f(1) - 2.1875\Delta^3 f(1)$$
$$= 7 - 1.5 \times 18 + 1.875 \times 18 - 2.1875 \times 6$$
$$= 0.625.$$

The answer may be confirmed by direct calculation.

Example 6.4.2. The area under the unit normal curve to the left of the point x is given in table 6.4.1 for values of x at intervals of 0.1. Find the area to the left of the point $x = 1.96$.

We begin by completing the difference table as shown. Third differences are small; so we shall ignore fourth and higher-order differences. The area to the left of the point $x = 1.96$ can be obtained by applying (6.4.1) with $a = 1.9$, $h = 0.1$ and $r = 0.6$.

$$f(1.96) = f(1.9) + 0.6\Delta f(1.9) - 0.12\Delta^2 f(1.9) + 0.056\Delta^3 f(1.9)$$
$$= 0.9713 + 0.6 \times 0.0060 - 0.12 \times (-0.0012) + 0.056 \times 0.0004$$
$$= 0.9750.$$

When we compare this result with the value in the table of the normal distribution on p. 323, we find that we have obtained the answer correct to four decimal places.

> *Further reading*: Freeman [27] 29–35, 81–90; Hartree [41] 60–2, 64, 92–4; Hildebrand [43] 94–5; Ralston [84] 52; Sokolnikoff and Sokolnikoff [92] 550–2.

6.5 Bessel's interpolation formula

Newton's advancing-difference interpolation formula is exact whenever the function is a polynomial. If the function is not a polynomial, the method is only approximate and the approximation may not be as accurate as we would like. To obtain the value at the point $x = a + \frac{1}{2}h$, for example, we use the tabulated values at the points $a, a+h, a+2h, a+3h, \dots$. All these points (except for the first) lie to the right of the interpolation point. A more satisfactory result will normally be obtained if we use a formula which is more or less symmetric about the interpolation point. Bessel's interpolation

Table 6.4.1. *The area under the unit normal curve to the left of the point x.*

x	Area $f(x)$	$\Delta f(x)$	$\Delta^2 f(x)$	$\Delta^3 f(x)$
1.9	0.9713			
		0.0060		
2.0	0.9773		−0.0012	
		0.0048		0.0004
2.1	0.9821		−0.0008	
		0.0040		0.0000
2.2	0.9861		−0.0008	
		0.0032		
2.3	0.9893			

formula is one such formula, and it should provide an accurate answer whenever r lies in the interval $(0,1)$:

■ $$f(a + rh) = f(a) + r\Delta f(a)$$
$$+ \tfrac{1}{4}(r^2 - r)\,(\Delta^2 f(a) + \Delta^2 f(a - h))$$
$$+ \tfrac{1}{6}(r^2 - r)\,(r - \tfrac{1}{2})\,\Delta^3 f(a - h)$$
$$+ \tfrac{1}{48}(r^3 - r)\,(r - 2)\,(\Delta^4 f(a - h) + \Delta^4 f(a - 2h))$$
$$+ \dots . \tag{6.5.1}$$

Example 6.5.1. The following values of a function $f(x)$ are given:

x	$f(x)$	x	$f(x)$
0.7	1.428 5714	1.1	0.909 0909
0.8	1.250 0000	1.2	0.833 3333
0.9	1.111 1111	1.3	0.769 2308
1.0	1.000 0000		

Calculate $f(0.95)$.

Let us use Newton's advancing-difference formula (6.4.1) and the difference table 6.5.1. We choose $a = 0.9, h = 0.1$ and $r = 0.5$, and obtain the value 1.052 6541.

When we use Bessel's formula with $a = 0.9, h = 0.1$ and $r = 0.5$, we obtain the value 1.052 6372.

This example was chosen for demonstration purposes and it is known that $f(x) = 1/x$. The true value of $f(0.95)$ is therefore 1.052 6316. We see that both formulae have given reasonably accurate results, but the Bessel result is more accurate than that obtained via Newton's formula.

Further reading: Freeman [27] 61–72; Hartree [41] 67–72; Henrici [42] 222–7; Hildebrand [43] 97–110.

Table 6.5.1. *The difference table for example 6.5.1*

$f(x)$	$\Delta f(x)$	$\Delta^2 f(x)$	$\Delta^3 f(x)$	$\Delta^4 f(x)$
1.428 5714				
	$-0.178\ 5714$			
1.250 0000		0.039 6825		
	$-0.138\ 8889$		$-0.011\ 9047$	
1.111 1111		0.027 7778		0.004 3289
	$-0.111\ 1111$		$-0.007\ 5758$	
1.000 0000		0.020 2020		0.002 5253
	$-0.090\ 9091$		$-0.005\ 0505$	
0.909 0909		0.015 1515		0.001 5541
	$-0.075\ 7576$		$-0.003\ 4964$	
0.833 3333		0.011 6551		
	$-0.064\ 1025$			
0.769 2308				

6.6   Numerical differentiation

Situations which give rise to numerical differentiation are outlined in section 6.1. It is important to remember that when we use a numerical method of differentiation, we are actually evaluating the slope of an approximating polynomial arc. This arc will oscillate above and below the actual function. It follows that numerical differentiation is a rather hazardous occupation and should be avoided whenever possible.

Let us imagine that the function $f(x)$ can be represented satisfactorily by a polynomial of degree two or less in the neighbourhood of the point $x = a$, and that the values $f(a-h)$, $f(a)$ and $f(a+h)$ are known. We shall denote these values by f_{-1}, f_0 and f_1 respectively. Then

■ $$f'_{-1} = \frac{1}{2h}(-3f_{-1} + 4f_0 - f_1),$$ (6.6.1)

■ $$f'_0 = \frac{1}{2h}(-f_{-1} + f_1),$$ (6.6.2)

■ $$f'_1 = \frac{1}{2h}(f_{-1} - 4f_0 + 3f_1).$$ (6.6.3)

Formulae (6.6.1) and (6.6.3) should only be used at the beginning and end of a table.

When the function $f(x)$ can be represented satisfactorily by a polynomial of degree four or less, the following formulae apply:

■ $$f'_{-2} = \frac{1}{12h}(-25f_{-2} + 48f_{-1} - 36f_0 + 16f_1 - 3f_2),$$ (6.6.4)

■ $$f'_{-1} = \frac{1}{12h}(-3f_{-2} - 10f_{-1} + 18f_0 - 6f_1 + f_2),$$ (6.6.5)

■ $f_0' = \dfrac{1}{12h}(f_{-2} - 8f_{-1} + 8f_1 - f_2),$ (6.6.6)

■ $f_1' = \dfrac{1}{12h}(-f_{-2} + 6f_{-1} - 18f_0 + 10f_1 + 3f_2),$ (6.6.7)

■ $f_2' = \dfrac{1}{12h}(3f_{-2} - 16f_{-1} + 36f_0 - 48f_1 + 25f_2).$ (6.6.8)

All these formulae can be proved by expanding the right-hand sides in Taylor series about the left-hand side subscript point. Formulae (6.6.4), (6.6.5), (6.6.7) and (6.6.8) should only be used at the beginning and end of the table.

It must be emphasised that these formulae can only be applied to smooth data. When doubtful about the degree of the appropriate approximating polynomial, use the five-point formulae rather than the three-point.

The calculation of the second derivative is even more hazardous than the calculation of the first derivative. The above formulae can be applied twice, but it is usually easier to employ one of the following:

■ $f_0'' = \dfrac{1}{h^2}(f_{-1} - 2f_0 + f_1),$ (6.6.9)

■ $f_0'' = \dfrac{1}{12h^2}(-f_{-2} + 16f_{-1} - 30f_0 + 16f_1 - f_2).$ (6.6.10)

The three-point formula assumes an approximating polynomial of degree two, and the five-point a polynomial of degree four.

Finite-difference methods can also be used for numerical differentiation, although the above methods are usually preferable. Our starting point will be the Taylor series (1.2.1), and we shall write D instead of d/dx, D^2 instead of d^2/dx^2, etc. Then

$$f(a + rh) = f(a) + \frac{rh}{1!}\,Df(a) + \frac{(rh)^2}{2!}\,D^2f(a) + \dots$$

$$= \left(1 + \frac{rhD}{1!} + \frac{(rhD)^2}{2!} + \dots\right)f(a)$$

$$= e^{rhD}f(a).$$

This formula is now compared with (6.4.2). It would seem that e^{hD} is equivalent to $1 + \Delta$, and

■ $Df(a) = \left(\dfrac{1}{h}\ln(1 + \Delta)\right)f(a)$

$$= \frac{1}{h}(\Delta f(a) - \tfrac{1}{2}\Delta^2 f(a) + \tfrac{1}{3}\Delta^3 f(a) - \dots),$$ (6.6.11)

■ $D^2f(a) = \left(\dfrac{1}{h}\ln(1 + \Delta)\right)^2 f(a)$

$$= \frac{1}{h^2}(\Delta^2 f(a) - \Delta^3 f(a) + \tfrac{11}{12}\Delta^4 f(a) + \dots).$$ (6.6.12)

These formulae are exact for polynomials, and approximate for other functions. They suffer from the same disadvantage as Newton's advancing-difference interpolation formula (see section 6.5). The method, however, is sometimes useful when a derivative is required at a non-tabular point (example 6.6.3).

Example 6.6.1. Values of $\tan^{-1}(e^x)$ are given in table 6.6.1. Calculate the first and second derivatives of the function at $x = 1.2$.

Explicit formulae for these two derivatives are somewhat complicated. Let us use formula (6.6.6) to calculate the first derivative:

$$f'(1.2) = \frac{1}{12 \times 0.1} (1.218\,283 - 8 \times 1.249\,462 \\ + 8 \times 1.304\,726 - 1.329\,023)$$

$$= 0.276\,143,$$

and this answer happens to be correct in all six places!

The second derivative (the slope of the slope curve) may be obtained via formula (6.6.10):

$$f''(1.2) = \frac{1}{12 \times (0.1)^2} (-1.218\,283 + 16 \times 1.249\,462 - 30 \times 1.278\,244 \\ + 16 \times 1.304\,726 - 1.329\,023)$$

$$= -0.230\,15.$$

The actual value is $-0.230\,21$; so our answer is correct to four decimal places.

Table 6.6.1. *Values of* $\tan^{-1}(e^x)$

x	$\tan^{-1}(e^x)$
1.0	1.218 283
1.1	1.249 462
1.2	1.278 244
1.3	1.304 726
1.4	1.329 023

Example 6.6.2. Find the slope of the empirical function in table 6.6.2 at the point $x = 60$.

According to (6.4.4), the slope is -1491. We can also use the finite-difference method. Third differences are effectively constant, and they are relatively small anyway.

$$f'(60) = \Delta f(60) - \tfrac{1}{2}\Delta^2 f(60) + \tfrac{1}{3}\Delta^3 f(60)$$

$$= -1536 - \tfrac{1}{2}(-97) + \tfrac{1}{3}(-8)$$

$$= -1490.$$

Table 6.6.2. *The difference table for example 6.6.2*

x	$f(x)$	Δf	$\Delta^2 f$	$\Delta^3 f$
60	63 620			
		−1536		
61	62 084		− 97	
		−1633		−8
62	60 451		−105	
		−1738		−6
63	58 713		−111	
		−1849		−5
64	56 864		−116	
		−1965		0
65	54 899		−116	
		−2081		−1
66	52 818		−117	
		−2198		
67	50 620			

Example 6.6.3. Find the slope of the empirical function in table 6.6.2 at the point $x = 61.5$.

We make use of formulae (6.6.11) and (6.4.2):

$$
\begin{aligned}
f'(61.5) &= (\ln(1 + \Delta))\, f(61.5) \\
&= (\ln(1 + \Delta))\,(1 + \Delta)^{\frac{1}{2}}\, f(61) \\
&= (\Delta - \tfrac{1}{2}\Delta^2 + \tfrac{1}{3}\Delta^3 - \ldots)\,(1 + \tfrac{1}{2}\Delta - \tfrac{1}{8}\Delta^2 + \ldots)\, f(61) \\
&= (\Delta - \tfrac{1}{24}\Delta^3 + \ldots)\, f(61) \\
&= \Delta f(61) - \tfrac{1}{24}\Delta^3 f(61) \\
&= -1633 - \tfrac{1}{24}(-6) \\
&= -1633.
\end{aligned}
$$

Further reading: Freeman [27] 126–7; Hildebrand [43] 82, 134; Nielsen [73] 56; Ralston [84] 83–5.

6.7 Unequally-spaced ordinates – divided-difference interpolation

All the finite-difference formulae we have used so far depend upon values of the function at equally-spaced points. This will not always be the case, and we therefore introduce the concept of a *divided difference*.

The construction of a divided-difference table is demonstrated in table 6.7.1 with the third-degree polynomial $f(x) = x^3 - x + 1$. We use Freeman's notation and denote, for example, the third-order divided-difference of $f(a)$ which involves $f(a), f(b), f(c)$ and $f(d)$ by $\underset{b,c,d}{\Delta}{}^3 f(a)$. The actual calculations in the

Table 6.7.1. *A divided-difference table*

x	$f(x)$	$\Delta f(x)$	$\Delta^2 f(x)$	$\Delta^3 f(x)$	$\Delta^4 f(x)$
0	1				
		$\dfrac{1-1}{1-0}=0$			
1	1		$\dfrac{20-0}{4-0}=5$		
		$\dfrac{61-1}{4-1}=20$		$\dfrac{10-5}{5-0}=1$	
4	61		$\dfrac{60-20}{5-1}=10$		$\dfrac{1-1}{6-0}=0$
		$\dfrac{121-61}{5-4}=60$		$\dfrac{15-10}{6-1}=1$	
5	121		$\dfrac{90-60}{6-4}=15$		
		$\dfrac{211-121}{6-5}=90$			
6	211				

Note: the sloping lines are inserted to demonstrate the calculation of $\underset{1,4,5}{\Delta^3} f(0)$.

table should be noted carefully; $\underset{1,4,5}{\Delta^3} f(0)$ for example, is computed in the following manner:

$$\underset{1,4,5}{\Delta^3} f(0) = (\underset{4,5}{\Delta^2} f(1) - \underset{1,4}{\Delta^2} f(0))/(5-0)$$
$$= (10-5)/(5-0)$$
$$= 1.$$

This number appears in the third-difference column at the tip of the triangle from $f(0), f(1), f(4)$ and $f(5)$. The reader should note that $f(x)$ is a polynomial of degree three and that third-order divided-differences are constant.

Newton's divided-difference interpolation formula takes the following form:

■ $\qquad f(x) = f(a) + (x-a) \underset{b}{\Delta} f(a) + (x-a)(x-b) \underset{b,c}{\Delta^2} f(a) + \dots .$ $\qquad$ (6.7.1)

It is worth noting that with a polynomial, function values can be listed in any order in the divided-difference table and the correct answer will always be obtained. For non-polynomial functions, best results will usually be obtained by listing the function values in their proper sequence and using divided differences as near as possible to the point in question. Newton's advancing-difference interpolation formula (6.4.1) is a special case of (6.7.1).

Example 6.7.1. Use the divided-difference table 6.7.1 to calculate the value of the polynomial at the point $x = 3$.

According to (6.7.1),

$$f(3) = f(0) + (3-0) \underset{1}{\Delta} f(0) + (3-0)(3-1) \underset{1,4}{\Delta^2 f(0)}$$
$$+ (3-0)(3-1)(3-4) \underset{1,4,5}{\Delta^2 f(0)}$$
$$= 1 + 3 \times 0 + 3 \times 2 \times 5 + 3 \times 2 \times (-1) \times 1$$
$$= 25.$$

This answer may be confirmed by direct calculation.

Further reading: Freeman [27] 39–45; Hildebrand [43] 35–45; Nielsen [73] 50, 82–7.

6.8   Derivatives using unequally-spaced ordinates

The most obvious approach is to use the divided-difference interpolation method to obtain values at equally-spaced points and then use a numerical differentiation formula from section 6.6. The spline-function method of section 4.3 could also be used to obtain the values at equally-spaced points. Either way, the amount of work involved is considerable. The best approach is to differentiate (6.7.1):

■ $$f'(x) = \underset{b}{\Delta} f(a) + ((x-a) + (x-b)) \underset{b,c}{\Delta^2 f(a)} + ((x-a)(x-b)$$
$$+ (x-a)(x-c) + (x-b)(x-c)) \underset{b,c,d}{\Delta^3 f(a)} + \dots . \qquad (6.8.1)$$

This formula will give exact results with a polynomial. Approximate results will be obtained when the function is not a polynomial, and the dangers inherent in numerical differentiation must be remembered (section 6.6).

Example 6.8.1. Find the slope of the polynomial in table 6.7.1 at the point $x = 2$.

Let us use (6.8.1) with $a = 0, b = 1, c = 4$ and $d = 5$:

$$f'(2) = 0 + ((2-0) + (2-1)) \times 5 + ((2-0)(2-1) + (2-0)(2-4)$$
$$+ (2-1)(2-4)) \times 1$$
$$= 11,$$

and this result may be confirmed by direct calculation.

6.9   The straight line through two points

Let (x_0, y_0) and (x_1, y_1) be two distinct points. The equation of the straight line through these two points is

■ $$y = Ax + B, \qquad (6.9.1)$$

where

■ $$A = (y_1 - y_0)/(x_1 - x_0), \Big\}$$
■ $$B = y_0 - Ax_0. \qquad\qquad (6.9.2)$$

Example 6.9.1. The straight line joining the points $(37, 324)$ and $(42, 584)$ is $y = 52x - 1600$.

6.10 The quadratic through three points

Let (x_0, y_0), (x_1, y_1) and (x_2, y_2) be three points with distinct abscissae.[3] The method of divided differences can be used to fit a quadratic through these points. We set $a = x_0$, $b = x_1$ and $c = x_2$, and form a small divided-difference table using $f(a) = y_0$, $f(b) = y_1$ and $f(c) = y_2$. The quadratic is given by the first three terms of (6.7.1). This method generalises to polynomials of any order.

The following method, however, requires no knowledge of divided differences and is quite straightforward. We compute

$$
\left.
\begin{aligned}
&a = x_1 - x_0, \\
&b = x_2 - x_0, \\
&\alpha = y_1 - y_0, \\
&\beta = y_2 - y_0, \\
&A = -(a\beta - b\alpha)/\{ab(a-b)\}, \\
&B = (a^2\beta - b^2\alpha)/\{ab(a-b)\}.
\end{aligned}
\right\}
\qquad (6.10.1)
$$

The quadratic equation takes the form:

$$
y = A(x - x_0)^2 + B(x - x_0) + y_0. \qquad (6.10.2)
$$

Note that the three points can be listed in any order.

Example 6.10.1. Let us fit a quadratic through the points $(0, 1)$, $(0.5, 0.7788)$ and $(2, 0.3679)$. We compute

$$
\begin{aligned}
a &= 0.5, \\
b &= 2.0, \\
\alpha &= -0.2212, \\
\beta &= -0.6321, \\
A &= 0.084\,2\dot{3}, \\
B &= -0.484\,51\dot{6}.
\end{aligned}
$$

The quadratic equation is therefore

$$
y = 0.084\,2\dot{3}x^2 - 0.484\,51\dot{6}x + 1.0.
$$

---

[3] Distinct x-values.

6.11   Exercises

1. (*a*) Form a difference table for the polynomial $f(x) = x^4$ using the function values at $x = 1, 2, 3, ..., 8$.
   (*b*) Use the fact that fourth differences are constant to complete an extra line in the table and hence obtain $f(9)$.
   (*c*) Use Newton's advancing-difference interpolation formula to obtain $f(1.5)$.
   (*d*) Use a numerical method to calculate $f'(1)$.
   (*e*) Use a numerical method to calculate $f''(5)$.
   (*f*) Use a method of example 6.6.3 to evaluate $f'(1.5)$.
   Confirm all these answers by direct calculation.

2. A function of five variables x_1, x_2, x_3, x_4 and x_5 is defined by

$$y = \sum_{j=2}^{4} \frac{(x_{j+1} - 2x_j + x_{j-1})^2}{(x_{j+1} - x_j)^2 + (x_j - x_{j-1})^2}.$$

Calculate the partial derivative of y with respect to x_3 at the point $x_1 = 1$, $x_2 = 4$, $x_3 = 10$, $x_4 = 17$, $x_5 = 24$.

3. Use (6.7.1) and the polynomial values in table 6.7.1 at the points $x = 1, 4, 5$ and 6 to calculate $f(3)$.

7
Some other numerical techniques

Summary   In this chapter we describe a variety of useful numerical techniques. We begin with the problem of regrouping grouped data, and then move on to Hardy's formula, which allows us to estimate the central ordinate of an area. The next important topic is a technique for fusing together two smooth curves to form a single curve. The method of steepest descent for minimising a function of several variables is described in section 7.5; this technique finds applications in many areas, including non-linear least squares (chapter 18). The chapter ends with a simple trick for increasing the data storage capacity of a programmable calculator.

7.1   Regrouping grouped data

Data are often grouped. This may be done to produce a concise table or because the data are sparse or some other reason. The user of a particular table may find that the grouping in the table is not suitable and he needs to regroup the data. This can be a very difficult task, particularly when the numbers in certain groups considerably exceed those in neighbouring groups.

The best approach is to apply an interpolation method to the *cumulative total.*

Example 7.1.1. Births to unmarried mothers in Australia in 1968 totalled 18980. The ages of these mothers are shown in table 7.1.1, in broad age groups. Estimate the number of births to unmarried mothers in each of the quinquennial age groups 12-16, 17-21, 22-26, ..., 47-51.

We begin by calculating the cumulative totals as follows:
births to unmarried mothers under age 15 =   121,
$$\text{under age } 20 = \quad 7\,290,$$
$$\text{under age } 30 = 16\,147,$$
$$\text{under age } 40 = 18\,574,$$
$$\text{under age } 50 = 18\,980.$$
We now fit a smooth curve through the cumulative points $(15, 121)$, $(20, 7290)$, $(30, 16\,147)$, $(40, 18\,574)$ and $(50, 18\,980)$. Any reasonable interpolation method can be used although, in this situation, the graphical method seems to be the most popular. We shall use a spline function (section 4.3) and work backwards from age 50.

There are very few births to unmarried mothers near age 50; so the initial polynomial may be chosen passing through $(50, 18\,980)$ and $(40, 18\,574)$ with

Table 7.1.1. *Births to unmarried mothers, Australia, 1968*

Age last birthday of mother	Number of births
14 and under	121
15–19	7 169
20–29	8 857
30–39	2 427
40–49	406
50 and over	0
Total	18 980

zero first- and second-derivatives at age 50. From the resulting spline curve, we obtain the following estimates:

births to unmarried mothers under age 17 = 1 782,
under age 22 = 10 415,
under age 27 = 14 843,
under age 32 = 16 825,
under age 37 = 18 099,
under age 42 = 18 772,
under age 47 = 18 969,
under age 50 = 18 980.

By subtraction, we obtain the results shown in table 7.1.2. (Zero births are assumed under age 12.)

Table 7.1.2. *Births to unmarried mothers, Australia, 1968. Estimated numbers in quinquennial age groups*

Age last birthday of mother	Number of births
12–16	1 782
17–21	8 633
22–26	4 428
27–31	1 982
32–36	1 274
37–41	673
42–46	197
47–51	11
Total	18 980

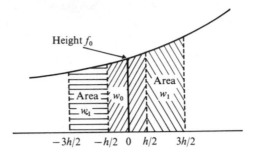

Fig. 7.2.1. The use of Hardy's formula to estimate the height of a curve.

7.2   The central ordinate of an area – Hardy's formula

The situation sometimes arises where the areas under a curve between consecutive equally-spaced ordinates are known and the height of the curve is required midway between two ordinates. The answer is provided by *Hardy's formula,* which may be regarded as a special method of numerical differentiation.

Consider the curve of an unknown function $f(x)$ over the range $-3h/2$ to $3h/2$. All we know about this function is that the areas under the curve and between $x = -3h/2$, $x = -h/2$, $x = h/2$ and $x = 3h/2$ are respectively w_{-1}, w_0 and w_1. These areas are depicted in fig. 7.2.1. An approximation to the height of the curve at the point $x = 0$ is given by[1]

■ $$f(0) = \frac{1}{h}\left(w_0 - \tfrac{1}{24}\Delta^2 w_{-1}\right). \tag{7.2.1}$$

This formula is exact when $f(x)$ is a polynomial of degree three.

Example 7.2.1. The figures in table 7.2.1 give the average weight at different points of time of groups of male rats on different dosages of monotertiary butyl hydroquinone. Calculate the daily rate of weight increase on day 135 for each dosage.

The weight of a rat at time x is the integral of the rate of increase in weight of the rat from conception up to time x. We need to calculate the rate of increase at time 135, midway between 128 days and 142 days. Hardy's formula (7.2.1) can be used.

The area under the rate-of-increase-in-weight curve between days 128 and 142 is equal to the change in weight over that period. For the control group therefore:

---

[1] The finite-difference operator Δ is defined in section 6.2. It is sufficient to note however that $\Delta^2 w_{-1} = w_1 - 2w_0 + w_{-1}$.

Table 7.2.1. *The effect of monotertiary butyl hydroquinone on the weights of male rats*

| Days on experiment | Average weight (g) | | | |
| | Dosage (%) | | | |
	0.02	0.1	0.5	Control
86	535	504	475	503
100	568	529	508	535
114	575	545	528	558
128	614	588	548	586
142	630	608	567	599
156	649	622	582	601
170	667	627	595	616

Source: personal communication from K. D. Cairncross, School of Biological Sciences, Macquarie University, North Ryde 2113, Australia.

$$w_{-1} = 586 - 558 = 28,$$
$$w_0 = 599 - 586 = 13,$$
$$w_1 = 601 - 599 = 2,$$

and the daily rate of increase at the mid-point (day 135) is

$$f(0) = \tfrac{1}{14}(13 - \tfrac{1}{24}(2 - 2 \times 13 + 28))$$
$$= 0.917 \text{ g/day.}$$

The equivalent figures for the 0.02 per cent, 0.1 per cent and 0.5 per cent dosages of monotertiary butyl hydroquinone are 1.065, 1.378 and 1.366 respectively.

Further reading: Tetley [96] 21-2.

7.3 The central ordinate of a sum of ordinates

Hardy's formula (7.2.1) gives the height of a curve in terms of areas under that curve. It is possible to derive an analogous formula for the middle ordinate of a group of ordinates added together.

Consider an unknown function $f(x)$. Let us denote the value of $f(x)$ at the point $x = a + rh$ by f_r $(r = \ldots, -2, -1, 0, 1, 2, \ldots)$. All we know about the function is that the sum of the n central ordinates is w_0, the sum of the preceding n ordinates is w_{-1} and the sum of the succeeding n ordinates is w_1. The integer n is odd. An approximation to the middle ordinate is given by[2]

---

[2] See footnote 1 in section 7.2.

$$\blacksquare \qquad f_0 = \frac{1}{n}\left(w_0 - \frac{n^2-1}{24n^2}\Delta^2 w_{-1}\right). \qquad (7.3.1)$$

This formula is exact when $f(x)$ is a polynomial of degree three.

Example 7.3.1. The smoothing effect of a running-average was noted in section 4.2. The simple running-average method, however, had a tendency to distort values already smooth. It is possible to get around the distortion problem, however. The sum of n consecutive observations will provide a non-distorted estimate of the sum of the n underlying smooth values, and (7.3.1) can then be used to estimate the central ordinate.

Let us use (7.3.1) with $n = 5$ to estimate smoothed ordinates at $x = 42$, 52 and 62 for the data in table 4.2.1.

At the point $x = 42$,

$$w_{-1} = 95 + 638 + 191 + 419 + 278 = 1621,$$
$$w_0 = 381 + 384 + 665 + 477 + 1015 = 2922,$$
$$w_1 = 1093 + 860 + 779 + 862 + 951 = 4545,$$

and

$$\Delta^2 w_{-1} = 4545 - 2 \times 2922 + 1621 = 322.$$

The smoothed value at $x = 42$ is therefore

$$\tfrac{1}{5}\left\{2922 - \frac{25-1}{24 \times 25}(322)\right\} = 581.82.$$

In a similar manner, the smoothed values at $x = 52$ and $x = 62$ are found to be 1203.78 and 2648.93 respectively.

The method of section 6.10 can be used to fit a quadratic through these three points, and the resulting curve is given in fig. 7.3.1. Over the range $x = 42$ to $x = 62$, the fit appears to be reasonable, but outside that range, it is not.

Formula (7.3.1) can also be used to obtain smoothed values at $x = 47$ and $x = 57$, and the spline method (section 4.3) can then be used to join all five points. The smoothed values at $x = 47$ and $x = 57$ are, however, particularly awkward ones to incorporate into a smooth curve.

Further reading: Tetley [96] 20–2; Whittaker and Robinson [100] 57–9.

7.4 The fusing together of two smooth curves

Suppose that a curve has been fitted to data over the range $x = 0$ to $x = s$ and a second curve has been fitted to data from $x = r$ onwards ($r < s$). The two curves overlap in the range $r \leqslant x \leqslant s$. The problem is to fuse the values that overlap in such a way that the final values pass smoothly from the first curve to the second.

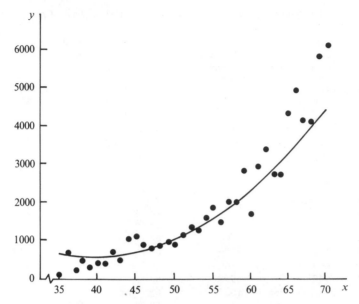

Fig. 7.3.1. A quadratic curve near 36 points.

Let us denote the first curve by $g(x)$, the second curve by $h(x)$ and the blending function by $\kappa(x)$. In the range $r \leqslant x \leqslant s$, we seek a function

■ $$f(x) = \kappa(x) g(x) + \{1 - \kappa(x)\} h(x), \qquad (7.4.1)$$

and for a smooth transition we must have

$$\kappa(r) = 1, \qquad \kappa'(r) = 0, \qquad \kappa''(r) \text{ small};$$
$$\kappa(s) = 0, \qquad \kappa'(s) = 0, \qquad \kappa''(s) \text{ small}.$$

It is usual, though not essential, to make the blending curve $\kappa(x)$ symmetric about $(r + s)/2$.

The so-called *curve of sines* is often used as a blending function:

■ $$\kappa(x) = \tfrac{1}{2}\left[1 + \cos\left\{\frac{(x - r)\pi}{s - r}\right\}\right]. \qquad (7.4.2)$$

Better results are sometimes obtained using the *curve of squares*:

■ $$\kappa(x) = \begin{cases} 1 - 2\left(\dfrac{x - r}{s - r}\right)^2, & r < x < \dfrac{r + s}{2}, \\ 2\left(\dfrac{x - s}{s - r}\right)^2, & \dfrac{r + s}{2} < x < s. \end{cases} \qquad (7.4.3)$$

Both (7.4.2) and (7.4.3) satisfy our requirements. The curve of squares has a slightly smaller second derivative at $x = r$ and $x = s$.

Table 7.4.1. *The blending of two quadratics by the curve of the sines*

x	First quad-ratic	Second quad-ratic	Blended curve	x	First quad-ratic	Second quad-ratic	Blended curve
35	281	–	281	53	1359	1134	1281
36	288	–	288	54	1477	1266	1372
37	302	–	302	55	1601	1420	1483
38	322	–	322	56	1732	1596	1624
39	349	–	349	57	1868	1793	1800
40	381	–	381	58	2011	2011	2011
41	419	–	419	59	–	2251	2251
42	464	–	464	60	–	2513	2513
43	515	–	515	61	–	2796	2796
44	572	–	572	62	–	3100	3100
45	635	–	635	63	–	3426	3426
46	704	–	704	64	–	3774	3774
47	779	–	779	65	–	4143	4143
48	860	–	860	66	–	4534	4534
49	948	–	948	67	–	4946	4946
50	1041	866	1037	68	–	5380	5380
51	1141	934	1121	69	–	5835	5835
52	1247	1023	1201	70	–	6312	6312

Example 7.4.1. Let us again return to the data in table 4.2.1
(fig. 4.2.1). A reasonably smooth curve might be obtained by fitting
a quadratic curve through the points (40, 381), (47, 779) and (58, 2011) and
another quadratic through (50, 866), (58, 2011) and (69, 5835), and blending
these two curves together by means of the curve of sines.

The quadratic curves in table 7.4.1 were obtained using the method of
section 6.10. The blended curve is also given in that table. It can be seen from
fig. 7.4.1 that the transition from the first quadratic to the second is not quite
as smooth as we would like. The pivotal points for the quadratics and the
amount of overlap of the two curves can be varied, and with a little skill and
patience, we should end up with a reasonably smooth curve.

Further reading: Tetley [96] 239–47.

7.5 Optimising a function of several variables – the method of steepest descent

The problem of minimising (or maximising) a function of several
variables often arises. In theory, all we need to do is equate the partial
derivatives[3] to zero and solve the resulting equations. But this can be very
tedious, and numerous numerical methods have been devised to solve the
problem. We now describe such a method – the method of steepest descent.

[3] Section 1.6.

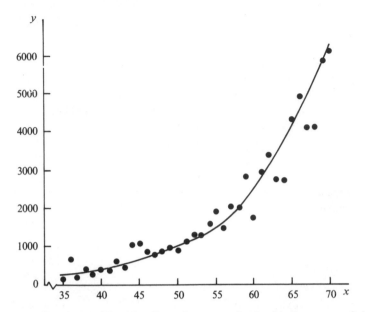

Fig. 7.4.1. The blending of two quadratics by the curve of sines.

Let us imagine that we wish to find the minimum point of the function $f(x, y, z)$ involving three variables x, y and z. We begin at an initial point (x_0, y_0, z_0) and move in the direction of steepest descent, which happens to be the direction with components proportional to the three partial derivatives at (x_0, y_0, z_0). In fact, we move to a point

■ $$(x_1, y_1, z_1) = \left(x_0 - k\frac{\partial f}{\partial x}, y_0 - k\frac{\partial f}{\partial y}, z_0 - k\frac{\partial f}{\partial z}\right). \tag{7.5.1}$$

The quantity k is chosen so that the point (x_1, y_1, z_1) gives the minimum value of the function for all points in that particular direction. The process is then repeated to obtain even better estimates of the minimum point: (x_2, y_2, z_2), (x_3, y_3, z_3), etc.

Numerical differentiation methods (chapter 6) may be used in preference to direct differentiation when the partial derivatives are complicated.

The search for the value of k which gives the minimum value of the function in a particular direction can be quite tedious, and the *quadratic minimum*[4] is normally used instead of the true minimum. To obtain the quadratic minimum, we calculate the function values K_0, K_1 and K_2 for three equally-spaced values of k (k_0, k_1 and k_2 say) near the minimum, preferably with k_0 on one side of the minimum and k_2 on the other. The value of k for the quadratic minimum is

_____

[4] Derived by assuming that $f(x_1, y_1, z_1)$ is a quadratic function of k.

$$k_{\min} = k_0 + (k_1 - k_0)\left(\frac{1}{2} - \frac{K_1 - K_0}{K_2 - 2K_1 + K_0}\right). \tag{7.5.2}$$

The steepest-descent method outlined above is straightforward to program. Convergence is sometimes slow. This may occur, for example, in the case of a long thin valley. If a starting point P_0 is chosen near one end of the valley, the process may zigzag from side to side across that end of the valley and move only slowly down the valley towards the actual minimum point. The optimisation process may be improved by the following modification.

1. Use the steepest-descent formulae (7.5.1) and (7.5.2) to obtain the point P_1 from the initial point P_0.
2. Use the steepest-descent formulae to obtain the even points P_2, P_4, P_6, etc., from the immediately preceding odd points (P_1, P_3, P_5, etc.).
3. Choose point P_3 by applying the quadratic-minimum formula along the line connecting P_0 and P_2. Thus

$$(x_3, y_3, z_3) = (x_2 + k(x_2 - x_0), y_2 + k(y_2 - y_0), z_2 + k(z_2 - z_0)). \tag{7.5.3}$$

4. Choose the odd points P_{2n+1} ($n > 1$) using the quadratic-minimum formula along the line connecting P_{2n} and P_{2n-3}. The appropriate formula is obtained from (7.5.3) by writing $2n + 1$ instead of 3, $2n$ instead of 2 and $2n - 3$ instead of 0.

The even points tend to zigzag across the valley and the odd points tend to shoot the process along the valley (fig.7.5.1).

Example 7.5.1. Find the minimum point of the function

$$f(x, y) = \exp(x^2 + y^2) + \exp\{(x - 1)^2\} + \exp\{(y - 2)^2\}.$$

The partial derivatives of this function are

$$\partial f / \partial x = 2x \exp(x^2 + y^2) + 2(x - 1) \exp\{(x - 1)^2\},$$
$$\partial f / \partial y = 2y \exp(x^2 + y^2) + 2(y - 2) \exp\{(y - 2)^2\}.$$

Let us use as our initial point

$$x_0 = 0.4, \qquad \partial f / \partial x = 0.831\,951\,3234, \qquad f = 7.341\,544\,519;$$
$$y_0 = 1.0, \qquad \partial f / \partial y = 0.943\,302\,8953.$$

If we try $k = 0.1$ and $k = 0.05$ in (7.5.1), we obtain the function values 7.417 637 066 and 7.322 035 734 respectively. Using the values at $k = 0$ (7.341 544 519), $k = 0.05$ (7.322 035 734) and $k = 0.1$ (7.417 637 066), we find $k_{\min} = 0.033\,473\,9662$. We therefore choose

$$x_1 = 0.372\,151\,2894, \qquad \partial f / \partial x = 0.321\,292\,9978, \qquad f = 7.315\,486\,331;$$
$$y_1 = 0.968\,423\,9106, \qquad \partial f / \partial y = -0.297\,214\,523.$$

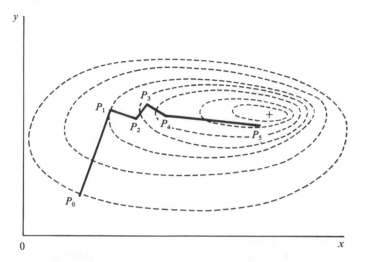

Fig. 7.5.1. The use of the modified steepest-descent method to obtain the minimum point of a function of two variables. The contours of the function are shown and the minimum point is indicated by a cross.

When we try $k = 0.1$ and $k = 0.05$, we obtain the function values 7.314 395 113 and 7.310 418 708 respectively. Using the function values at $k = 0$, $k = 0.05$ and $k = 0.1$, we find $k_{min} = 0.053\,016\,4063$, whence

$$x_2 = 0.355\,117\,4893, \qquad x_2 - x_0 = -0.044\,882\,5107,$$
$$y_2 = 0.984\,181\,1565, \qquad y_2 - y_0 = -0.015\,818\,8435,$$
$$f = 7.310\,401\,834.$$

We now use (7.5.3) to obtain P_3. When we try $k = 0.5$ and $k = 0.25$, we obtain the function values 7.310 290 573 and 7.309 080 967 respectively. Using these function values and the value at $k = 0$ (7.310 401 834), we obtain $k_{min} = 0.255\,496\,0575$, whence

$$x_3 = 0.343\,650\,1848, \qquad \partial f/\partial x = 0.001\,826\,8161, \qquad f = 7.309\,080\,443;$$
$$y_3 = 0.980\,139\,5044, \qquad \partial f/\partial y = -0.006\,188\,0223.$$

After a few more steps, we obtain the solution correct to five decimal places:

$$x = 0.343\,45, \qquad y = 0.980\,34, \qquad f = 7.309\,08.$$

The amount of work involved in this process is considerable, and it is best performed on a computer or at least on a sophisticated programmable calculator. Note that, for a minimisation process, k_{min} must be positive in formula (7.5.1), but it may have either sign in (7.5.3). The same formulae are used to maximise a function, and (7.5.2) will give the quadratic maximum point k_{max}, which must be negative in (7.5.1).

Further reading: Draper and Smith [19] 270–2; Scheid [89] 328–30.

7.6 A trick for increasing the data-storage capacity of a programmable calculator

Some of the programmable calculators and mini-computers currently available have quite adequate storage for program instructions, but rather limited data-storage facilities. The machine in the School of Biological Sciences at Macquarie University, for example, allows up to 512 program steps, but the data memory is limited to 51 registers. The memories, however, do allow the recording of numbers with large numbers of digits (twelve decimal digits and even more).

More often than not, the scientist only needs a small number of significant figures (less than six say), and it is by no means unreasonable for him to store two or more numbers in a single memory. The technique was well known to users of first-generation computers. Instructions are necessary to arrange the data for storage and to rearrange them later for computation. As a result, the method may not work on a machine with very limited program storage space.

Example 7.6.1. Consider a machine storing twelve significant digits plus sign together with a two-digit exponent and sign. In this machine, the number $-0.000\,123$, for example, is normally stored as $-1.230\,000\,000\,00-04$.

Let us imagine that all the numbers we need to store compactly lie between 300 and 900 and that six-digit accuracy is quite adequate for our purposes. We can store a pair of numbers 379.6548 and 436.0917 correct to six digits as one large integer: $3.796\,554\,360\,92+11$. This can be achieved using a simple subroutine which operates in the following manner:

take first number:	$3.796\,548\,000\,00+02$
multiply by 10^3:	$3.796\,548\,000\,00+05$
take nearest integer:	$3.796\,550\,000\,00+05$
multiply by 10^6:	$3.796\,550\,000\,00+11$ (A)
take second number:	$4.360\,917\,000\,00+02$
multiply by 10^3:	$4.360\,917\,000\,00+05$
take nearest integer:	$4.360\,920\,000\,00+05$ (B)
add (A) and (B):	$3.796\,554\,360\,92+11$

When we recall the number pair from memory, the two numbers are separated using a subroutine which operates in the following manner:

recall stored number	3.796 554 360 92 + 11 (a)
divide by 10^6	3.796 554 360 92 + 05
take integer part	3.796 550 000 00 + 05 (b)
multiply by 10^6	3.796 550 000 00 + 11 (c)
subtract (c) from (a)	4.360 920 000 00 + 05
divide by 10^3	4.360 920 000 00 + 02 (d)
divide (b) by 10^3	3.796 550 000 00 + 02 (e)

The first number is (e) and the second number is (d).

The process we have described is perhaps the simplest one. Greater sophistication can be achieved by suitable scaling, reorientation and the use of more complicated multiple-storage conventions.

7.7   Exercises

1. (*a*) Use (6.6.6) to find the rate of increase in weight on day 128 for the control group of rats in table 7.2.1.
   (*b*) Use (7.2.1) to find the rate of increase in weight on day 128 for the control group of rats in table 7.2.1.
   (*c*) Which answer do you believe to be the more reliable?

2. Find smoothed values at $x = 47$ and $x = 57$ for the data in table 4.2.1 using the method of example 7.3.1. Fit a spline curve through the smoothed points corresponding to $x = 42, 47, 52, 57$ and 62.

3. Use the curve of squares to blend the two quadratics in example 7.4.1.

4. Describe the procedures you would use to store and retrieve numbers, three per memory, in the machine described in example 7.6.1. Each number is known to lie in the range -90 to 360.

Part II
Basic statistical techniques

8

Probability, statistical distributions and moments

Summary   In this chapter we summarise the basic results of probability and
statistics needed in subsequent chapters. First we outline the axioms and
operating rules of probability; we continue with a discussion of univariate
statistical distributions, moments, and measures of central tendency,
dispersion, skewness and kurtosis. Bivariate distributions and moments are
described in the final two sections of the chapter.

8.1   The axioms and operating rules of probability

A gambler[1] tosses a coin ten times. This process may be referred to
as an *experiment*. At each toss, the gambler will obtain either a 'head' (H)
or a 'tail' (T). The particular sequence of 'heads' and 'tails' obtained by the
gambler is called the outcome of the experiment, and it is clear that the
experiment we have described has $2^{10} = 1024$ possible outcomes. The sequence
H H T H T T T T H H H is one possible outcome.

The gambler may count the number of 'heads' he obtains. This number is a
random variable which may take any one of the non-negative integer values
0, 1, 2, ..., 10.

Let us imagine that the gambler obtains a sequence containing six 'heads'
and four 'tails'. Then, we can say that the *event* 'six "heads" and four "tails"'
has occurred, or, for example, the event ' an even number of " heads "' has
occurred, or the event ' more " heads " than " tails "' has occurred.

Each event A has a number $P(A)$ associated with it which is non-negative
and less than or equal to one. This number is called the *probability* of the
event. The sum of the probabilities over all the possible mutually exclusive
events associated with an experiment is one. Thus, in our gambling example,
the probability of an even number of 'heads' plus the probability of an odd
number of 'heads' is one. If the event A has probability $P(A)$ of occurring
in a particular experiment and we were to repeat that experiment an infinite
number of times, a proportion $P(A)$ of the experiments will result in event A.

We now note the following axioms and operating rules of probability:

---

[1]   Statistical theory finds application in almost every area of human endeavour.
The basic concepts, however, are readily described in terms of games of chance.

- $0 \leqslant P(A) \leqslant 1,$ (8.1.1)
- $P(\text{not } A) = 1 - P(A),$ (8.1.2)
- $P(A_1 \text{ or } A_2 \text{ or } ... \text{ or } A_n) = P(A_1) + ... + P(A_n)$
  if $A_1, A_2, ..., A_n$ are mutually exclusive,   (8.1.3)
- $P(A_1 \text{ and } A_2 \text{ and } ... \text{ and } A_n) = P(A_1)P(A_2) ... P(A_n)$
  if $A_1, A_2, ..., A_n$ are mutually independent.   (8.1.4)

The *conditional probability* of event B given that event A has occurred is denoted by $P(B\,|\,A)$, and we note that

- $P(A \text{ and } B) = P(A)P(B\,|\,A).$ (8.1.5)

If A and B are independent,

- $P(B\,|\,A) = P(B).$ (8.1.6)

From (8.1.6), we see that (8.1.5) reduces to an example of (8.1.4) when A and B are independent.

Example 8.1.1. A gambler makes five independent throws of an unbiased coin. What is the probability that he obtains exactly two 'heads'?

The following ten sequences are the only outcomes resulting in exactly two 'heads':

$$\text{HHTTT} \quad \text{HTHTT} \quad \text{HTTHT} \quad \text{HTTTH} \quad \text{THHTT}$$
$$\text{THTHT} \quad \text{THTTH} \quad \text{TTHHT} \quad \text{TTHTH} \quad \text{TTTHH}$$

The probability of obtaining a 'head' and the probability of obtaining a 'tail' from a particular throw are both equal to $\frac{1}{2}$. The five throws of the coin are independent. According to (8.1.4), therefore, the probability of obtaining H at the first throw, H at the second throw, T at the third throw, T at the fourth throw and T at the fifth throw is $(\frac{1}{2})^5 = \frac{1}{32}$. The same probability is obtained for each of the other nine sequences. Each one of these ten sequences results in exactly two 'heads', and the sequences are mutually exclusive. It follows from (8.1.3) that the required probability is the sum of these ten equal probabilities, namely $\frac{10}{32}$.

Example 8.1.2. A gambler makes five independent throws of an unbiased coin. We shall see later[2] in section 10.1 that

$$P(0 \text{ 'heads'}) = \tfrac{1}{32}, \qquad P(3 \text{ 'heads'}) = \tfrac{10}{32},$$
$$P(1 \text{ 'head'}) = \tfrac{5}{32}, \qquad P(4 \text{ 'heads'}) = \tfrac{5}{32},$$
$$P(2 \text{ 'heads'}) = \tfrac{10}{32}, \qquad P(5 \text{ 'heads'}) = \tfrac{1}{32}.$$

Note that each of these probabilities satisfies inequality (8.1.1). Furthermore, the six events are mutually exclusive and exhaustive, and their probabilities

---

[2] See also example 8.1.1 above.

sum to one. What is the probability that the gambler obtains at least two 'heads'?

The events 'two "heads"', 'three "heads"', 'four "heads"' and 'five "heads"' are mutually exclusive, and so by (8.1.3), the probability of at least two 'heads' is

$$\tfrac{10}{32} + \tfrac{10}{32} + \tfrac{5}{32} + \tfrac{1}{32} = \tfrac{26}{32}.$$

Alternatively, we note that the probability of obtaining at least two 'heads' is the same as the probability of not obtaining zero or one 'head'. According to (8.1.3), the probability of obtaining zero or one 'head' is $\tfrac{1}{32} + \tfrac{5}{32} = \tfrac{6}{32}$. So, by (8.1.2), the probability of at least two heads is

$$1 - \tfrac{6}{32} = \tfrac{26}{32}.$$

Example 8.1.3. Find the probability that the gambler in example 8.1.2 obtains an even number of 'heads' given that he obtains at least two 'heads'

Let us denote the event 'at least two "heads"' by A and the event 'an even number of "heads"' by B. We need to find $P(B|A)$, and according to (8.1.5),

$$P(B|A) = P(A \text{ and } B)/P(A).$$

From example 8.1.2, we know that $P(A) = \tfrac{26}{32}$. For events A and B both to be true, the gambler must obtain either two or four 'heads'. The events 'two "heads"' and 'four "heads"' are mutually exclusive, and so by (8.1.3) we add the probabilities to obtain the probability of either event. Thus,

$$\begin{aligned} P(A \text{ and } B) &= P(\text{two 'heads' or four 'heads'}) \\ &= \tfrac{10}{32} + \tfrac{5}{32} \\ &= \tfrac{15}{32}. \end{aligned}$$

Finally,

$$P(B|A) = \tfrac{15}{32} / \tfrac{26}{32} = \tfrac{15}{26}.$$

Further reading: Balaam [3] 41–8; Brownlee [7] 1–11; Fraser [26] 4–29; Hoel [45] 5, 33–52; Hoel [46] 4–15; Mood and Graybill [70] 6–52; Spiegel [93] 99–121; Wilks [101] 1–26; Wilks [102] 58–92.

8.2 Discrete and continuous distributions

In many experimental situations, the events of interest are mutually exclusive and may be represented by the non-negative integers 0, 1, 2, ... (for example, the number of 'heads' in a coin-tossing experiment). Let us imagine that the probability of event j is p_j ($j = 0, 1, 2, ...$). We can define a random variable X which assumes the value j when event j occurs. Then

■ $$P(X = r) = p_r, \qquad (8.2.1)$$

■ $$P(X \leqslant r) = \sum_{j=0}^{r} p_j, \qquad (8.2.2)$$

■ $$P(r < X \leqslant s) = \sum_{j=r+1}^{s} p_j, \qquad (8.2.3)$$

■ $$\sum_{j=0}^{\infty} p_j = 1. \qquad (8.2.4)$$

The probabilities $\{p_j\}$ define a *probability distribution* on the non-negative integers. The distribution is *discrete* because the random variable can only assume certain discrete values (the non-negative integers). The sum (8.2.2) is sometimes denoted by $F(r)$ and referred to as the *distribution function*.

Discrete distributions on the non-negative integers are perhaps the most common, but many other types of discrete distribution exist; for example, distributions over the positive integers and distributions over all the integers (positive and negative). The common discrete distributions are described in chapter 10.

Some random variables like height and weight are essentially continuous and have *continuous distributions*. For such a random variable X, we define a cumulative *distribution function $F(x)$* such that

■ $$P(X \leqslant x) = F(x). \qquad (8.2.5)$$

Clearly $F(x)$ cannot decrease as x increases, and $F(\infty) = 1$. When $F(x)$ is differentiable,

■ $$P(x < X \leqslant x + \mathrm{d}x) = f(x)\,\mathrm{d}x, \qquad (8.2.6)$$
where
■ $$f(x) = \frac{\mathrm{d}}{\mathrm{d}x} F(x), \qquad (8.2.7)$$
and
■ $$F(x) = \int_{-\infty}^{x} f(x)\,\mathrm{d}x. \qquad (8.2.8)$$

The function $f(x)$ is called the *probability-density function* of the distribution, and it is always non-negative. We also note that

■ $$P(a < X \leqslant b) = \int_{a}^{b} f(x)\,\mathrm{d}x = F(b) - F(a). \qquad (8.2.9)$$

In the integral (8.2.8) the lower limit is $-\infty$. A random variable like height is normally restricted to positive values, and for such a random variable $f(x)$ and $F(x)$ must both be zero when x is negative.

Some of the more important continuous distributions are described in chapter 9.

Example 8.2.1. A coin is tossed until the first 'head' appears. If X denotes the number of tosses required, X is a discrete random variable over the positive integers.

Example 8.2.2. A biologist wishes to compare the concentrations of two types of bacteria A and B. He counts the number of bacteria of type A on a plate and the number of bacteria of type B on the plate. Both these random variables have distributions over the non-negative integers. He then computes the ratio of these two random variables. The result is a random variable with a discrete distribution over the non-negative rational points (that is, those non-negative points which may be expressed as the ratio of two integers).

Further reading: Balaam [3] 55; Fraser [26] 38–40, 59–64; Hoel [45] 73–85; Hoel [46] 26–46, 52–89; Mood and Graybill [70] 53–102; Wilks [101] 30–48; Wilks [102] 98–121.

8.3 Mean and variance

The *mean* or *expected value* of a random variable X distributed over the non-negative integers is defined by

$$\mu = \sum_{j=0}^{\infty} j p_j, \tag{8.3.1}$$

and for a continuous distribution over the range $(-\infty, \infty)$ with probability-density function $f(x)$,

$$\mu = \int_{-\infty}^{\infty} x f(x)\, dx. \tag{8.3.2}$$

The mean is a measure of location (Balaam [3] 21–2; Spiegel [93] 45–68).

The *variance* of a distribution provides a measure of the spread of the distribution (Balaam [3] 23; Spiegel [93] 69–88). In the case of a discrete distribution over the non-negative integers, the variance is

$$\sigma^2 = \sum_{j=0}^{\infty} (j-\mu)^2 p_j, \tag{8.3.3}$$

and for a continuous distribution over the range $(-\infty, \infty)$,

$$\sigma^2 = \int_{-\infty}^{\infty} (x-\mu)^2 f(x)\, dx. \tag{8.3.4}$$

When all the likely outcomes are concentrated near the mean, the variance σ^2 will be small. The following alternative (but mathematically equivalent) formulae for variance are often useful:

$$\sigma^2 = \sum_{j=0}^{\infty} j^2 p_j - \mu^2 \text{ (discrete)}, \tag{8.3.5}$$

$$\sigma^2 = \int_{-\infty}^{\infty} x^2 f(x)\, dx - \mu^2 \text{ (continuous)}. \tag{8.3.6}$$

The square root of the variance is called the *standard deviation* and it is denoted by σ.

Further reading: Brownlee [7] 34–42; Chakravarti *et al.* [9] 100–2; Colquhoun [16] 24–32; Hoel [45] 16–20, 80–3; Mood and Graybill [70] 103–21; Pollard [81] 32–54; Spiegel [93] 45–88; Zar [105] 19–21, 31–5.

8.4   The sample mean and variance

In most experimental situations, we do not know the mean and variance of our underlying distribution. We need to estimate them from our experimental data, and the sample mean and variance are usually used for this purpose (chapter 13). For a sample $x_1, x_2, ..., x_n$, of size n, the *sample mean* $\bar{x}$ and the *sample variance* s^2 are calculated as follows:

■ $$\bar{x} = \frac{1}{n} \sum_{j=1}^{\infty} x_j, \tag{8.4.1}$$

■ $$m_2 = \frac{1}{n} \sum_{j=1}^{n} x_j^2 - \bar{x}^2, \tag{8.4.2}$$

■ $$s^2 = \{n/(n-1)\} \, m_2. \tag{8.4.3}$$

The quantity m_2 is a sample second-order moment (section 8.6), and the similarity of (8.3.5) and (8.4.2) should be noted. On average, m_2 slightly underestimates σ^2, and the adjustment $n/(n-1)$ in s^2 corrects for this. The *sample standard deviation s* is the square root of the sample variance.

Data from discrete distributions and grouped data from continuous distributions may be given in frequency form. If the observed frequency of the value x is f_x, the calculation formulae become

■ $$\bar{x} = \frac{1}{n} \sum_{\text{all } x} x f_x, \tag{8.4.4}$$

■ $$m_2 = \frac{1}{n} \sum_{\text{all } x} x^2 f_x - \bar{x}^2, \tag{8.4.5}$$

where

■ $$n = \sum_{\text{all } x} f_x \tag{8.4.6}$$

is the sample size.

Further reading: Balaam [3] 21–4; Mood and Graybill [70] 144–7; Wilks [102] 34–57.

8.5   Median and mode

The *median* value is that value of the variate which divides the total distribution into two halves, and like the mean it is a measure of location. In the case of a continuous distribution over the range $(-\infty, \infty)$, the median is the value m such that

■ $\int_{-\infty}^{m} f(x)\, dx = \int_{m}^{\infty} f(x)\, dx = \frac{1}{2}.$ (8.5.1)

With a discrete distribution on the non-negative integers, certain conventions are necessary. We take as the median the integer m such that

■ $\sum_{j=0}^{m-1} p_j < \frac{1}{2}$ and $\sum_{j=0}^{m} p_j > \frac{1}{2}.$ (8.5.2)

We occasionally find that an integer M exists such that

■ $\sum_{j=0}^{M} p_j = \frac{1}{2}.$ (8.5.3)

We then choose $m = M + \frac{1}{2}$. In the case of a symmetric distribution, the mean and median coincide.

A *sample median* also exists, and again certain conventions are necessary. In a sample of size $2N + 1$, the median is the $(N + 1)$th largest observation. If a sample is of size $2N$, the median is chosen half-way between the Nth and $(N + 1)$th largest observations.

The *mode* is that value of the variate with the greatest probability. In the case of a continuous distribution with a density function, it corresponds to the highest point of the probability-density function. The common distributions of statistics are unimodal; i.e. they each have a single maximum point.

It is worth observing that in the case of a unimodal distribution, the mean, median and mode always occur in the same order as in the dictionary (or in the reverse order), and, like the words in the dictionary, the median is closer to the mean than to the mode!

Further reading: Balaam [3] 21–2; Chakravarti *et al.* [9] 100–1; Kendall and Stuart [53] 38–40; Pollard [81] 1–55; Spiegel [93] 45–88; Zar [105] 19–25.

8.6 Higher-order moments, skewness and kurtosis

The *r*th *moment about the origin* for a discrete distribution over the non-negative integers is defined by

■ $\mu'_r = \sum_{j=0}^{\infty} j^r p_j,$ (8.6.1)

and for a continuous distribution over the range $(-\infty, \infty)$, the *r*th-order moment is

■ $\mu'_r = \int_{-\infty}^{\infty} x^r f(x)\, dx.$ (8.6.2)

These moments are sometimes referred to as *non-central moments*. It will be noted that the means (8.3.1) and (8.3.2) are actually first-order moments about the origin, and therefore

■ $\mu \equiv \mu'_1.$ (8.6.3)

Furthermore, the variances (8.3.5) and (8.3.6) are obtained by subtracting the square of the mean from the second-order non-central moment.

Moments about the mean are also defined:

■ $\mu_r = \sum_{j=0}^{\infty} (j-\mu)^r \, p_j$ (discrete),   (8.6.4)

■ $\mu_r = \int_{-\infty}^{\infty} (x-\mu)^r \, f(x) \, dx$ (continuous).   (8.6.5)

These moments are also referred to as *central moments*. We see that the variances (8.3.3) and (8.3.4) are actually second-order central moments, and

■ $\sigma^2 \equiv \mu_2.$ (8.6.6)

Moments above fourth order are seldom used, but the third- and fourth-order central moments do find some applications (for example, for fitting Pearson-type curves (chapter 11)). The odd central moments are all zero for a symmetric distribution; they are positive for a skew distribution with a long tail to the right, and negative for a skew distribution with a long tail to the left. The moment function

■ $\sqrt{\beta_1} = \mu_3 / (\mu_2)^{\frac{3}{2}}$ (8.6.7)

is therefore commonly used as a measure of *skewness*. The even central moments must always be positive, and an indication of *kurtosis* (or peakedness) is provided by

■ $\beta_2 = \mu_4 / \mu_2^2.$ (8.6.8)

For the normal distribution (section 9.1) $\sqrt{\beta_1} = 0$ and $\beta_2 = 3$. Distributions with higher values of β_2 are usually more peaked, and those with lower β_2 values are usually flatter than the normal curve.

The following formulae connect the central and non-central moments:

■ $\sigma^2 = \mu_2 = \mu_2' - (\mu_1')^2,$ (8.6.9)

■ $\mu_3 = \mu_3' - 3\mu_2'\mu_1' + 2(\mu_1')^3,$ (8.6.10)

■ $\mu_4 = \mu_4' - 4\mu_3'\mu_1' + 6\mu_2'(\mu_1')^2 - 3(\mu_1')^4;$ (8.6.11)

and it is clear that (8.3.5) and (8.3.6) are particular examples of these formulae.

For a sample $x_1, x_2, \ldots, x_n$ of size n, the sample rth-order non-central moment m_r' is calculated by means of the formula

■ $m_r' = \frac{1}{n} \sum_{i=1}^{n} x_i^r.$ (8.6.12)

Data from discrete distributions and grouped data from continuous distributions may be given in frequency form. If the observed frequency of the value x is f_x, the sample moment calculation formulae become

■ $$m'_r = \frac{1}{n} \sum_{\text{all } x} x^r f_x,$$ (8.6.13)

where

■ $$n = \sum_{\text{all } x} f_x$$ (8.6.14)

is the sample size (example 8.6.1).

To calculate the central moments of a sample, we merely use formulae (8.6.9), (8.6.10) and (8.6.11) with each μ replaced by an m. It should be noted that

■ $$\bar{x} \equiv m'_1,$$ (8.6.15)

■ $$s^2 \equiv \{n/(n-1)\} \, m_2.$$ (8.6.16)

Adjustments are sometimes made for the effects of grouping (section 11.3). The sample measures of skewness and kurtosis are

■ $$\sqrt{b_1} = m_3/(m_2)^{\frac{3}{2}},$$ (8.6.17)

■ $$b_2 = m_4/m_2^2.$$ (8.6.18)

Example 8.6.1. Some grouped frequency data are given in the first two columns of table 11.2.1. The skewness measure $\sqrt{b_1}$ is calculated in the following manner:

$m'_1 = (17 \times 34 + 22 \times 145 + 27 \times 156 + ...)/1000 = 37.875,$
$m'_2 = (17^2 \times 34 + 22^2 \times 145 + 27^2 \times 156 + ...)/1000 = 1626.075,$
$m'_3 = (17^3 \times 34 + 22^3 \times 145 + 27^3 \times 156 + ...)/1000 = 77\,986.575,$ $(8.6.9)$
$m_2 = 1626.075 - (37.875)^2 = 191.559\,375,$ $(8.6.10)$
$m_3 = 77\,986.575 - 3 \times 1626.075 \times 37.875 + 2 \times (37.875)^3$

$\quad = 1888.361\,72,$
$\sqrt{b_1} = 1888.361\,72/(191.559\,375)^{\frac{3}{2}} = 0.712\,246\,085.$ $(8.6.17)$

a tail to right $(8.6.7)$

Further reading: Chakravarti *et al.* [9] 105–10; Kendall and Stuart [53] 85–6; Mather [66] 24–5; Spiegel [93] 89–98; Wilks [101] 72–86.

8.7   Bivariate distributions

Consider two discrete random variables X and Y taking non-negative integral values (for example, the ages last birthday of husband and wife of a random Australian couple). The probability

$$P(X = i \quad and \quad Y = j)$$

may be denoted by p_{ij}, and the values $\{p_{ij}\}$ ($i, j = 0, 1, 2, ...$) define a bivariate distribution. The sum over all i and all j of p_{ij} is of course one. The following results should be noted:

- $$P(X = i \quad and \quad Y = j) = p_{ij}, \tag{8.7.1}$$

- $$P(X = i) = \sum_{j=0}^{\infty} p_{ij}, \tag{8.7.2}$$

- $$P(X = i \mid Y = j) = p_{ij} \bigg/ \left(\sum_{k=0}^{\infty} p_{kj} \right). \tag{8.7.3}$$

The expected value of X given that $Y = j$ is obtained by multiplying (8.7.3) by i and summing for all values of i. When X and Y are *independent*,

- $$p_{ij} = P(X = i)P(Y = j). \tag{8.7.4}$$

For continuous random variables X and Y distributed over the range $(-\infty, \infty)$, we define a *joint probability-density function* $f(x, y)$ such that

- $$P(x < X \leqslant x + dx \quad and \quad y < Y \leqslant y + dy) = f(x, y)\, dx\, dy. \tag{8.7.5}$$

The double integral of $f(x, y)$ over the complete range of x and y is one, and we note that

- $$P(x < X \leqslant x + dx) = \left(\int_{-\infty}^{\infty} f(x, y)\, dy \right) dx, \tag{8.7.6}$$
- $$P(x < X \leqslant x + dx \mid y < Y \leqslant y + dy)$$
  $$= (f(x, y)\, dx) / \left(\int_{-\infty}^{\infty} f(x, y)\, dx \right). \tag{8.7.7}$$

The expected value of X given that Y lies between y and $y + dy$ is obtained by multiplying (8.7.7) by x and integrating over the complete range of x. When X and Y are *independent*

- $$f(x, y)\, dx\, dy = P(x < X \leqslant x + dx) P(y < Y \leqslant y + dy). \tag{8.7.8}$$

The particularly important bivariate normal distribution is described in section 9.6.

Example 8.7.1. Table 8.7.1 gives the relative ages of husbands and wives in a hypothetical population. The figures in parentheses give the proportions of couples in the various cells (for example, $0.193\,79 = 21\,643/111\,682$). A couple is chosen at random from the population. Let us denote the ages of husband and wife by H and W respectively. Then

(i) $P(H \text{ is } 45+ \quad and \quad W \text{ is } 15\text{–}29) = 0.008\,74,$
(ii) $P(W \text{ is } 15\text{–}29) = 0.250\,02,$
(iii) $P(H \text{ is } 45+) = 0.363\,75,$
(iv) $P(W \text{ is } 15\text{–}29 \mid H \text{ is } 45+) = 0.008\,74/0.363\,75 = 0.024\,03,$
(v) $P(H \text{ is } 45+ \mid W \text{ is } 15\text{–}29) = 0.008\,74/0.250\,02 = 0.034\,96.$

Further reading: Brownlee [7] 45–53; Fraser [26] 41, 72–5.

Table 8.7.1. *The relative ages of husband and wife in a hypothetical population*

Age of wife (years)	Age of husband (years)			
	15–29	30–44	45+	Total
15–29	21 643	5 304	976	27 923
	(0.193 79)	(0.047 49)	(0.008 74)	(0.250 02)
30–44	3 905	35 219	6 742	45 866
	(0.034 97)	(0.315 35)	(0.060 37)	(0.410 69)
45+	391	4 596	32 906	37 893
	(0.003 50)	(0.041 15)	(0.294 64)	(0.339 29)
Total	25 939	45 119	40 624	111 682
	(0.232 26)	(0.403 99)	(0.363 75)	(1.000 00)

8.8 Covariance and correlation

Moments of univariate distributions were described in sections 8.3 and 8.6. Bivariate and multivariate moments also exist; we shall confine our attention to a particular bivariate moment, the covariance.

Consider two discrete random variables X and Y distributed on the non-negative integers. Let the mean values of X and Y be μ_x and μ_y respectively. The *covariance* of X and Y is defined as

$$\text{cov}(X, Y) = \sum_{i=0}^{\infty} \sum_{j=0}^{\infty} (i - \mu_x)(j - \mu_y) p_{ij}. \qquad (8.8.1)$$

If the values assumed by X tend to be large when Y is large, and conversely, the covariance is positive. This would occur, for example, with data like those in table 8.7.1. If large values of X tend to be associated with small values of Y, and conversely, the covariance will be negative. The covariance of two independent random variables is zero, but it is *not* true that random variables with zero covariance are necessarily independent.[3] Zero covariance merely implies the lack of any *linear* connection between two random variables.

The following alternative (but mathematically identical) expression for the covariance is often of greater practical use:

$$\text{cov}(X, Y) = \sum_{i=0}^{\infty} \sum_{j=0}^{\infty} ij p_{ij} - \mu_x \mu_y. \qquad (8.8.2)$$

---

[3] A special result holds in the case of the bivariate normal distribution (section 9.6).

Similar definitions are adopted for continuous random variables:

■ $$\text{cov}(X, Y) = \int_{-\infty}^{\infty} \int_{-\infty}^{\infty} (x - \mu_x)(y - \mu_y) f(x,y) \, dx \, dy, \qquad (8.8.3)$$

or

■ $$\text{cov}(X, Y) = \int_{-\infty}^{\infty} \int_{-\infty}^{\infty} xyf(x,y) \, dx \, dy - \mu_x \mu_y. \qquad (8.8.4)$$

It should be noted that $\text{cov}(X, Y) = \text{cov}(Y, X)$, and the covariance of a random variable X with itself is equal to $\text{var}(X)$.

A *covariance matrix* displays the various covariances of a set of random variables. The three random variables X, Y and Z, for example, have a covariance matrix of the form

$$\begin{pmatrix} \text{var}(X) & \text{cov}(X,Y) & \text{cov}(X,Z) \\ \text{cov}(Y,X) & \text{var}(Y) & \text{cov}(Y,Z) \\ \text{cov}(Z,X) & \text{cov}(Z,Y) & \text{var}(Z) \end{pmatrix}.$$

The diagonal elements are always positive, and the matrix is symmetric.

The *correlation coefficient* of X and Y is usually denoted by ρ, and it is obtained by dividing the covariance of X and Y by the product of their standard deviations. Thus

$$\rho = \text{cov}(X, Y)/(\sigma_X \sigma_Y). \qquad (8.8.5)$$

This coefficient must lie between -1 and 1, and it measures the degree of linearity in the connection between X and Y. If X and Y are completely linearly related (for example, $Y = 10X + 3$), the absolute value of ρ is one. If large values of X are associated with large values of Y, the correlation will be positive; if large values of X are associated with small values of Y, the correlation is negative. Zero correlation does *not* imply independence,[4] but rather the lack of any *linear* connection between X and Y (Campbell [8] 271; Zar [105] 237).

For a sample $(x_1, y_1), ..., (x_n, y_n)$ of size n, we calculate a *sample product moment* m_{11} as follows:

$$m_{11} = \frac{1}{n} \sum_{i=1}^{n} x_i y_i - \bar{x}\bar{y}. \qquad (8.8.6)$$

The similarity between this formula and (8.8.2) should be noted. On average, m_{11} slightly underestimates the population covariance. The bias is removed by multiplying m_{11} by $n/(n-1)$ and defining the *sample covariance* as follows:

■ $$\text{sample covariance} = \{n/(n-1)\} m_{11}. \qquad (8.8.7)$$

Sometimes bivariate data are given in frequency form (example 8.8.2), and

---

[4] See footnote 3.

the outcome (x, y) is observed $f_{x,y}$ times. The appropriate formula for calculating the sample product moment m_{11} is then

■
$$m_{11} = \frac{1}{n} \sum_x \sum_y xy f_{x,y} - \overline{x}\overline{y},$$
(8.8.8)

where

■
$$n = \sum_x \sum_y f_{x,y}$$
(8.8.9)

is the total frequency.

To compute the *sample correlation coefficient r*, we divide the sample covariance by the product of the sample standard deviations. Thus,

$$r = (\text{sample covariance})/(s_X s_Y).$$
(8.8.10)

This coefficient must lie between -1 and 1. The following equivalent formula is usually more convenient for computational purposes:

■
$$r = \frac{m_{11}}{\sqrt{(m_2(X) m_2(Y))}};$$
(8.8.11)

$m_2(X)$ denotes the second central moment of X (sections 8.4 and 8.6).

Example 8.8.1. An unbiased six-sided die is tossed, and X denotes the numerical value of the outcome. If the outcome is odd, Y is set equal zero; otherwise, it assumes the value one. Calculate the mean and variance of X and Y, their covariance and their correlation coefficient.

The bivariate distribution is shown in table 8.8.1. We calculate

$$\mu_X = 1 \times \tfrac{1}{6} + 2 \times \tfrac{1}{6} + \dots + 6 \times \tfrac{1}{6} = 3.5,$$
$$\mu_Y = 0 \times \tfrac{1}{2} + 1 \times \tfrac{1}{2} = 0.5,$$
$$\sigma_X^2 = (1^2 \times \tfrac{1}{6} + 2^2 \times \tfrac{1}{6} + \dots) - (3.5)^2 = 2.91\dot{6},$$
$$\sigma_Y^2 = (0^2 \times \tfrac{1}{2} + 1^2 \times \tfrac{1}{2}) - (0.5)^2 = 0.25.$$

Furthermore, using (8.8.2),

$$\begin{aligned}
\text{cov}(X, Y) = {}& (1 \times 0) \times \tfrac{1}{6} + (1 \times 1) \times 0 \\
& + (2 \times 0) \times 0 + (2 \times 1) \times \tfrac{1}{6} \\
& + (3 \times 0) \times \tfrac{1}{6} + (3 \times 1) \times 0 \\
& + (4 \times 0) \times 0 + (4 \times 1) \times \tfrac{1}{6} \\
& + (5 \times 0) \times \tfrac{1}{6} + (5 \times 1) \times 0 \\
& + (6 \times 0) \times 0 + (6 \times 1) \times \tfrac{1}{6} - 3.5 \times 0.5 \\
= {}& 0.25.
\end{aligned}$$

It follows from (8.8.5) that

$$\rho = 0.25/(2.91\dot{6} \times 0.25)^{\frac{1}{2}} = 0.292\,77.$$

Table 8.8.1. *A bivariate distribution (example 8.8.1)*

y	x 1	2	3	4	5	6	Total
0	$\frac{1}{6}$	0	$\frac{1}{6}$	0	$\frac{1}{6}$	0	$\frac{1}{2}$
1	0	$\frac{1}{6}$	0	$\frac{1}{6}$	0	$\frac{1}{6}$	$\frac{1}{2}$
Total	$\frac{1}{6}$	$\frac{1}{6}$	$\frac{1}{6}$	$\frac{1}{6}$	$\frac{1}{6}$	$\frac{1}{6}$	1

Example 8.8.2. In a certain bivariate experiment, the following (X, Y) outcomes are possible: $(1,1)$, $(1,2)$, $(1,3)$, $(2,1)$, $(2,2)$ and $(2,3)$. Fifty-seven observations are made. The first outcome occurs 12 times, the second 6 times, the third 5 times, the fourth 15 times, the fifth 11 times, and the sixth 8 times. These results are shown in table 8.8.2. Calculate the sample correlation coefficient.

The sample means $\bar{x}$ and $\bar{y}$ are obtained by applying formula (8.4.4) to the marginal totals for x and y respectively:

$$\bar{x} = (1 \times 23 + 2 \times 34)/57 = 1.5965,$$
$$\bar{y} = (1 \times 27 + 2 \times 17 + 3 \times 13)/57 = 1.7544.$$

The same marginal totals are used to obtain the second-order central moments $m_2(X)$ and $m_2(Y)$ (formula (8.4.5)):

$$m_2(X) = (1^2 \times 23 + 2^2 \times 34)/57 - \bar{x}^2 = 0.240\,69,$$
$$m_2(Y) = (1^2 \times 27 + 2^2 \times 17 + 3^2 \times 13)/57 - \bar{y}^2 = 0.641\,43.$$

Formula (8.8.8) is used to obtain the sample product moment:

$$
\begin{aligned}
m_{11} = \{&(1 \times 1) \times 12 + (2 \times 1) \times 15 \\
+ &(1 \times 2) \times 6 + (2 \times 2) \times 11 \\
+ &(1 \times 3) \times 5 + (2 \times 3) \times 8\}/57 - \bar{x}\bar{y} \\
= &\,0.023\,700.
\end{aligned}
$$

Table 8.8.2. *The results of a bivariate experiment. The entries in the table give the observed frequencies of the various (X, Y) outcomes*

x	y 1	2	3	Total
1	12	6	5	23
2	15	11	8	34
Total	27	17	13	57

The sample correlation coefficient is then calculated using (8.8.11):

$$r = 0.023\,700/(0.240\,69 \times 0.641\,43)^{\frac{1}{2}} = 0.0603.$$

Further reading: Brownlee [7] 55–7; Campbell [8] 268–72; Hoel [45] 187–92; Pollard [81] 132–44.

8.9   Exercises

1. A friend tosses an unbiased six-sided die, and tells us that the outcome is an even number. What is the chance that the outcome is a 'six'?

2. A gambler makes five independent throws of an unbiased coin. What is the probability that he obtains at least two 'heads' given that the number of 'heads' is even. *Hint*: refer to examples 8.1.2 and 8.1.3.

3. A die is biased and we are told that $P(\text{one}) = \frac{1}{21}$, $P(\text{two}) = \frac{2}{21}$, $P(\text{three}) = \frac{3}{21}$, $P(\text{four}) = \frac{4}{21}$, $P(\text{five}) = \frac{5}{21}$ and $P(\text{six}) = \frac{6}{21}$. The die is tossed and we are told that the outcome is not less than three. What is the chance that the outcome is greater than four?

4. The score associated with side j of an unbiased six-sided die is j. A gambler tosses such a die. Calculate the mean, variance, skewness and kurtosis of the random variable representing his score.

5. A bivariate distribution is shown in table 8.8.1. Find the conditional mean value of X given that $Y = 0$. Find also the conditional mean value of Y given that $X = 3$.

9
The normal and related distributions

Summary   The normal, chi-square, *t*- and *F*-distributions all play prominent roles in statistical theory. The importance of the normal distribution is a direct result of the central limit theorem, and the other distributions are defined in terms of the normal. In this chapter, we summarise the important properties of each of these distributions. We also describe the log-normal distribution and the multivariate normal distribution.

9.1   The normal distribution

Random errors or fluctuations in the results of scientific experiments frequently follow a distribution which approximates the normal (or Gaussian) form. There are good reasons to expect this to be so. The *central limit theorem*[1] states that the distribution function of a random variable which is the sum of n independent identically-distributed random variables with means μ and variances σ^2 approaches the normal distribution function with mean $n\mu$ and variance $n\sigma^2$ as n becomes large. A particular variable which we measure is often the result of combining a large number of variables which we cannot or do not measure; the deviation of an observation from the mean value or expected value is thus a sum of a number of deviations, some positive and some negative, and often of comparable magnitude.

The normal distribution is a continuous distribution completely defined by its mean μ and variance σ^2. The distribution is over the range $-\infty$ to $+\infty$, and the probability density function takes the form

$$\blacksquare \qquad \frac{1}{\sigma\sqrt{(2\pi)}} \exp\left\{-\tfrac{1}{2}(x-\mu)^2/\sigma^2\right\}. \qquad\qquad (9.1.1)$$

The probability that a normal random variable X with mean μ and variance σ^2 lies between a and b is obtained by integrating (9.1.1) over the interval (a, b). The integral does not have an explicit mathematical form, and tables are therefore necessary. At first sight, it would appear that tables are needed for all possible values and combinations of values of μ and σ^2. This is not so, however, because the standardised normal random variable $Y = (X-\mu)/\sigma$ has

---

[1]   This form of the central limit theorem is probably the simplest; there are many more complicated versions.

zero mean and unit variance, and a probability-density function

■ $\phi(y) = \dfrac{1}{\sqrt{(2\pi)}} \exp\left(-\tfrac{1}{2}y^2\right),$ (9.1.2)

which requires only one basic table of probabilities. It follows that

■ $P(a < X \leqslant b) = \Phi\left(\dfrac{b-\mu}{\sigma}\right) - \Phi\left(\dfrac{a-\mu}{\sigma}\right),$ (9.1.3)

where

■ $\Phi(x) = \int_{-\infty}^{x} \phi(y)\, dy.$ (9.1.4)

A table of the unit normal distribution is given on pp. 321–4, and a graph of the unit normal probability-density curve appears in fig. 9.2.2. The $\phi(x)$ curve is symmetric about $x = 0$, and values of $\Phi(x)$ for negative x can be found by subtracting $\Phi(|x|)$ from one.

     Example 9.1.1. Adult males in a certain country are known to have heights which are approximately normally distributed with mean 175.6 cm and standard deviation 7.63 cm. An adult male will be chosen at random. What is the probability that his height lies between 175 and 185 cm?
     We calculate

$$\Phi((185 - 175.6)/7.63) - \Phi((175 - 175.6)/7.63).$$

Look up
$p = -0.079$
Then
$1 - \phi$

That is, $\Phi(1.232) - \Phi(-0.079) = 0.891\,02 - 0.468\,52 = 0.422\,50.$ The probability is approximately 0.423.

> *Further reading*: Balaam [3] 78–84; Colquhoun [16] 64–74; Fraser [26] 68–71; Johnson and Kotz [48] 40–94; Johnson and Leone [50] 100–8; Mather [65] 25–30; Mather [66] 10–19; Mood and Graybill [70] 123–5; Pollard [81] 68–75; Remington and Schork [86] 124–47; Wilks [101] 156–7; Wilks [102] 144–59; Zar [105] 70–84.

9.2   The chi-square distribution

     Let us consider a set of mutually independent normal random variables $\{X_j\}$ ($j = 1, 2, \dots, \nu$), each with zero mean and variance one. We shall define another random variable

$$Y = \sum_{i=1}^{\nu} X_i^2.$$

Then Y has the chi-square distribution with ν degrees of freedom.
     From this definition, we see that a chi-square random variable must always be positive. It is also clear that if Y_1 has the chi-square distribution with ν_1 degrees of freedom and Y_2 has the chi-square distribution with ν_2 degrees of freedom, and Y_1 and Y_2 are independent, $Y = Y_1 + Y_2$ has the chi-square distribution with $\nu = \nu_1 + \nu_2$ degrees of freedom.

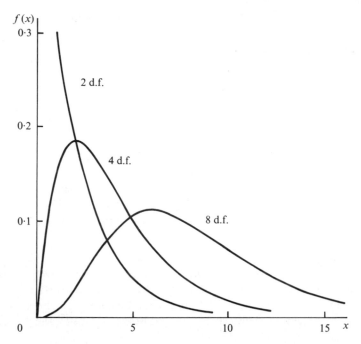

Fig. 9.2.1. The probability-density functions of the chi-square distributions with 2, 4 and 8 degrees of freedom. The χ_2^2 density function meets the y-axis at the point $(0, 0.5)$. The χ_1^2 density function approaches the y-axis asymptotically.

It is possible to prove that the chi-square distribution with ν degrees of freedom has probability-density function

$$f(x) = \frac{1}{2^{\frac{1}{2}\nu}\,\Gamma(\frac{1}{2}\nu)} x^{(\frac{1}{2}\nu)-1} e^{-\frac{1}{2}x} \quad (0 \leqslant x < \infty). \tag{9.2.1}$$

The gamma function $\Gamma(n)$ is defined in footnote 1 of section 11.2. Graphs of the chi-square density function for certain degrees of freedom are given in fig. 9.2.1 and chi-square tables are given on p. 325. The entries in this table are values of x such that

$$P(\chi_\nu^2 > x) = \alpha,$$

where ν (the number of degrees of freedom) is specified in the row heading, and α (the probability, or area to the right of x) is specified in the column heading.

The expected value of a chi-square random variable with ν degrees of freedom is ν and the variance is 2ν. The distribution is not symmetric. It is worth noting, however, that if Y has the chi-square distribution with ν degrees of freedom

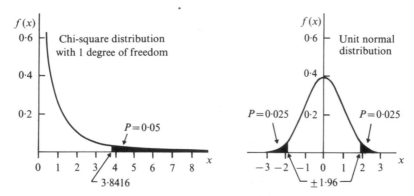

Fig. 9.2.2. The connection between the χ_1^2 and unit normal distributions. The shaded area under the chi-square curve is equal to the sum of the two shaded areas under the normal curve and 3.8416 is the square of 1.96.

and v is large (greater than 30, say), $Z = \sqrt{(2Y)} - \sqrt{(2v-1)}$ is approximately a unit normal variate, and normal tables can be used.

The chi-square distribution becomes important when we study the sample variance s^2 from a normal population. It is also used in an accurate approximate test of association in contingency tables.

Example 9.2.1. This example demonstrates the connection between the unit normal and χ_1^2 distributions. According to the above definition, a χ_1^2 random variable is merely the square of a unit normal variable. It follows that

$$P(\chi_1^2 \text{ variable} > a^2) = P(\text{unit normal variable} < -a)$$
$$+ P(\text{unit normal variable} > a)$$
$$= 2P(\text{unit normal variable} > a).$$

When $a = 1.96$, the right-hand side is equal to 0.05 and $a^2 = 3.8416$. From the chi-square table we confirm that the left-hand side is also equal to 0.05. This situation is depicted in fig. 9.2.2.

Further reading: Balaam [3] 90–2, 111; Brownlee [7] 82; Fraser [26] 190–3; Hoel [45] 234–45; Hoel [46] 252–6; Johnson and Kotz [48] 166–99; Johnson and Leone [50] 118; Kendall and Stuart [53] 369–74; Mood and Graybill [70] 226–7; Parzen [75] 14.

9.3   The *t*-distribution

Let X_1 be a unit normal random variable and X_2 an independent chi-square random variable with v degrees of freedom. Then

$$t = \frac{\sqrt{v}\, X_1}{\sqrt{X_2}}$$

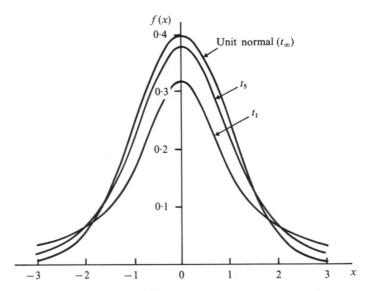

Fig. 9.3.1. The probability-density functions of the t distributions with one, five and infinity degrees of freedom. The latter is the same as the unit normal density function.

is defined as a random variable having the t-distribution with ν degrees of freedom.

It is possible to prove that the t_ν-distribution has probability-density function

$$\blacksquare \qquad f(x) = \frac{\Gamma(\tfrac{1}{2}\nu + \tfrac{1}{2})}{\nu^{\frac{1}{2}}\,\Gamma(\tfrac{1}{2})\,\Gamma(\tfrac{1}{2}\nu)}\left(1 + \frac{x^2}{\nu}\right)^{-\frac{1}{2}(\nu+1)} \qquad (-\infty < x < \infty). \qquad (9.3.1)$$

The gamma function $\Gamma(n)$ is defined in footnote 1 of section 11.2. Graphs of the t_ν probability-density function for various degrees of freedom are given in fig. 9.3.1 and t-tables are to be found on p. 327.

The expected value of a t_ν random variable is zero and the variance is $\nu/(\nu-2)$. The distribution is symmetric.

The t-distribution becomes important when we study sample means and we do not know the population variance. In this situation, we need to make use of the sample variance, and we then encounter the t-distribution.

For large values of ν, the variance of t is approximately one; the mean is still zero. In fact, for large values of ν (> 30, say) the distribution of t_ν is effectively normal with zero mean and unit variance (X_2/ν is close to one with a high probability, and t is then almost identical to X_1).

Example 9.3.1. Find the probability that a t-variable with ten degrees of freedom lies between 1.5 and 2.0.

According to our t-tables,

$$P(1.372 < t_{10} < 1.812) = 0.05;$$
$$P(1.812 < t_{10} < 2.228) = 0.025.$$

The required probability is approximately

$$0.05 \times (1.812 - 1.5)/(1.812 - 1.372)$$
$$+ 0.025 \times (2.0 - 1.812)/(2.228 - 1.812) = 0.047.$$

Further reading: Balaam [3] 121–44; Colquhoun [16] 75–7; Hoel [45] 146–8; Hoel [46] 257–60; Johnson and Kotz [49] 94–124; Johnson and Leone [50] 119; Kendall and Stuart [53] 374–7; Mood and Graybill [70] 233.

9.4 The F-distribution

Let X_1 be a chi-square random variable with m degrees of freedom and X_2 an independent chi-square random variable with n degrees of freedom. Then

$$F_{m,n} = \frac{(X_1/m)}{(X_2/n)}$$

is defined as an F-variate with m and n degrees of freedom. The random variable must of course be positive.

It is possible to prove that the $F_{m,n}$-distribution has probability-density function

$$\blacksquare \qquad f(x) = \frac{\Gamma(\tfrac{1}{2}m + \tfrac{1}{2}n)}{\Gamma(\tfrac{1}{2}m)\,\Gamma(\tfrac{1}{2}n)} \left(\frac{m}{n}\right)^{\tfrac{1}{2}m} \frac{x^{(\tfrac{1}{2}m)-1}}{\left(1 + \dfrac{m}{n}x\right)^{\tfrac{1}{2}(m+n)}} \qquad (0 \leqslant x < \infty).$$

(9.4.1)

The gamma function $\Gamma(n)$ is defined in footnote 1 of section 11.2. The probability-density function of the $F_{10,\,12}$-distribution is shown in fig. 9.4.1, and F-tables may be found on pp. 328–31.

The expected value of an $F_{m,n}$ random variable is $n/(n-2)$ and the variance is

$$\frac{2n^2\,(m+n-2)}{m(n-2)^2\,(n-4)} \qquad (n > 4).$$

The distribution is not symmetric.

When one considers the limit as $n \to \infty$, the expected value of $F_{m,n}$ becomes one and the variance becomes $2/m$. These are the moments of a χ_m^2/m random variable, and in fact $mF_{m,\infty}$ is a chi-square variable with m degrees of freedom.

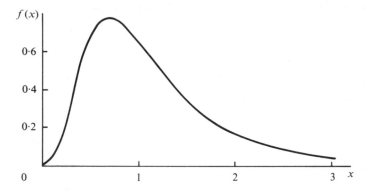

Fig. 9.4.1. The probability-density function of the $F_{10,12}$ distribution.

The F-distribution becomes important when we wish to compare sample variances from normal populations. It also finds extensive use in regression analysis and the analysis of variance.

Example 9.4.1. This example demonstrates the connection between the χ^2_m- and $F_{m,\infty}$-distributions. We have seen that

$$P\left(F_{m,\infty} > \frac{a}{m}\right) = P(\chi^2_m > a).$$

Let us select $m = 10$ and $a = 18.31$. The right-hand side is equal to 0.05, and we can soon verify that the left-hand side is also equal to 0.05.

Example 9.4.2. This example demonstrates the connection between the t_n- and $F_{1,n}$-distributions. According to the definition of t_n (section 9.3), t_n is the ratio of a unit normal random variable to the square root of an independent χ^2_n/n random variable. If we square a t_n random variable, therefore, we have the ratio of a $\chi^2_1/1$ random variable to an independent χ^2_n/n random variable, and this ratio is an $F_{1,n}$ random variable. Thus,

$$P(F_{1,n} > a^2) = P(t_n > a) + P(t_n < -a).$$

Let us select $n = 10$ and $a = 2.23$ so that $a^2 = 4.97$. The right-hand side is equal to 0.05, and this is also the value of the left-hand side.

Example 9.4.3. The tables in the Appendix only give the right-hand tail points of an F-distribution, and the distribution is not symmetric. In this important example, we demonstrate how the left-hand tail points can be obtained from the tables. Let us find x such that $P(F_{7,9} < x)$ is 0.05.

From the definition of $F_{m, n}$, we see that

$$P(F_{7,9} < x) = P\left(F_{9,7} > \frac{1}{x}\right).$$

According to our tables, $P(F_{9,7} > 3.68) = 0.05$, and it is soon apparent that the value we require is $x = 1/3.68 = 0.272$. We deduce the following rule for finding the lower 100α per cent point of an $F_{m,n}$-distribution: take the reciprocal of the upper 100α per cent point of the $F_{n,m}$-distribution.

Further reading: Balaam [3] 145–50; Fraser [26] 202–13; Hoel [45] 269–70; Johnson and Kotz [49] 75–89; Johnson and Leone [50] 123; Kendall and Stuart [53] 377–82; Mood and Graybill [70] 231–2; Parzen [75] 14.

9.5 The log-normal distribution

Let X be a random variable. If there is a number θ such that the random variable $Z = \ln(X - \theta)$ is normally distributed, X is said to have the log-normal distribution. Since logarithms of positive numbers only are defined, X must always be greater than θ. The distribution of X involves three parameters: θ, μ and σ^2, the last two being the mean and variance of the related normal distribution. The probability-density function takes the form

$$\blacksquare \qquad f(x) = \frac{1}{(x - \theta)\,\sigma\sqrt{(2\pi)}} \exp[-\tfrac{1}{2}\{\ln(x - \theta) - \mu\}^2/\sigma^2], \qquad (9.5.1)$$

and we note that the mean and variance are given by

$$\blacksquare \qquad \text{mean} = \theta + \exp(\mu + \tfrac{1}{2}\sigma^2), \qquad (9.5.2)$$

$$\blacksquare \qquad \text{variance} = e^{2\mu}\,e^{\sigma^2}(e^{\sigma^2} - 1). \qquad (9.5.3)$$

The distribution is unimodal and positively skew. When σ^2 is small, the shape of the distribution is close to normal. The probability-density function of the log-normal distribution with parameters $\theta = 3$, $\mu = 1.5$ and $\sigma^2 = 0.36$ is shown in fig. 9.5.1.

In many applications, θ is known to be zero and X then has the two-parameter log-normal distribution. This two-parameter distribution is in at least one respect a more realistic representation of distributions of quantities like weight, height, etc., which must be positive, than is the normal distribution, which ascribes positive probability to negative values.

Parameter estimation methods are given in section 13.25.

Example 9.5.1. Find the lower $2\tfrac{1}{2}$ per cent point of the log-normal distribution with parameters θ, μ and σ^2.

We need the find x such that $P(X < x) = 0.025$, or

$$P\left(\frac{\ln(X - \theta) - \mu}{\sigma} < \frac{\ln(x - \theta) - \mu}{\sigma}\right) = 0.025.$$

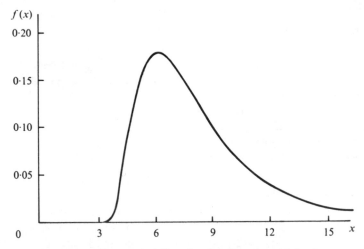

Fig. 9.5.1. The probability-density function of the log-normal distribution with parameters $\theta = 3$, $\mu = 1.5$ and $\sigma^2 = 0.36$. The mean is 8.366 and the variance is 12.475.

The random variable $\{\ln (X - \theta) - \mu\}/\sigma$ has the unit normal distribution with lower $2\frac{1}{2}$ per cent point -1.96. We therefore equate $\{\ln (x - \theta) - \mu\}/\sigma$ to -1.96 and obtain

$$x = \theta + \exp(\mu - 1.96\sigma).$$

Further reading: Chakravarti *et al.* [9] 153; Colquhoun [16] 78–9; Johnson and Kotz [48] 112–36; Williams [103] 5–16.

9.6 The bivariate and multivariate normal distributions

Two random variables X_1 and X_2 are said to have the bivariate normal distribution if all linear combinations of X_1 and X_2 of the form $a_1 X_1 + a_2 X_2$ are normal. The joint probability-density function of X_1 and X_2 takes the form

$$f(x_1, x_2) = \frac{1}{2\pi\sigma_1\sigma_2(1 - \rho^2)^{\frac{1}{2}}} \exp\left[\frac{-1}{2(1 - \rho^2)}\left\{\left(\frac{x_1 - \mu_1}{\sigma_1}\right)^2\right.\right.$$
$$\left.\left. - 2\rho\left(\frac{x_1 - \mu_1}{\sigma_1}\right)\left(\frac{x_2 - \mu_2}{\sigma_2}\right) + \left(\frac{x_2 - \mu_2}{\sigma_2}\right)^2\right\}\right]$$

$$(9.6.1)$$

The parameters μ_1 and μ_2 are the means of X_1 and X_2, σ_1^2 and σ_2^2 are the variances, and ρ is the correlation coefficient. The covariance of X_1 and X_2 is $\rho\sigma_1\sigma_2$. When the correlation coefficient ρ is zero, (9.6.1) is the product of two univariate normal probability-density functions, and it follows that two normal random variables are independent if they are uncorrelated.

The expected value of X_2 given X_1 is a linear function of X_1. Denoting this expectation by $E(X_2|X_1)$,

■ $$E(X_2|X_1) = \mu_2 + \rho(\sigma_2/\sigma_1)(X_1 - \mu_1). \tag{9.6.2}$$

The covariance matrix of X_1 and X_2 displays the covariance of X_1 with itself, the covariance of X_1 and X_2, the covariance of X_2 and X_1, and the covariance of X_2 with itself. Let us denote this matrix by $\mathbf{V}$. Then

$$\mathbf{V} = \begin{pmatrix} \sigma_1^2 & \rho\sigma_1\sigma_2 \\ \rho\sigma_1\sigma_2 & \sigma_2^2 \end{pmatrix} \text{ and the determinant } |\mathbf{V}| = \sigma_1^2\sigma_2^2(1-\rho^2).$$

The inverse matrix $\mathbf{V}^{-1}$ can be written down readily using the method of cofactors (section 1.10).

If we list the $\{X_i\}$, $\{x_i\}$ and $\{\mu_i\}$ in column vectors $\mathbf{X}$, $\mathbf{x}$ and μ respectively, the probability-density function of $\mathbf{X}$, given above as (9.6.1), may be written

■ $$f(\mathbf{x}) = \frac{1}{(2\pi)^{\frac{1}{2}n}|\mathbf{V}|^{\frac{1}{2}}} \exp\left\{-\tfrac{1}{2}(\mathbf{x}-\mu)'\mathbf{V}^{-1}(\mathbf{x}-\mu)\right\} \tag{9.6.3}$$

with $n = 2$. Formula (9.6.3) in fact gives the joint density of the n-variate normal distribution. The covariance matrix $\mathbf{V}$ is then $n \times n$ and the column vectors $\mathbf{X}$, $\mathbf{x}$ and μ are of dimension n.

Further reading: Brownlee [7] 397–416; Fraser [26] 75–7; Johnson and Leone [50] 129–32; Mood and Graybill [70] 198–218; Wilks [101] 158–69.

9.7   Exercises

1. Adult males in a certain country are known to have heights which are approximately normally distributed with mean 175.6 cm and standard deviation 7.63 cm. Three adult males are chosen at random. What is the probability that all three are taller than 185 cm?

2. Adult males in a certain country are known to have heights which are approximately normally distributed with mean 175.6 cm. The variance is unknown. We do know however that 10 per cent of adult males are taller than 187 cm. Calculate the variance in adult male height.

3. What is the probability that a χ^2_{10} random variable is greater than 20?

4. What is the probability that a χ^2_{80} random variable is greater than 100?

5. Find the probability that a t_{10} random variable lies between -1 and $+2$.

6. Find the upper and lower $2\frac{1}{2}$ per cent points of the $F_{19,\ 59}$-distribution.

IO

The common discrete distributions

Summary In this chapter, we summarise the common discrete distributions (binomial, Poisson, geometric, negative binomial, logarithmic and hypergeometric). We also describe certain relationships between the distributions and some useful approximations.

10.1   The binomial distribution

When a certain coin is tossed, the outcome is a 'head' with probability p and a 'tail' with probability $q = 1 - p$. The probability of obtaining r 'heads' and $n - r$ 'tails' in n independent tosses of the coin is

■ $$P(r \text{ 'heads' from } n \text{ tosses}) = \binom{n}{r} p^r q^{n-r}. \tag{10.1.1}$$

This probability defines the *binomial distribution*. A particular case of (10.1.1) was proved in example 8.1.1, and the general proof is similar. The combinatorial notation is explained in section 1.5. The binomial distribution has

■ $$\text{mean} = np, \tag{10.1.2}$$

■ $$\text{variance} = npq. \tag{10.1.3}$$

Fig. 10.1.1 depicts the binomial distribution with parameters $n = 10$ and $p = 0.3$.

The calculation of individual binomial probabilities and cumulative sums of probabilities tends to be rather tedious. Extensive tables have been prepared by the National Bureau of Standards [71], the U.S. Army Material Command [98] and Weintraub [99]. The reader's attention is also drawn to the normal approximation of section 10.2 and the F-method of section 10.3.

Parameter estimation procedures are given in section 13.5.

Example 10.1.1. Six cows are allowed to roam at random in a large field with uniform vegetation. The cows have no inclination to congregate, and each cow is equally likely to be at any point in the field. If the area of the field is A and the cows act completely independently of one another, find the probability that at least four of the cows are inside a quadrat of area a within the field.

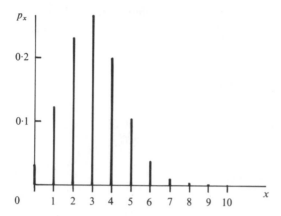

Fig. 10.1.1. The binomial distribution with parameters $n = 10$ and $p = 0.3$.

The probability that a particular cow is inside the quadrat is $p = a/A$. The probability that four or more cows are in the quadrat is

$$\binom{6}{4} p^4 q^2 + \binom{6}{5} p^5 q^1 + \binom{6}{6} p^6 q^0.$$

Example 10.1.2. In a certain country, the heights of adult males are known to be approximately normally distributed with mean 175.6 cm and standard deviation 7.63 cm. Six adult males are chosen at random. What is the probability that at least four have heights between 175 and 185 cm?

From example 9.1.1, we know that the height of a random adult male lies between 175 and 185 cm with probability 0.423. The chance that at least four males out of six have heights between 175 and 185 cm is

$$\binom{6}{4} (0.423)^4 (0.577)^2 + \binom{6}{5} (0.423)^5 (0.577)^1$$

$$+ \binom{6}{6} (0.423)^6 (0.577)^0 = 0.212.$$

Further reading: Bailey [2] 6–20; Balaam [3] 63–9; Brownlee [7] 30–1; Campbell [8] 295–9; Feller [21] 135–41; Fraser [26] 42–4; Hoel [45] 91–9; Johnson and Kotz [47] 50–86; Kendall and Stuart [53] 120–5; Mather [66] 26–31; Mood and Graybill [70] 64–9; Pollard [81] 84–94; Remington and Schork [86] 112–18; Wilks [102] 122–32.

10.2   The normal approximation to the binomial

If a coin is tossed n times, the probability of obtaining a number of 'heads' between r_1 and r_2 (inclusive) is given by

$$\sum_{r=r_1}^{r_2} \binom{n}{r} p^r q^{n-r}. \tag{10.2.1}$$

This sum is tedious to evaluate for large n. When n is greater than about twenty, however, the central limit theorem[1] allows us to approximate the sum in (10.2.1) by

$$■ \quad \int_{\alpha_1}^{\alpha_2} \frac{1}{\sqrt{(2\pi)}} \exp(-\tfrac{1}{2}x^2) \, dx, \qquad (10.2.2)$$

where

$$■ \quad \alpha_1 = (r_1 - np - \tfrac{1}{2})/(npq)^{\frac{1}{2}}, \qquad (10.2.3)$$

$$■ \quad \alpha_2 = (r_2 - np + \tfrac{1}{2})/(npq)^{\frac{1}{2}}. \qquad (10.2.4)$$

The area implied by (10.2.2) is readily available from normal tables.

The corrections of $\pm\tfrac{1}{2}$ in α_1 and α_2 are often omitted, particularly when n is very large. The rationale behind the corrections is evident in fig. 10.2.1: we require the sum of the binomial ordinates; this sum is equal to the polygon area, which is approximated by the area under the normal curve between α_1 and α_2.

Let us use the random variable X to denote the number of 'heads' obtained in n tosses. When we use (10.2.2), we are in effect saying that $(X-np)/\sqrt{(npq)}$ is approximately a unit normal variable. It follows from section 9.2 that $(X-np)^2/(npq)$ is approximately a χ_1^2 random variable. We now note that $(X-np)^2$, which represents the square of the difference between the observed number of 'heads' and the expected number of 'heads', is also equal to the square of the difference between the observed number of 'tails' and the expected number of 'tails'. Furthermore

$$1/(npq) \equiv 1/(np) + 1/(nq).$$

It follows that

$$■ \quad \frac{(\text{observed 'heads'} - \text{expected 'heads'})^2}{\text{expected 'heads'}}$$

$$+ \frac{(\text{observed 'tails'} - \text{expected 'tails'})^2}{\text{expected 'tails'}} \qquad (10.2.5)$$

is a chi-square random variable with one degree of freedom.

---

[1] See section 9.1. The outcome of a binomial experiment involving n trials is the sum of the outcomes in each of the n independent trials. The central limit approximation for the binomial was one of the earliest forms of the central limit theorem.

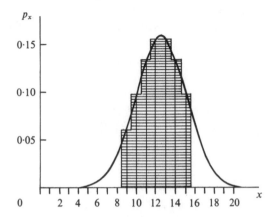

Fig. 10.2.1. The normal approximation to the binomial in the case $n = 25, p = 0.5$. The probability that the random variable lies between 9 and 15 inclusive is equal to the sum of the ordinates from 9 to 15 inclusive. This sum is equal to the shaded histogram area, which can be approximated by the area under the normal curve between 8.5 and 15.5.

Example 10.2.1. An unbiased coin is tossed 1000 times. What is the chance that the number of 'heads' obtained will deviate from the expected number of 500 by more than 31?

Using the symmetry of the normal curve and the normal approximation to the binomial without the $\pm \frac{1}{2}$ adjustments, the required probability is

$$2 \left\{ 1 - \Phi \left(\frac{531 - 500}{\sqrt{(250)}} \right) \right\} = 0.05.$$

Alternatively, we can calculate

$$\chi_1^2 = \frac{31^2}{500} + \frac{31^2}{500} = 3.84,$$

and according to our table, the chance of obtaining a χ_1^2 value this size or larger is 0.05.

Note that by its very definition, the chi-square method gives the probability of a deviation of more than 31 in either direction. When the normal curve is used to obtain the answer to this question, both tails of the distribution must be included.

Further reading: Brownlee [7] 136–9; Feller [21] 164–79; Hoel [45] 105–11; Johnson and Kotz [47] 61–70; Kendall and Stuart [53] 106–7; Pollard [81] 84–94.

10.3   The critical points of the binomial distribution

Let X be a binomial random variable with parameters n and p. The upper 100α per cent critical value of the binomial (n, p) distribution is an integer r such that

$$P(X \geqslant r) \leqslant \alpha < P(X \geqslant r-1). \tag{10.3.1}$$

Binomial tables can be used to determine such critical values, and approximate critical points can be calculated using the normal approximation of section 10.2. It is worth noting, however, that

■ $$P(X \geqslant r) = P\left(Y \geqslant \frac{r}{n+1-r} \frac{1-p}{p}\right), \tag{10.3.2}$$

where Y is an F-variate with $2(n+1-r)$ and $2r$ degrees of freedom (Brownlee [7] 148). The following procedure can therefore be used to find the upper 100α per cent point of the binomial distribution.

1. Compute r_0, the normal approximation to the 100α per cent point, and round r_0 to the nearest integer.
2. Equate $\{r/(n+1-r)\}\{(1-p)/p\}$ to the upper 100α per cent point of the F-distribution with $2(n+1-r_0)$ and $2r_0$ degrees of freedom.
3. Solve for r and round *up* to the next integer.

If r differs from the previous estimate r_0 by more than one, steps 2 and 3 should be repeated using r instead of r_0.

To obtain the lower 100α per cent critical point, we note that

■ $$P(X \leqslant r) = P\left(Z \geqslant \frac{n-r}{r+1} \frac{p}{1-p}\right), \tag{10.3.3}$$

where Z is an F-variate with $2(r+1)$ and $2(n-r)$ degrees of freedom. The following procedure can therefore be used to find the lower 100α per cent point of the binomial distribution.

1. Compute r_0, the normal approximation to the 100α per cent point, and round r_0 to the nearest integer.
2. Equate $\{(n-r)/(r+1)\}\{p/(1-p)\}$ to the upper 100α per cent point of the F-distribution with $2(r_0+1)$ and $2(n-r_0)$ degrees of freedom.
3. Solve for r and round *down* to the next integer.

If r differs from the previous estimate r_0 by more than one, repeat steps 2 and 3 using r instead of r_0.

Example 10.3.1. Find the upper 5 per cent point of the binomial $(10, 0.3)$ distribution.

This distribution has mean 3 and variance 2.1. The normal approximation to the upper 5 per cent point is therefore

$$r_0 = 3 + 1.65 \times \sqrt{(2.1)} \doteqdot 5.$$

When we solve the equation

$$\frac{r}{11-r}\frac{0.7}{0.3} = 2.91$$

and round up, we obtain the value $r = 7$. This differs from r_0 by more than one; so we repeat the process with $r_0 = 7$. We then obtain the critical value $r = 6$, which may be confirmed by direct calculation or by reference to fig. 10.1.1.

Further reading: Brownlee [7] 148–50; Johnson and Kotz [47] 58–9.

10.4   The multinomial distribution

The multinomial distribution is a natural multivariate generalisation of the binomial distribution.[2] We shall describe it in terms of tossing a k-sided die. If the chance of obtaining side i at a single toss is p_i ($i = 1$, $2, ..., k$) and n independent tosses are made, the chance of obtaining n_1 of the first side, n_2 of the second side, etc. is

■ $$P(n_1, n_2, ..., n_k) = \frac{n!}{n_1!\, n_2! \,...\, n_k!}\, p_1{}^{n_1} p_2{}^{n_2} \,...\, p_k{}^{n_k}, \qquad (10.4.1)$$

where

■ $$\sum_{i=1}^{k} p_i = 1, \qquad\qquad (10.4.2)$$

■ $$\sum_{i=1}^{k} n_i = n. \qquad\qquad (10.4.3)$$

At each toss the outcome is either side i with probability p_i or not side i with probability $1-p_i$ and this is a binomial situation. It follows that the number of side i outcomes has

■   mean $= np_i,$ $(10.4.4)$

■   variance $= np_i(1-p_i).$ $(10.4.5)$

In the binomial situation, the number of 'heads' and the number of 'tails' are linearly related (they sum to n). It follows that their correlation coefficient is minus one and because the variance of each is npq, their covariance must be $-npq$. In the multinomial situation, the covariance of the number of side i outcomes X_i and the number of side j outcomes X_j is given by

■   cov $(X_i, X_j) = -np_i p_j.$ $(10.4.6)$

---

[2] Even the binomial distribution can be thought of as a bivariate distribution, but the number of 'tails' is fully determined once we know the number of 'heads'.

The correlation coefficent is

■ $$\rho_{ij} = \left\{ \frac{p_i p_j}{(1-p_i)(1-p_j)} \right\}^{\frac{1}{2}}. \tag{10.4.7}$$

Parameter estimation methods are given in section 13.7.

> **Example 10.4.1.** Twenty-one unbiased dice are rolled. Find the probability of obtaining the following result:
>
> | | |
> |---|---|
> | exactly 1 one, | exactly 4 fours, |
> | exactly 2 twos, | exactly 5 fives, |
> | exactly 3 threes, | exactly 6 sixes. |

The probability of obtaining this result is

$$\frac{21!}{1!\,2!\,3!\,4!\,5!\,6!} \, (\tfrac{1}{6})^{21} = 0.000\,094.$$

> *Further reading*: Brownlee [7] 206–20; Feller [21] 157; Fraser [26] 48–52; Hoel [46] 225; Johnson and Kotz [47] 281–5; Johnson and Leone [50] 86; Kendall and Stuart [53] 141; Mood and Graybill [70] 69; Wilks [101] 138–40.

10.5 The chi-square approximation to the multinomial

The chi-square (or normal) approximation to the binomial distribution was discussed in section 10.2. We now describe a more general result applicable to the multinomial distribution.

Let us define X_i to be the number of side i outcomes in n tosses of a k-sided die. As a generalisation of (10.2.5), it may be shown that

■ $$\sum_{i=1}^{k} (X_i - np_i)^2/(np_i) \tag{10.5.1}$$

is approximately a χ^2_{k-1} random variable. For the approximation to be accurate, the expectations $\{np_i\}$ must not be too small. Many authors recommend a minimum expectation of about five, although Cochran [13] suggests a minimum expectation as low as one. Formula (10.5.1) may be written less formally as

■ $$\sum_{\text{all cells}} (\text{observed} - \text{expected})^2/\text{expected}. \tag{10.5.2}$$

The reduction from k to $(k-1)$ in the number of degrees of freedom is caused by the constraint whereby the total number of observations and the total expected observations are equal. A large chi-square value is obtained when some of the cells contain values markedly different from their expected values.

The results of this section become important when we study contingency tables in section 12.9.

Further reading: Brownlee [7] 207–10; Hoel [45] 225–9; Johnson and Kotz [47] 285–8; Lancaster [57] 36–8.

10.6   The Poisson distribution

The Poisson distribution is a non-negative integer-valued distribution which plays a prominent role in statistical theory. Traditional examples of the use of the Poisson distribution include the number of alpha particles emitted from a radioactive source in a given time, the number of bacteria visible under the microscope on a fixed area of a plate, blood counts, mutations caused by radiation, flying bomb hits on London[3] and deaths of Prussian soldiers from horse-kicks.[4] Other examples include defects in materials, the distribution of animal litters in a field, stars in space and even the distribution of raisins in a cake. The distribution is also used for studying telephone exchange problems.

A random variable X taking non-negative integral values is said to have the Poisson distribution[5] if

■ $$P(X = r) = e^{-\lambda} \lambda^r / r!,$$ (10.6.1)

where $r = 0, 1, 2, \ldots$ and λ is a positive constant. It is not difficult to prove that the Poisson distribution has

■   mean $= \lambda$,   (10.6.2)
■   variance $= \lambda$.   (10.6.3)

Poisson distributions with means 0.9 and 5.0 are shown in fig. 10.6.1. Poisson tables are given, for example, by Molina [69] and Pearson and Hartley [76] 185–204.

Parameter estimation methods are given in section 13.8.

Example 10.6.1. A baker produces 160 cakes for a child's party by adding 300 raisins to 10 kg of dough. What is the probability that a particular cake contains no raisins?

One approach to this problem is to use the binomial distribution. Each raisin has chance 159/160 of not being in the particular cake. The chance that all the raisins are not in that cake is $(159/160)^{300}$ or 0.152.

The Poisson distribution can also be used to solve the problem. On average each cake has $300/160 = 1.875$ raisins. Let us assume that the distribution of the number of raisins in a cake is Poisson with mean 1.875. From (10.6.1), we see that

$$P(\text{no raisins}) = e^{-1.875} (1.875)^0 / 0! = e^{-1.875} = 0.153.$$

---

[3] Clarke [12].   [4] Bortkiewicz [4].   [5] Note that $\lambda^0 = 1$ and $0! = 1$.

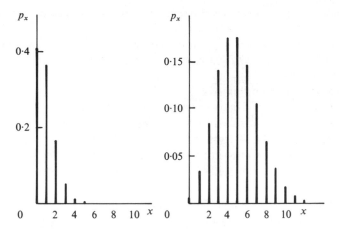

Fig.10.6.1. Poisson distributions with means 0.9 and 5.0.

This is almost exactly the same as the binomial answer. The connection between the two approaches is explained in section 10.7.

> *Further reading*: Bailey [2] 6–20; Balaam [3] 70–2; Campbell [8] 303–6;
> Colquhoun [16] 53–4; Feller [21] 146–54; Fraser [26] 45–7; Johnson and
> Kotz [47] 87–121; Kendall and Stuart [53] 125–30; Mather [66] 32–6;
> Mood and Graybill [70] 70; Remington and Schork [86] 119–21;
> Spiegel [93] 122–40; Wilks [102] 133–43.

10.7 The Poisson distribution as a limiting case of the binomial

The Poisson distribution can be arrived at in many different ways, although it is possible to define it as a distribution in its own right, as we have done in section 10.6. It is useful to know, however, that it can be derived from the binomial distribution.

Let us consider the *large* number of soldiers in the nineteenth-century Prussian army and assume that this number is n. Each of these soldiers is exposed to a *very small* risk of being kicked to death by a horse, and we shall assume that the probability for a particular soldier during a given year is p. The expected number of horse-kick deaths during a particular year is given by the binomial mean np, and this mean will be of moderate size, say λ. Because p is very small, q will be very close to one, and the binomial variance is very close to the mean λ. Intuition might even suggest the use of the Poisson distribution!

The probability of r horse-kick deaths during a given year is equal to

$$\binom{n}{r} p^r q^{n-r},$$

where n is large, p is small and $np = \lambda$ is moderate. An approximation to this probability can be obtained by considering the limit as $n \to \infty$ and $p \to 0$ such that $np = \lambda$. The limit is $e^{-\lambda}\lambda^r/r!$ which is a Poisson probability. For this reason, the Poisson distribution is sometimes regarded as the distribution of rare events.

Example 10.7.1. A type-setter produces on average 1.5 misprints per page. Calculate the probability that a given page contains more than seven misprints, and the probability that, in a book of 200 pages, no page contains more than seven misprints.

Each page contains a large number of symbols, and the chance that a particular symbol is wrong is rather remote. The Poisson distribution should be applicable. The probability that a particular page contains more than seven misprints is

$$1 - e^{-1.5}\left\{\frac{(1.5)^0}{0!} + \frac{(1.5)^1}{1!} + \frac{(1.5)^2}{2!} + \ldots + \frac{(1.5)^7}{7!}\right\} = 0.000\,170.$$

The probability that a particular page contains seven or less misprints is 0.999 830 and the probability that none of the 200 pages contain more than seven misprints is $(0.999\,830)^{200} = 0.967$.

The Poisson distribution can also be used for this latter calculation. Let us call a page defective if it contains more than seven misprints. The chance that a particular page is defective is remote (0.000 170), but we look at a large number of pages (200). The average number of defective pages for a book of this length is $200 \times 0.000\,170$ or 0.0340. From (10.6.1), the probability that the book contains no defective pages is

$$e^{-0.0340}\frac{(0.0340)^0}{0!} = 0.967.$$

Example 10.7.2. The square in fig. 10.7.1 is divided into 900 small squares. Using a table of random numbers,[6] each small square has been filled with an object with probability $\frac{1}{25}$ and left empty with probability $\frac{24}{25}$. Larger 5×5 squares containing 25 small squares are also marked in the diagram. (The large 30×30 square might be thought of as a section of a forest containing trees as objects, or as part of a plate which is covered with bacteria.)

Each small square has a relatively small probability ($p = \frac{1}{25}$) of being occupied, but a 5×5 square is composed of a relatively large number ($n = 25$) of such small squares. The expected number of objects per 5×5 square is one, and according to our theory the probability that a 5×5 square contains r objects ($r = 0, 1, 2, \ldots$) is approximately $e^{-1}1^r/r!$. For a particular 5×5 square therefore

[6] Section 14.1.

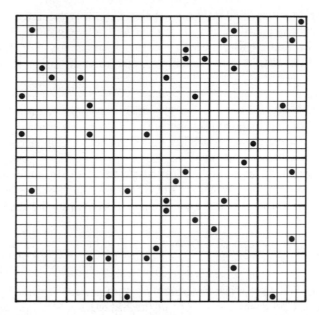

Fig. 10.7.1. Random points on a plate.

$P(0$ objects$) = 0.3679,$	$P(3$ objects$) = 0.0613,$
$P(1$ object$) = 0.3679,$	$P(4$ or more$) = 0.0190.$
$P(2$ objects$) = 0.1839,$	

But we have thirty-six 5×5 squares. Of these thirty-six squares, we should expect about $0.3679 \times 36 = 13.2$ to be empty, $0.3679 \times 36 = 13.2$ to contain only one object, $0.1839 \times 36 = 6.7$ to contain two objects, 0.0613×36 to contain three objects, and 0.0190×36 to contain four or more objects. These expected numbers and the actual observed numbers are given in table 10.7.1. Similar results are also given for the 900 small squares. The differences between the observed and expected values are caused by sampling fluctuations.

Example 10.7.3. Consider an infinite forest in which the trees are distributed at random with a uniform density of λ per unit area. A forestry official wishes to estimate λ. He selects a random point in the forest and notes the distance to the nearest tree. If the forest is dense, this distance will tend to be rather small and if the forest is sparse, the converse is likely to be true.

Let us denote the distance random variable by R, and consider the situation depicted in fig. 10.7.2. If the nearest tree is at distance r, the circle of radius r around the random point with area πr^2 must contain no trees. According to

Table 10.7.1. *An example of the Poisson limit to the binomial distribution*

n	Number of squares containing exactly *n* objects			
	5 × 5 squares		1 × 1 squares	
	Observed	Expected	Observed	Expected
0	10	13.2	859	864.7
1	15	13.2	41	34.6
2	7	6.7	0	0.7
3	4	2.2	0	0.0
4 or more	0	0.7	0	0.0
Total	36	36.0	900	900.0

the Poisson distribution, the probability of this event is $\exp(-\pi r^2 \lambda)$. The ring of width dr and area $2\pi r\, dr$, on the other hand, must contain one tree, and the Poisson probability of this event is

$$\exp\left(-2\pi r\lambda\, dr\right)(2\pi r\lambda\, dr)^1/1! = 2\pi r\lambda\, dr$$

(ignoring terms of order $(dr)^2$ and smaller). These two events are independent and it follows that the probability-density function of the distance R from a random point to the nearest tree is

$$f(r) = 2\pi r\lambda \exp\left(-\pi r^2\lambda\right).$$

The average distance to the nearest tree is

$$\int_0^\infty 2\pi r^2\lambda \exp\left(-\pi r^2\lambda\right) dr = \tfrac{1}{2}\lambda^{-\frac{1}{2}}.$$

Variations of this type of model appear in the ecological literature, and

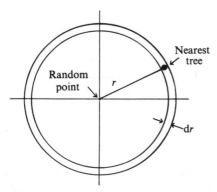

Fig. 10.7.2. The nearest tree to a random point.

the method can also be applied to defects in cloth and other materials, and even extended to three dimensions to study stars in space (Parzen [75] 32-3).

Further reading: Balaam [3] 73-7; Feller [21] 142-5; Kendall and Stuart [53] 125-6; Pielou [79] 111-14; Pollard [82]; Skellam [90].

10.8   The normal approximation to the Poisson

The probability of obtaining an outcome between r_1 and r_2 (inclusive) from a Poisson random variable with mean λ is given by

$$\sum_{r=r_1}^{r_2} e^{-\lambda} \lambda^r / r!. \tag{10.8.1}$$

This sum is often tedious to evaluate. When λ is greater than about 10, the central limit theorem[7] allows us to approximate the sum in (10.8.1) by

■ $\int_{\alpha_1}^{\alpha_2} \dfrac{1}{\sqrt{(2\pi)}} \exp\left(-\tfrac{1}{2}x^2\right) dx,$ ⁣ (10.8.2)

where

■ $\alpha_1 = (r_1 - \lambda - \tfrac{1}{2})/\sqrt{\lambda},$ (10.8.3)

■ $\alpha_2 = (r_2 - \lambda + \tfrac{1}{2})/\sqrt{\lambda}.$ (10.8.4)

The area implied by (10.8.2) is readily available from normal tables.

The corrections $\pm\tfrac{1}{2}$ in α_1 and α_2 are often omitted, particularly when λ is very large. The rationale behind these corrections is explained (in connection with the binomial distribution) in section 10.2.

Let us use the random variable X to denote the outcome of a Poisson experiment. When we use (10.8.2), we are in effect saying that $(X-\lambda)/\sqrt{\lambda}$ is approximately a unit normal random variable. It follows from section 9.2 that $(X-\lambda)^2/\lambda$ is approximately a χ_1^2 random variable. This quantity may be written in the form

■ $\dfrac{(\text{observed} - \text{expected})^2}{\text{expected}}.$ (10.8.5)

Further reading: Johnson and Kotz [47] 98-101.

10.9   The critical points of the Poisson distribution

Let X be a Poisson random variable with parameter λ. The upper 100α per cent critical value of the Poisson distribution with mean λ is an integer r such that

$$P(X \geqslant r) \leqslant \alpha < P(X \geqslant r-1). \tag{10.9.1}$$

Poisson tables can be used to determine such critical values, and approximate

---

[7] See section 9.1. The sum of a number of independent Poisson random variables is also Poisson. It follows that a Poisson random variable with $\lambda = 10$ may be regarded, for example, as the sum of ten independent Poisson variables each with $\lambda = 1$.

critical points can be calculated using the normal approximation of section 10.8. It is worth noting, however, that

■ $P(X \geqslant r) = P(Y \leqslant 2\lambda)$, (10.9.2)

where Y has the chi-square distribution with $2r$ degrees of freedom (Brownlee [7] 173). If we find r such that 2λ lies between the lower 100α per cent points of the chi-square distributions with $2r$ and $2(r-1)$ degrees of freedom, r is the upper 100α per cent point of the Poisson distribution with mean λ.

To obtain the lower 100α per cent point, we note that

■ $P(X \leqslant r) = P(Z \geqslant 2\lambda)$, (10.9.3)

where z has the chi-square distribution with $2(r+1)$ degrees of freedom. If we find r such that 2λ lies between the upper 100α per cent points of the chi-square distributions with $2r$ and $2(r+1)$ degrees of freedom, r is the lower 100α per cent point of the Poisson distribution with mean λ.

Example 10.9.1. Find the upper 5 per cent point of the Poisson distribution with mean 0.9.

We need to find r such that the lower 5 per cent point of the χ^2_{2r} distribution is greater than 1.8 and the lower 5 per cent point of the $\chi^2_{2(r-1)}$ distribution is less than 1.8. From the chi-square tables we see that $r = 4$, and this value may be confirmed by direct calculation or by reference to fig. 10.6.1.

Example 10.9.2. Find the lower 10 per cent point of the Poisson distribution with mean 5.0.

We need to find r such that the upper 10 per cent point of the χ^2_{2r} distribution is less than 10.0 and the upper 10 per cent point of the $\chi^2_{2(r+1)}$ distribution is greater than 10.0. From the chi-square tables we see that $r = 2$, and this result may be confirmed by direct calculation or by reference to fig. 10.6.1.

Further reading: Brownlee [7] 173–4; Johnson and Kotz [47] 98–9.

10.10 The geometric (Pascal) distribution

Like the Poisson distribution, the geometric distribution is a distribution over the non-negative integers. It may be derived by considering a man who tosses a coin until he obtains a 'head'. If the chance of obtaining a 'head' at a single toss is p and $q = 1-p$, the probability of obtaining j 'tails' before the first 'head' is

■ $p_j = q^j p$ $(j = 0, 1, 2, ...)$. (10.10.1)

This probability defines the *geometric distribution*. The mean and variance are as follows:

■   mean $= q/p$, (10.10.2)
■   variance $= q/p^2$. (10.10.3)

The geometric distribution with $p = 0.3$ is shown in fig. 10.11.1. It should be noted that the geometric distribution is a special case of the negative binomial distribution.

Parameter estimation methods are given in section 13.9.

Further reading: Johnson and Kotz [47] 123–4.

10.11 The negative binomial distribution

The negative binomial distribution is also a distribution over the non-negative integers, and it is defined in terms of two parameters: k and p. We also use $q = 1-p$. The distribution is in fact the distribution of the sum of k mutually-independent geometric variables, or the distribution of the number of 'tails' obtained before the kth 'head' in a coin-tossing experiment, and it takes the form

∎
$$p_j = \binom{k+j-1}{j} p^k q^j \quad (j = 0, 1, 2, ...).$$
(10.11.1)

The combinatorial notation is explained in section 1.5.

The parameters of the negative binomial distribution are p and k. The parameter p must lie between zero and one. The parameter k, which we restricted to positive integral values in our derivation of (10.11.1), may in fact assume any positive value because (10.11.1) defines a valid distribution for all positive k. The mean and variance are as follows:

∎
$$\text{mean} = \frac{kq}{p},$$
(10.11.2)

∎
$$\text{variance} = \frac{kq}{p^2}.$$
(10.11.3)

Two examples of the negative binomial distribution are given in fig. 10.11.1. When $k = 1$, the distribution becomes the geometric distribution.

A negative binomial random variable with parameters k and p may be regarded as the sum of k independent geometric variables each with parameter p. When k is large, the central limit theorem of section 9.1 allows us to use the following approximation for the sum of consecutive negative binomial probabilities $\{p_j\}$:

∎
$$\sum_{j=r_1}^{r_2} p_j \doteq \int_{\alpha_1}^{\alpha_2} \frac{1}{\sqrt{(2\pi)}} \exp\left(-\tfrac{1}{2}x^2\right) dx,$$
(10.11.4)

where

∎
$$\alpha_1 = (r_1 - kq/p - \tfrac{1}{2})/(kq/p^2)^{\frac{1}{2}},$$
(10.11.5)

∎
$$\alpha_2 = (r_2 - kq/p + \tfrac{1}{2})/(kq/p^2)^{\frac{1}{2}}.$$
(10.11.6)

The corrections $\pm \tfrac{1}{2}$ are explained (in connection with the binomial distribution) in section 10.2. They are often omitted, particularly when k is very large.

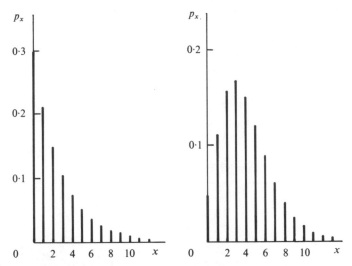

Fig. 10.11.1. Negative-binomial distributions with parameters $k = 1$, $p = 0.3$ and $k = 6$, $p = 0.6$ respectively. The first distribution is also called a geometric distribution.

In section 10.6, we observed that the mean and variance of the Poisson distribution were equal. The variance of the binomial distribution, on the other hand, is less than the binomial mean (section 10.1), and for this reason the binomial distribution is sometimes said to be *under-dispersed*. The variance of the negative binomial distribution is clearly greater than the mean, and the distribution is said to be *over-dispersed*. This property is often useful when we need to fit a non-negative integer distribution to over-dispersed data.

The mean and variance of a negative binomial distribution with a large parameter k and a moderate mean are approximately equal, because the parameter p must be close to one. In fact, it is possible to prove that the limiting distribution is the Poisson distribution.

Tables of the negative binomial distribution have been prepared by Williamson and Bretherton [104]. Parameter estimation methods are given in section 13.10.

Example 10.11.1. Fit a discrete distribution to the observed frequency data in table 10.11.1.

The mean number of eggs is $6.7\dot{3}$. The Poisson distribution with mean $6.7\dot{3}$ has its modal value at $j = 6$, and the distribution is quite the wrong shape to fit these data. Let us try fitting a negative binomial distribution by the method of moments (section 13.3). We equate the negative binomial mean and variance to the observed sample values. Thus,

$$kq/p = 6.7\dot{3} \quad \text{and} \quad kq/p^2 = 116.19\dot{5}.$$

Table 10.11.1. *Distribution of eggs of parasitic nematodes in sheep in Scotland, together with numbers calculated on the basis of the negative binomial and Poisson distributions*

		Fitted frequencies				
		Negative binomial distribution with mean 6.73 and parameter k equal to				Poisson with mean 6.73
Number of eggs	Observed frequency	0.414 19	0.5	0.6	100	
0	20	27.7	23.7	20.0	0.1	0.1
1	12	10.8	11.0	11.0	0.8	0.7
2	14	7.2	7.7	8.1	2.7	2.4
3	7	5.5	6.0	6.5	5.7	5.5
4	3	4.4	4.9	5.3	9.3	9.2
5	6	3.6	4.1	4.5	12.2	12.4
6	3	3.1	3.5	3.9	13.5	13.9
7	3	2.7	3.0	3.3	12.9	13.3
8	2	2.3	2.6	2.9	10.9	11.2
9	3	2.1	2.3	2.6	8.2	8.4
10–14	4	7.4	8.2	9.0	13.0	12.5
15–19	2	4.5	4.8	5.1	0.5	0.4
20+	11	8.7	8.2	7.8	0.2	0.0
Total	90	90.0	90.0	90.0	90.0	90.0

Source: Williams [103], table 119.

When we divide the first equation by the second, we obtain $p = 0.057\,948\,286$, whence $q = 0.942\,051\,714$ and $k = 0.414\,186\,524$. Because

$$\binom{k-1}{0} = 1 \quad \text{and} \quad p_{j+1} = \{(k+j)\,q/(j+1)\}p_j,$$

the fitted frequencies $\{f_j\}$ may be calculated in the following manner:

$$f_0 = 90 \times (0.057\,948\,286)^{0.414\,186\,524}(0.942\,051\,714)^0 = 27.7,$$
$$f_1 = f_0 \times (0.414\,186\,524) \times (0.942\,051\,714)/1 = 10.8,$$
$$f_2 = f_1 \times (1.414\,186\,524) \times (0.942\,051\,714)/2 = 7.2,$$
$$f_3 = f_2 \times (2.414\,186\,524) \times (0.942\,051\,714)/3 = 5.5.$$

The fitted frequency f_0 appears to be too high, while many of the other frequencies are too low. Negative binomial distributions with lower f_0 values can be obtained using the same mean 6.73 but larger values of k (and correspondingly lower variances). Such distributions are given in table 10.11.1. A reasonable fit[8] seems to have been obtained when $k = 0.6$.

The similarity between the Poisson distribution with mean 6.73 and the negative binomial with mean 6.73 and $k = 100$ should be noted. Both are unsuitable for fitting the given data.

> *Further reading*: Feller [21] 155–6; Fraser [26] 55–8; Johnson and Kotz [47] 122–42; Kendall and Stuart [53] 130–1; Williams [103] 5–16.

10.12 Fisher's logarithmic series distribution

R. A. Fisher, in a co-operative study with A. S. Corbet and C. B. Williams [24], developed a mathematical theory which describes with some success the relative numbers of different species obtained when sampling at random from a heterogeneous population. Williams has shown that the logarithmic series can also be applied to a great variety of biological problems, in which the integer n is variously the number of species per genus, the number of genera per sub-family, the number of parasites per host and even the number of research papers per biologist (published in a particular year). It is hard to believe that a single mechanism will be found to explain the relevance of the logarithmic series to all these problems, but the distribution is of frequent use to the biologist.

The distribution is over the positive integers (excluding zero):

■
$$P(X = j) = \{-\ln(1-\alpha)\}^{-1} \alpha^j / j. \tag{10.12.1}$$

The parameter α lies strictly between zero and one, and j takes the values 1, 2, 3, ... We note that the distribution has

■
$$\text{mean} = \{-\ln(1-\alpha)\}^{-1} \alpha/(1-\alpha), \tag{10.12.2}$$

■
$$\text{variance} = \frac{\text{mean}}{1-\alpha} - (\text{mean})^2. \tag{10.12.3}$$

The distribution may be derived from the negative binomial distribution. When the negative binomial parameter k is very small, the conditional distribution of the negative binomial random variable, given that it is non-zero, becomes the logarithmic series distribution.

Parameter estimation methods are given in section 13.11.

Example 10.12.1. A biologist uses a light trap to obtain a large number of moths. He knows that there are very many different species in the area, but the actual number of species is not known. He also knows that the probability of catching a moth of any particular species during a certain length

---

[8] The chi-square goodness-of-fit test of section 12.11 (or possibly the Kolmogorov-Smirnov test of section 12.12) may be used to test the adequacy of the fitted distribution.

of time is constant and independent of the species. He continues trapping until he has a total of n moths, where n is fairly large. What distribution would he expect the number of moths of the different species captured to have?

The biologist continues trapping until he has a certain total number of moths. The number of species is very large; so a particular species may not be represented at all with a high probability. In effect, the biologist is sampling until he has an average of k of each species, where k is less than one and rather small. So for each species, the number of moths caught follows a negative binomial distribution with k small. We deduce that for each species represented in the sample, the number of moths caught is a random observation from Fisher's logarithmic series distribution. All the species are assumed to have the same abundance and the same probabilities of capture. We conclude therefore that the numbers of moths of the different captured species will follow the logarithmic series distribution.

In practice, many of the above assumptions are not strictly correct. Nevertheless, C. B. Williams has shown that the logarithmic series distribution provides a good fit in many different circumstances.

It may seem that we can perform such an experiment and then work backwards to the negative binomial distribution to determine the total number of species including those represented by zero captures. This is not possible, because in our assumptions we include one which effectively says that the total number of species is infinite.

Example 10.12.2. Fit a logarithmic series distribution to the moth data in table 10.12.1. The mean number of individuals per species is $6814/197 = 34.588\,832$ and we equate (10.12.2) to this value. By trial and error we find that $\alpha = 0.994\,47$, and the fitted frequencies in table 10.12.1 are calculated in the following manner:

$$f_1 = 197 \times 0.994\,47/\{-\ln(0.005\,53)\} = 37.7,$$
$$f_2 = f_1 \times 0.994\,47 \times 1/2 = 18.7,$$
$$f_3 = f_2 \times 0.994\,47 \times 2/3 = 12.4,$$
$$f_4 = f_3 \times 0.994\,47 \times 3/4 = 9.3.$$

The chi-square goodness-of-fit test of section 12.11 (or possibly the Kolmogorov-Smirnov test of section 12.12) can be used to test the adequacy of the fit.

Further reading: Johnson and Kotz [47] 166–82; Kendall and Stuart [53] 131–3; Williams [103] 5–16.

10.13   The hypergeometric distribution

Consider a population containing R red elements and B black elements. The total population size is $R + B$. From this population n elements are chosen

Table 10.12.1. *Frequency distribution of lepidoptera captured in a light trap at Rothamsted Experimental Station during 1935. The total number of individuals captured is 6814*

Individuals per species j	Number of species with j individuals	
	Observed	Fitted
1	37	37.7
2	22	18.7
3	12	12.4
4	12	9.3
5	11	7.4
6	11	6.1
7	6	5.2
8	4	4.5
9	3	4.0
10	5	3.6
11	2	3.2
12	4	3.0
13	2	2.7
14	3	2.5
15	2	2.3
16+	61	74.4
Total	197	197.0

Source: Williams [103] 25, table 8.

at random without replacement. The distribution of the number of red elements in the sample is known as the *hypergeometric distribution*:

■
$$P(r \text{ red elements}) = \binom{R}{r}\binom{B}{n-r}\Big/\binom{R+B}{n}. \qquad (10.13.1)$$

The sample size n must be less than $R + B$, and $0 < r < \min (R, n)$. We note that the mean and variance are given by

■ $\quad\text{mean} = nR/(R+B),$ $\qquad\qquad(10.13.2)$

■ $\quad\text{variance} = \dfrac{nRB}{(R+B)^2}\left(1 - \dfrac{n-1}{R+B-1}\right).$ $\qquad(10.13.3)$

Table 10.14.1. *Counts of parasite infestations*

Parasite count	Number of animals
Under 8	0
8	2
9	0
10	2
11	2
12	5
13	2
14	4
15	6
16	3
17	1
18	3
over 18	0
Total	30

When the population size is large and the sample size is small relative to the population, the distribution is effectively binomial with parameters n and $p = R/(R + B)$. Furthermore, when n is large, $R + B$ is very large, $p = B/(R + B)$ is small and np is moderate, the distribution is effectively Poisson (section 10.7). Normal approximations can also be used in certain circumstances (sections 10.2 and 10.8). Tables of the hypergeometric distribution have been calculated by Liebermann and Owen [59].

The Pearson system of frequency curves has its origins in the hypergeometric distribution.

> *Further reading*: Feller [21] 41–4; Fraser [26] 53–5; Johnson and Kotz [47] 143–65; Johnson and Leone [50] 82–5; Kendall and Stuart [53] 133–5.

10.14 Exercises

1. Table 10.14.1 contains some parasite infestation data. Calculate the mean and variance of the distribution of counts. The sample variance happens to be smaller than the mean. Fit a binomial distribution to the data by equating np and npq to the sample mean and sample variance respectively.

2. Fit a Poisson distribution to the data in table 10.14.1 by equating λ to the sample mean.

3. An unbiased six-sided die is tossed 1000 times. Calculate the probability of obtaining 200 or more sixes. *Hint*: at each toss the outcome is either a six with probability $\frac{1}{6}$ or not a six with probability $\frac{5}{6}$.

4. An unbiased six-sided die is tossed 10 times. Calculate the probability of obtaining exactly 4 sixes and exactly 2 threes. *Hint*: at each toss the outcome is either a six with probability $\frac{1}{6}$ or a three with probability $\frac{1}{6}$ or something else with probability $\frac{2}{3}$.

5. The south of London has been divided into 576 small areas of equal size. R. D. Clarke [12] reports that 229 of the areas experienced zero flying-bomb hits during World War II, 211 experienced exactly one hit, 93 experienced two hits, 35 experienced three hits, 7 experienced four hits and one experienced five or more hits. Fit a Poisson distribution to these data. *Hint*: equate the Poisson mean (10.6.2) to the mean of the observed distribution.

6. Equate the geometric distribution mean (10.10.2) and variance (10.10.3) to the sample values of these moments in the numerical example 10.11.1 and hence fit a geometric frequency distribution to the data of table 10.11.1.

7. A box contains 10 000 red objects and 20 000 black objects. One hundred objects are chosen at random from the box without replacement. Find the probability that 40 or more of the selected objects are red.

8. Use the method of section 10.3 to find the upper and lower $2\frac{1}{2}$ per cent points of the binomial distribution with parameters $n = 18$, $p = 0.4$.

9. Use the method of section 10.9 to find the upper and lower 5 per cent points of the Poisson distribution with mean $\lambda = 1.5$.

11

The Pearson system of probability-density functions

Summary   The Pearson system of probability-density functions is defined by a differential equation similar in form to the difference equation of the hypergeometric distribution. By varying the parameters in the equation, it is possible to produce continuous distributions with different levels of skewness and kurtosis. The parameters are completely defined in terms of the first four moments, and the first four sample moments are used for curve-fitting purposes. The technique is convenient if one just wants to fit a curve without worrying about the justification of the functional form.

11.1   Introduction

Many different shapes of discrete distribution can be obtained by varying the parameters R, B and n in the hypergeometric distribution (section 10.13). When, for example, $R = B$ and both are very large, and a reasonably large random sample of size n is drawn, the distribution of the number of red objects chosen is effectively binomial with parameters n (large) and $p = \frac{1}{2}$. We then have a symmetric discrete distribution closely resembling the normal. By varying the hypergeometric parameters, we are able to produce distributions with different levels of skewness and kurtosis (section 8.6). Karl Pearson observed this, and he devised the Pearson system of probability-density functions by using a differential equation similar to the difference equation of the discrete hypergeometric distribution.

The process of fitting a Pearson probability-density function usually involves a large amount of arithmetic, but the work is nowhere near as onerous as previously now that electronic desk calculators are available with exponential, logarithmic and trigonometric keys. The technique is convenient if one just wants to fit a curve without worrying about the justification of the functional form.

Let us denote the hypergeometric probability (10.13.1) by p_r. Then it is not difficult to show that the difference $\Delta p_r = p_{r+1} - p_r$ obeys the difference equation

$$\frac{1}{p_r} \Delta p_r = -\left\{ \frac{-(nR - B + n - 1) + (R + B + 2)r}{(B - n + 1) + (B - n + 2)r + r^2} \right\}. \tag{11.1.1}$$

Under the Pearson system, a probability-density function $f(x)$ obeys a differential equation of the form

■ $$\frac{1}{f(x)}\frac{df}{dx} = -\frac{a+x}{c_0+c_1x+c_2x^2},$$ (11.1.2)

where a, c_0, c_1 and c_2 are suitable constants.

Further reading: Chakravarti *et al.* [9] 155–9; Elderton and Johnson [20] 35–46; Johnson and Kotz [48] 9–14.

11.2 Fitting a Pearson curve

The differential equation (11.1.2) defines the Pearson system of curves, and the constants a, c_0, c_1 and c_2 are fully determined in terms of the first four moments. To fit a Pearson curve to a set of data, we use the method of moments (section 13.3). We calculate the first four non-central moments m'_1, m'_2, m'_3 and m'_4, the central moments m_2, m_3 and m_4, and the moment functions $\sqrt{b_1}$ and b_2 (section 8.6). If the origin is chosen to lie at the mean, and we define

■ $$d = 2(5b_2 - 6b_1 - 9),$$ (11.2.1)

the constants in the differential equation are given by

■ $a = \{\sqrt{m_2}(b_2+3)\sqrt{b_1}\}/d,$ (11.2.2)
■ $c_0 = m_2(4b_2-3b_1)/d,$ (11.2.3)
■ $c_1 = a,$ (11.2.4)
■ $c_2 = (2b_2-3b_1-6)/d.$ (11.2.5)

The type of curve obtained depends upon the roots of the quadratic in the denominator of (11.1.2). Pearson distinguished twelve different types. (These are summarised, for example, on p. 45 of Elderton and Johnson [20].) The main types are numbered I, IV and VI. In the limiting cases when one main type changes into another, we reach simpler forms of transition curves. The normal curve is obtained, for example, when $c_1 = c_2 = 0$ ($\sqrt{b_1} = 0$; $b_2 = 3$).

Detailed hand-calculation procedures for fitting the various types of curve are described elsewhere (for example, Elderton and Johnson [20]). These procedures involve some cumbersome formulae which tend to mask the actual simplicity of the method. In our examples we shall obtain the fitted curves from first principles, and avoid these formulae. It is perhaps worth noting, however, that Pearson's criterion

■ $$\kappa = c_1^2/(4c_0c_2)$$ (11.2.6)

distinguishes the different types of curve because it distinguishes the different types of root for the quadratic in the denominator of (11.1.2).

Once a probability-density curve has been fitted, the chi-square goodness-of-fit test of section 12.11 (or possibly the Kolmogorov–Smirnov test of section 12.12) can be used to test the adequacy of the fit.

The Pearson method can also be used to fit curves to frequency data which are not independent observations on a single random variable (for example, the age distribution of a population); the chi-square and Kolmogorov–Smirnov goodness-of-fit tests cannot then be used.

> **Example 11.2.1.** Fit a Pearson curve to the data in table 11.2.1. We begin by calculating the moment functions
>
> $m'_1 = 37.875,$
> $m'_2 = 1626.075,$ $\qquad m_2 = 191.559\,375,$ $\qquad \sqrt{b_1} = 0.712\,246\,085,$
> $m'_3 = 77\,986.575,$ $\qquad m_3 = 1888.361\,72,$ $\qquad b_1 = 0.507\,294\,486,$
> $m'_4 = 4100\,394.675,$ $\quad m_4 = 107\,703.2971,$ $\quad b_2 = 2.935\,095\,088.$

The relevant formulae are given in section 8.6. We then use (11.2.2) to (11.2.5) to calculate

$$a = 11.115\,824\,33,$$
$$c_0 = 371.\,896\,975,$$
$$c_1 = 11.115\,824\,33,$$
$$c_2 = -0.313\,806\,272.$$

We now know that the differential equation (11.1.2) takes the form

$$\frac{1}{f}\frac{df}{dx} = -\frac{11.115\,824\,33 + x}{371.896\,975 + 11.115\,824\,33x - 0.313\,806\,272x^2}.$$

The roots of the denominator quadratic are real and take the values $-21.003\,132\,03$ and $56.425\,701\,06$. It follows that the right-hand side of the differential equation may be written in the form

$$\frac{A}{21.003\,132\,03 + x} + \frac{B}{56.425\,701\,06 - x}.$$

The constants A and B may be found by selecting two particular numerical values of x and equating the two different forms of the right-hand side. Using $x = 0$ and $x = 10$, for example, we obtain

$$0.047\,611\,946A + 0.017\,722\,420B = -0.029\,889\,526,$$
$$0.032\,254\,805A + 0.021\,539\,793B = -0.046\,750\,082,$$

whence $A = 0.406\,924\,333$ and $B = -2.779\,754\,990$. This solution can be checked by substituting another value of x, $x = 100$, say, into the two forms of the right-hand side. We obtain the almost identical values of $0.067\,156\,379$ and $0.067\,156\,376$.

Table 11.2.1. *Some data of W. P. Elderton.* * *Fitted values are also given*

Central value of quinquennial group	Observed frequency	Observed frequency divided by five	Fitted value at central point × 1000
17	34	6.8	6.7
22	145	29.0	27.8
27	156	31.2	30.1
32	145	29.0	28.6
37	123	24.6	25.4
42	103	20.6	21.6
47	86	17.2	17.6
52	71	14.2	13.7
57	55	11.0	10.2
62	37	7.4	7.2
67	21	4.2	4.7
72	13	2.6	2.8
77	7	1.4	1.4
82	3	0.6	0.6
87	1	0.2	0.1
Total	1000	—	—

Source: Elderton and Johnson [20] 53.

The left-hand side of the differential equation is the derivative of $\ln f$ with respect to x, and so we can write

$$\frac{d}{dx}\ln f = \frac{0.406\,924\,333}{21.003\,132\,03 + x} - \frac{2.779\,754\,990}{56.425\,701\,06 - x}.$$

Since $\frac{d}{dx}\ln(\alpha + \beta x) = \beta/(\alpha + \beta x)$, it follows that

$$\ln f = 0.406\,924\,333 \ln(21.003\,132\,03 + x)$$
$$+ 2.779\,754\,990 \ln(56.425\,701\,06 - x)$$
$$+ \text{constant}$$

and $f = f_0(\gamma + x)^u(\delta - x)^v,$

where $\gamma = 21.003\,132\,03,$ $u = 0.406\,924\,333,$
$\delta = 56.425\,701\,06,$ $v = 2.779\,754\,990.$

This Pearson Type I curve is a beta distribution defined for values of x in the range $-\gamma \leqslant x \leqslant \delta$.

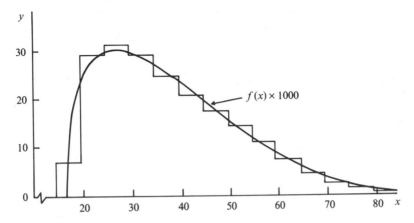

Fig. 11.2.1. The use of a Pearson type I curve to fit some frequency data of W. P. Elderton.

To ensure that the total area under the curve is one, we calculate the constant f_0 as follows:[1]

$$f_0 = \left\{ (\gamma + \delta)^{u+v+1} \frac{\Gamma(u+1)\Gamma(v+1)}{\Gamma(u+v+2)} \right\}^{-1}.$$

We find that $f_0 = 0.970\,08 \times 10^{-7}$. When we use the true origin rather than the mean and the true values of x, the frequency function takes the form

$$f_0(\gamma - m_1' + x)^u (\delta + m_1' - x)^v$$

and x lies in the range $m_1' - \gamma$ to $m_1' + \delta$. The fitted curve and the original data are compared in table 11.2.1 and fig. 11.2.1. Areas under sections of the curve may be found by approximate integration (chapter 5).

The data in table 11.2.1 are not independent observations on a single random variable; so the goodness-of-fit tests of sections 12.11 and 12.12 cannot be used.

Example 11.2.2. Fit a Pearson curve to the data in table 11.2.3.

As before, we begin by calculating the first four moments and related quantities. We find that

$$m_1' = 44.577\,233\,99,$$
$$m_2' = 2102.403\,320, \quad m_2 = 115.273\,5302, \quad \sqrt{b_1} = -0.071\,273\,462,$$

---

[1] Numerical values of the gamma function $\Gamma(n) = \int_0^\infty e^{-x} x^{n-1} \, dx$ are given in table 11.2.2. When $n > 2$, the relationship $\Gamma(n) = (n-1)\,\Gamma(n-1)$ is used to evaluate the function. (Note that $\Gamma(n) = (n-1)!$ when n is a positive integer.) For large positive values of n, an accurate approximation is provided by Stirling's formula:
$$\Gamma(n) \cong n^n e^{-n} (2\pi/n)^{\frac{1}{2}}.$$

Table 11.2.2. *The gamma function* $\Gamma(n)$*

n	$\Gamma(n)$	n	$\Gamma(n)$	n	$\Gamma(n)$	n	$\Gamma(n)$
1.00	1.00000	1.25	.90640	1.50	.88623	1.75	.91906
1.01	.99433	1.26	.90440	1.51	.88659	1.76	.92137
1.02	.98884	1.27	.90250	1.52	.88704	1.77	.92376
1.03	.98355	1.28	.90072	1.53	.88757	1.78	.92623
1.04	.97844	1.29	.89904	1.54	.88818	1.79	.92877
1.05	.97350	1.30	.89747	1.55	.88887	1.80	.93138
1.06	.96874	1.31	.89600	1.56	.88964	1.81	.93408
1.07	.96415	1.32	.89464	1.57	.89049	1.82	.93685
1.08	.95973	1.33	.89338	1.58	.89142	1.83	.93969
1.09	.95546	1.34	.89222	1.59	.89243	1.84	.94261
1.10	.95135	1.35	.89115	1.60	.89352	1.85	.94561
1.11	.94739	1.36	.89018	1.61	.89468	1.86	.94869
1.12	.94359	1.37	.88931	1.62	.89592	1.87	.95184
1.13	.93993	1.38	.88854	1.63	.89724	1.88	.95507
1.14	.93642	1.39	.88785	1.64	.89864	1.89	.95838
1.15	.93304	1.40	.88726	1.65	.90012	1.90	.96177
1.16	.92980	1.41	.88676	1.66	.90167	1.91	.96523
1.17	.92670	1.42	.88636	1.67	.90330	1.92	.96878
1.18	.92373	1.43	.88604	1.68	.90500	1.93	.97240
1.19	.92088	1.44	.88580	1.69	.90678	1.94	.97610
1.20	.91817	1.45	.88565	1.70	.90864	1.95	.97988
1.21	.91558	1.46	.88560	1.71	.91057	1.96	.98374
1.22	.91311	1.47	.88563	1.72	.91258	1.97	.98768
1.23	.91075	1.48	.88575	1.73	.91466	1.98	.99171
1.24	.90852	1.49	.88595	1.74	.91683	1.99	.99581
						2.00	1.00000

Source: Standard Mathematical Tables (Twelfth Edition), C. D. Hodgman (editor), The Chemical Rubber Co., 1963. Used by permission of The Chemical Rubber Co.

$m'_3 = 103\,908.2641,$ $m_3 = -88.210\,900,$ $b_1 = 0.005\,079\,906,$
$m'_4 = 5349\,410.913,$ $m_4 = 42\,074.106\,00,$ $b_2 = 3.166\,326\,603,$

$a = -0.346\,901\,821,$
$c_0 = 107.203\,7139,$
$c_1 = -0.346\,901\,821,$
$c_2 = 0.023\,335\,268.$

The roots of the quadratic in the denominator of the right-hand side of the differential equation are complex in this case. We therefore rearrange the quadratic as follows:

$$\begin{aligned}
\text{denominator} &= 0.023\,335\,268x^2 - 0.346\,901\,821x + 107.203\,7139 \\
&= 0.023\,335\,268(x^2 - 14.865\,988\,29x + 4594.063\,967) \\
&= 0.023\,335\,268\{(x - 7.432\,994\,15)^2 + 4538.814\,565\} \\
&= c_2\{(x - \alpha)^2 + \beta^2\}.
\end{aligned}$$

We also rearrange the numerator as follows:

$$\begin{aligned}
\text{numerator} &= x - 0.346\,901\,821 \\
&= x - 7.432\,994\,15 + 7.086\,092\,329 \\
&= x - \alpha + \gamma.
\end{aligned}$$

We are now in a position to solve the differential equation, which may be written

$$\frac{d}{dx}\ln f = \frac{-2(x-\alpha)}{2c_2\{(x-\alpha)^2+\beta^2\}} - \frac{\gamma}{c_2\beta}\,\frac{\beta}{\{(x-\alpha)^2+\beta^2\}}.$$

Integrating,[2] we have

$$\ln f = -\frac{1}{2c_2}\ln\{(x-\alpha)^2+\beta^2\} - \frac{\gamma}{c_2\beta}\tan^{-1}\left(\frac{x-\alpha}{\beta}\right) + \text{constant},$$

so that

$$f = f_0\{(x-\alpha)^2+\beta^2\}^{-1/(2c_2)}\left[\exp\left\{\tan^{-1}\left(\frac{x-\alpha}{\beta}\right)\right\}\right]^{-\gamma/(c_2\beta)}.$$

This is the second main type of Pearson curve (Type IV) and it is the most difficult. The range of the distribution is unlimited in both directions. When we use the true origin rather than the mean, and the true values of x, the frequency function takes the form

$$f_0\{(x-m_1'-\alpha)^2+\beta^2\}^{-1/(2c_2)}\left[\exp\left\{\tan^{-1}\left(\frac{x-m_1'-\alpha}{\beta}\right)\right\}\right]^{-\gamma/(c_2\beta)}.$$

All the constants in this formula are known, except f_0, and its determination is no trivial matter. One method is outlined briefly by Elderton and Johnson [20] 59, but the following numerical method seems preferable.

Use an approximate integration formula such as Simpson's rule (section 5.3) to find the area under the curve, assuming that f_0 is one. In our case, this

[2] Note that $\frac{d}{dx}\ln g(x) = g'(x)/g(x)$ and $\frac{d}{dx}\tan^{-1}\left(\frac{x-\alpha}{\beta}\right) = \beta/\{(x-\alpha)^2+\beta^2\}$.

Table 11.2.3. *Some further data of W. P. Elderton.* * Fitted values are also given*

Central value of quinquennial group	Observed frequency	Observed frequency divided by five	Fitted value at central point × 9154
5	10	2.0	0.9
10	13	2.6	3.0
15	41	8.2	9.3
20	115	23.0	25.9
25	326	65.2	62.9
30	675	135.0	129.6
35	1113	222.6	222.1
40	1528	305.6	309.9
45	1692	338.4	346.3
50	1530	306.0	306.7
55	1122	224.4	214.6
60	610	122.0	119.1
65	255	51.0	53.1
70	86	17.2	19.3
75	26	5.2	5.9
80	8	1.6	1.5
85	2	0.4	0.4
90	1	0.2	0.1
95	1	0.2	0.0
Total	9154	—	—

Source: Elderton and Johnson [20] 60.

area turns out to be 33.475 781 26. We require the total area to be one, so we choose $f_0 = 0.029\,87$. The fitted curve and the original data are compared in table 11.2.3 and fig. 11.2.2. Areas under sections of the curve may be found by approximate integration (chapter 5).

The data in table 11.2.3 are not independent observations on a single random variable; so the goodness-of-fit tests of sections 12.11 and 12.12 cannot be used.

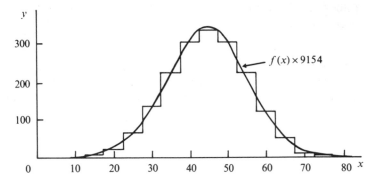

Fig. 11.2.2. The use of a Pearson Type IV curve to fit some frequency data of W. P. Elderton.

Example 11.2.3. Fit a Pearson curve to the data in table 11.2.4. As before, we calculate

$m_1' = 34.021\,739\,13$,
$m_2' = 1258.695\,652$, $\quad m_2 = 101.\,216\,9187$, $\quad \sqrt{b_1} = 0.877\,036\,700$,
$m_3' = 50\,603.260\,86$, $\quad m_3 = 893.094\,5420$, $\quad b_1 = 0.769\,193\,373$,
$m_4' = 2209\,891.304$, $\quad m_4 = 45\,657.182\,70$, $\quad b_2 = 4.456\,592\,092$,

$a = 3.795\,297\,465$,
$c_0 = 90.609\,144\,67$,
$c_1 = 3.795\,297\,465$,
$c_2 = 0.034\,934\,126$.

In this case the denominator quadratic has two real roots, namely $-35.425\,482\,16$ and $-73.216\,064\,07$. We know therefore that the right-hand side of the differential equation may be written in the form

$$\frac{A}{35.425\,482\,16 + x} + \frac{B}{73.216\,064\,07 + x}.$$

To find A and B, we use the method of example 11.2.1 with $x = 0$ and $x = 10$. We find that

$$0.028\,228\,267A + 0.013\,658\,204B = -0.041\,886\,472,$$
$$0.022\,014\,075A + 0.012\,016\,910B = -0.104\,465\,880,$$

whence $A = 23.958\,974\,99$ and $B = -52.584\,279\,41$. When we use the method of example 11.2.1 to check the answer and select $x = 100$, we obtain consistent values for the right-hand side: $-0.126\,659\,930$ and $-0.126\,659\,928$.

Table 11.2.4. *Some further data of W. P. Elderton.* * *Fitted values are also given*

Central value of decennial group	Observed frequency	Observed frequency divided by 10	Fitted value at central point × 368
10	1	0.1	0.03
20	56	5.6	6.06
30	167	16.7	16.09
40	98	9.8	10.01
50	34	3.4	3.44
60	9	0.9	0.91
70	2	0.2	0.22
80	1	0.1	0.05
Total 368		—	—

Source: Elderton and Johnson [20] 68.

We now see that the differential equation may be written in the form

$$\frac{d}{dx}\ln f = \frac{A}{\alpha+x} + \frac{B}{\beta+x},$$

whence

$$\ln f = A\ln(\alpha+x) + B\ln(\beta+x) + \text{constant} \left(\text{since } \frac{d}{dx}\ln(\alpha+\beta x) = \beta/(\alpha+\beta x)\right),$$

and

$$f = f_0(\alpha+x)^A(\beta+x)^B$$

This is the third main type (Type VI) of Pearson curve. The roots of the quadratic are both real and of the same sign. The range of the distribution is from $-\alpha$ to ∞ ($\alpha < \beta$). The constant f_0 must be chosen so that the total area under the curve is one. In fact,

$$f_0 = \Gamma(-B)/\{(\beta-\alpha)^{A+B+1}\Gamma(A+1)\Gamma(-B-A-1)\}.$$

Using the asymptotic formula for $\Gamma(n)$ in footnote 1, we find that $\ln f_0$ is equal to 142.988 9256.

When we use the true origin rather than the mean and the true values of x, the frequency function takes the form

$$f_0(\alpha - m_1' + x)^A(\beta - m_1' + x)^B,$$

and x lies in the range $m_1' - \alpha$ to $+\infty$. The final curve is compared with the original data in table 11.2.4 and fig. 11.2.3. Areas under sections of the curve may be found by approximate integration (chapter 5).

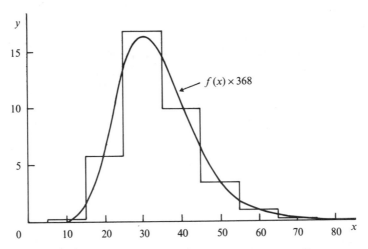

Fig. 11.2.3. The use of a Pearson Type VI curve to fit some frequency data of W. P. Elderton.

The data in table 11.2.4 are not independent observations on a single random variable; so the goodness-of-fit tests of sections 12.11 and 12.12 cannot be used.

Further reading: Chemical Rubber Company [11] 387; Elderton and Johnson [20] 47–109, 182–4; Feller [21] 50–2.

11.3 Adjustments for grouping

The data in each of the three examples of section 11.2 were grouped, and some authors believe that corrections should be made to the calculated moments to take account of this grouping. Sheppard's corrections are sometimes used when the tails of the distribution go gradually to zero in both directions (h denotes the width of each group):

- m_2 (adjusted) $= m_2 - h^2/12,$ (11.3.1)
- m_4 (adjusted) $= m_4 - h^2 m_2/2 + 7h^4/240.$ (11.3.2)

The moments m_1' and m_3 require no adjustment.

Further reading: Elderton and Johnson [20] 177–81; Kendall and Stuart [53] 75–81; Spiegel [93] 90.

11.4 Exercises

1. In example 11.2.2, the skewness measure $\sqrt{b_1}$ is close to zero and the kurtosis measure b_2 is close to three. Fit a normal curve to the data by equating μ and σ^2 to the sample moments m_1' and m_2. Compare the result with the Type IV fit.

2. Use Stirling's formula in footnote 1 to obtain approximate values for $n!$ ($n = 1$, 2, ..., 5). Calculate the percentage error in each answer.

12

Hypothesis testing

Summary   In this chapter we outline the general procedure for testing a statistical hypothesis and summarise a number of standard tests. These tests cover binomial and multinomial hypotheses (sections 12.6–12.8), association in contingency tables (sections 12.9–12.10), goodness-of-fit (sections 12.11–12.12), hypotheses about population means (sections 12.13–12.27), hypotheses about variances (sections 12.28–12.30), hypotheses about correlation (sections 12.31–12.34), and shape of distribution (sections 12.35–12.37).

12.1   Introduction

A scientist has recently been given a coin and he wishes to test whether it is unbiased. The obvious experiment for him to carry out is to toss the coin a number of times and observe the number of 'head' and 'tail' outcomes. He decides to toss the coin 1000 times. For an unbiased coin, $p = \frac{1}{2}$, and the expected number of 'heads' is 500 (section 10.1). If the number of heads turns out to be 511, the scientist will probably conclude that the coin is unbiased (511 is near 500); if on the other hand the number of heads turns out to be 897, he will be rather convinced that the coin is biased, and he will reject any suggestion that it is unbiased.

Both the probability of obtaining exactly 511 heads with an unbiased coin and the probability of obtaining exactly 897 heads with such a coin are small. Why then do we accept the null hypothesis that the coin is unbiased when we obtain 511 heads yet reject it when we obtain 897 heads? In the former case we accept the *null hypothesis* that the coin is unbiased because the probability of obtaining with an unbiased coin a number of heads differing from the expected number 500 by 11 or more is quite appreciable (about 0.5). The probability of obtaining with an unbiased coin a number of heads differing from 500 by 397 or more on the other hand is extremely remote, and we would reject the null hypothesis in this case.

This type of reasoning underlies all statistical tests. The probability level below which we reject the null hypothesis is known as the *significance level*. The general test procedure may be summarised as follows:

General procedure	*Coin-tossing example*
1. Choose a *level of significance* α.	$\alpha = 0.05$.
2. Formulate the *statistical model*.	The number of heads X has the binomial distribution with parameters $n = 1000$ and p.

3. Specify the *null hypothesis* H_0 and the *alternative hypothesis* H_1.

$H_0: p = \frac{1}{2}$;
$H_1: p \neq \frac{1}{2}$.

4. Select a *test statistic* whose behaviour is known.

$T = (X - np)/\sqrt{(npq)}$ is approximately a unit normal random variable when n is large (section 10.2).

5. Find the appropriate *critical region* for testing the null hypothesis. (If the test statistic falls in this region, the null hypothesis will be rejected.) The test statistic has probability α of falling in this region when H_0 is true.

From the normal tables, the critical region is where $|T| > 1.96$.

6. Compute the test statistic.

$T = (511 - 500)/\sqrt{(250)} = 0.696$.

7. Draw conclusions. If the test statistic lies in the critical region, reject the null hypothesis and accept the alternative. Otherwise accept the null hypothesis.

The test statistic is not in the critical region, so we accept the null hypothesis. We have no evidence that the coin is biased.

The fact that the test criterion is not significant does not prove that the null hypothesis is true; it merely tells us that the data do not contradict the null hypothesis. We accept the null hypothesis until evidence is brought forward disproving it. In other contexts this approach might be referred to as '*scientific method*'.

Further reading: Bailey [2] 27–32; Balaam [3] 12; Brownlee [7] 97–104; Colquhoun [16] 86–96; Guenther [39] 1–14; Hoel [45] 159–64; Johnson and Leone [50] 195–202; Mood and Graybill [70] 275–90; Pollard [81] 76–83; Remington and Schork [86] 193–9; Snedecor [91] 24–5; Spiegel [93] 167–87; Wilks [102] 216–19.

12.2   Types of error and the power of a test

We have seen that the *critical region* is the set of outcomes for an experiment which leads to the rejection of the null hypothesis. When the null hypothesis is true and a level of significance α is used, an experiment will fall in the critical region with probability α; we then wrongly reject the null hypothesis and commit a *Type I error*. Sometimes the outcome of an experiment will fail to fall in the critical region when the null hypothesis is untrue. We then wrongly accept the null hypothesis and commit a *Type II error*.

Naturally, we aim to minimise the probabilities of these types of error. We can reduce the probability of a Type I error merely by reducing the significance level α, but the probability of a Type II error is then increased. We therefore introduce the concept of the *power* of a test, which is defined as the probability of rejecting the null hypothesis. This probability depends upon the actual parameter value in the population, and as we do not know this, we consider the *power curve* which plots the power of the test against the possible values of the parameter. It is soon apparent that an ideal power curve would have an ordinate of one for all values of the parameter except those corresponding to the null hypothesis. It is impossible to achieve such a result in practice but the power of a test can usually be increased to any desired level by making the sample size sufficiently large.

A word of warning should be given about the dangers of applying more than one statistical test to the same set of data. If two different tests are applied to the same data to test the same null hypothesis (or very similar hypotheses) and a 5 per cent significance level is used in each case, the probability that at least one of the tests will lead to an incorrect rejection of a true null hypothesis is somewhat greater than 5 per cent. Only one test, preferably the more powerful one, should be used.

It sometimes happens that we need to test two very different hypotheses using the same data (for example, the values of the mean and variance of a normal population). If a 5 per cent significance level is used in both tests and both null hypotheses are true, the probability of an incorrect rejection of at least one null hypothesis is considerably greater than 5 per cent and often closer to 10 per cent.

Example 12.2.1. In the coin-tossing example of section 12.1, the null hypothesis is rejected whenever $|T| > 1.96$. That is, we reject the null hypothesis whenever the observed number of heads X lies outside the range 470 to 530 (inclusive).

Let us imagine that the coin is in fact slightly biased and that the unknown p is actually 0.52. Then, according to (10.2.2), the probability that the null hypothesis $p = 0.5$ will be rejected is

$$1 - \Phi\left(\frac{530.5 - 1000 \times 0.52}{\sqrt{(1000 \times 0.52 \times 0.48)}}\right) + \Phi\left(\frac{469.5 - 1000 \times 0.52}{\sqrt{(1000 \times 0.52 \times 0.48)}}\right) = 0.254.$$

This is the power of the test when $p = 0.52$, and the probability of a Type II error is 0.746. The complete power curve is shown in fig. 12.2.1.

Fig. 12.2.1 also shows the power curve of the scientist's test had he used a sample of size 10 000.

Further reading: Johnson and Leone [50] 195–202; Spiegel [93] 167–87.

12.3 One- and two-tail tests

A gambler has procured a coin which he intends using against unsuspecting opponents. He wishes to test the manufacturer's claim that the probability of a 'head' is somewhat less than one-half, and he tosses the coin 1000 times, obtaining 529 heads. Does this result contradict the manufacturer's claim?

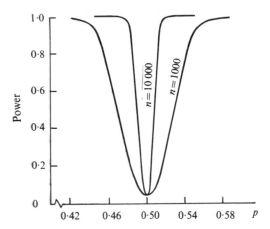

Fig. 12.2.1. Power curves corresponding to $n = 1000$ and $n = 10\,000$ for the test of the binomial hypothesis $p = 0.5$. The significance level is 5 per cent in both cases.

The manufacturer's claim will only be rejected if the experiment results in an excessively large number of 'heads' (a one-tail test), whereas the earlier null hypothesis about lack of bias is rejected if the observed number of 'heads' is either too large or too small (a two-tail test). Both tests are summarised in section 12.6.

12.4 Robustness; non-parametric tests

All statistical tests assume some form of mathematical model (for example a normal population; constant variance). In practice, the conditions of the mathematical model may not be met, and the probability that an incorrect conclusion will be drawn from the test is then increased. Some tests are affected more than others. A test which is little affected by moderate departures from the mathematical model is said to be *robust*.

Many of the tests in this chapter assume a normal population. The tests on means tend to be robust for moderate departures from normality, while the tests on variances are not.

Some of the tests we describe are *distribution-free* or *non-parametric*, and make no assumption about the distributional form of the underlying population. These tests have the advantage of not assuming normality when we are uncertain, but they are slightly less powerful than the corresponding normal-based test when the population is actually normal. A model which makes more assumptions usually leads to a more powerful test.

Further reading: Balaam [3] 219–21; Colquhoun [16] 96–9; Johnson and Leone [50] 255–8; Kendall and Stuart [54] 470–3.

12.5 How the tests are summarised in this chapter

Each test is summarised under six headings:

Usual form of data

This display gives the reader a mental picture of the underlying data to which the statistical test *may* be applicable.

Statistical model

The importance of this heading cannot be over-emphasised. Wrong conclusions are likely to be reached if the model assumed in the test is not appropriate to the data under consideration. Independence assumptions must always be met. Moderate departures from normality, on the other hand, can often be ignored (section 12.4).

Hypotheses

The usual null and alternative hypotheses are given.

Critical region

All the tests are summarised in terms of a 5 per cent significance level. The reader should have no problem adapting to other levels of significance.

Computation of test statistic

A description of the computational procedure.

Comments

Miscellaneous comments dealing with computation, alternative tests, other null hypotheses, robustness, etc.

12.6 Tests involving a single binomial probability p

Usual form of data

Number of trials resulting in the outcome		Total number of trials
Success	Failure	
r	$n-r$	n

Statistical model

The n trials are independent and each has probability p of success. The distribution of successes is therefore binomial.

Hypotheses

(a) Equality	(b) Inequality	(c) Inequality
$H_0: p = p_0$;	$H_0: p \leqslant p_0$;	$H_0: p \geqslant p_0$;
$H_1: p \neq p_0$.	$H_1: p > p_0$.	$H_1: p < p_0$.

Critical region

(a) *Equality*: greater than upper $2\frac{1}{2}$ per cent point or less than the lower $2\frac{1}{2}$ per cent point of the unit normal distribution.

(b) *Inequality*: greater than the upper 5 per cent point of the unit normal distribution.

(c) *Inequality*: less than the lower 5 per cent point of the unit normal distribution.

Computation of test statistic

Set $p = p_0$ in the statistic

■ $$T = (r - np)/\{np(1-p)\}^{\frac{1}{2}}.$$ (12.6.1)

Comments

1. The above test depends upon the normal approximation to the binomial (section 10.2) and should only be used when n is large (> 20, say). For smaller values of n, we can use r as the test statistic and the following critical regions:

(a) *Equality*: the upper and lower $2\frac{1}{2}$ per cent regions of the binomial (n, p_0) distribution.

(b) *Inequality*: the upper 5 per cent region of the binomial (n, p_0) distribution.

(c) *Inequality*: the lower 5 per cent region of the binomial (n, p_0) distribution.

2. The use of binomial tables tends to be inconvenient. It can be avoided by using the following test procedures, which are based on (10.3.2) and (10.3.3) and valid for all n.

(a) *Equality*. Compute

■ $$Y = \frac{r}{n+1-r} \frac{1-p_0}{p_0},$$ (12.6.2)

■ $$Z = \frac{n-r}{r+1} \frac{p_0}{1-p_0}.$$ (12.6.3)

The null hypothesis is rejected if Y exceeds the upper $2\frac{1}{2}$ per cent point of the F-distribution with $2(n+1-r)$ and $2r$ degrees of freedom, or if Z exceeds the upper $2\frac{1}{2}$ per cent point of the F-distribution with $2(r+1)$ and $2(n-r)$ degrees of freedom.

(b) *Inequality*. Use (12.6.2) to compute Y. If Y exceeds the upper 5 per cent point of the F-distribution with $2(n+1-r)$ and $2r$ degrees of freedom, the null hypothesis is rejected.

(c) *Inequality*. Use (12.6.3) to compute Z. The null hypothesis is rejected if Z exceeds the upper 5 per cent point of the F-distribution with $2(r+1)$ and $2(n-r)$ degrees of freedom.

Example 12.6.1. A population is known to have the normal distribution with variance one. The mean is unknown. A sample of size 35 is drawn from the population: twenty of the observations are greater than

10 and fifteen less than 10. Does a claim that the population mean is 9 seem reasonable?

If the population mean were 9, each observation would have the chance $p = \Phi((10-9)/\sqrt{1}) = 0.8413$ of being less than 10 and the chance 0.1587 of being greater than 10. We therefore test the null hypothesis H_0: $p = 0.8413$ against the alternative H_1: $p \neq 0.8413$. The sample size n is reasonably large; so we can use the test statistic (12.6.1) and the two-tail critical region derived from the normal curve. The 5 per cent critical region is $|T| > 1.96$. The value of the test statistic is

$$T = (15 - 35 \times 0.8413)/(35 \times 0.8413 \times 0.1587)^{\frac{1}{2}} = -6.68.$$

This value lies in the critical region; so we reject the null hypothesis and conclude that $p \neq 0.8413$.

The method of comment 2 can also be used. We compute

$$Y = \frac{15}{35+1-15} \frac{0.1587}{0.8413} = 0.13,$$

$$Z = \frac{35-15}{15+1} \frac{0.8413}{0.1587} = 6.63.$$

Z is much larger than the upper $2\frac{1}{2}$ per cent point of the $F_{32, 40}$-distribution; so we reject the null hypothesis and conclude that $p \neq 0.8413$.

> *Further reading*: Balaam [3] 113; Brownlee [7] 148–50; Hoel [45] 175–8; Zar [105] 287–90.

12.7 Testing a single set of multinomial probabilities

Usual form of data

Number of trials resulting in an outcome of type				Total number of trials
1	2	...	k	
n_1	n_2	...	n_k	$n = n_1 + ... + n_k$

Statistical model

The n trials are independent, and each trial has probability p_i of type i outcome ($i = 1, 2, ..., k$). The distribution is therefore multinomial.

Hypotheses

H_0: $p_1 = \pi_1, p_2 = \pi_2, ..., p_k = \pi_k$;
H_1: null hypothesis not true.

Critical region

Greater than the upper 5 per cent point of the chi-square distribution with $k-1$ degrees of freedom.

Computation of test statistic

Set $p_1 = \pi_1, ..., p_k = \pi_k$ in the test statistic

$$\chi^2 = (n_1 - np_1)^2/(np_1) + ... + (n_k - np_k)^2/(np_k). \qquad (12.7.1)$$

Comments

1. The test makes use of the chi-square approximation to the multinomial (section 10.5) and is only valid for large values of n ($n > 25$, say). The expected numbers $\{n\pi_i\}$ of outcomes in the various categories should not be too small. Many authors recommend a minimum expected number of five, although Cochran [13] would allow a minimum expectation of one, provided relatively few expectations are less than five (one cell out of five or two cells out of ten or more).

2. If the expected number in a particular cell is too small, combine it with another cell and reduce the number of degrees of freedom by one. (Add together the two expectations and the two observed numbers.) Unnecessary grouping will make the test less sensitive.

3. This test can be used instead of test (*a*) in section 12.6. The tests are equivalent in the binomial situation.

4. This test can be used to test whether a sample $x_1, x_2, ..., x_n$ of size n came from a certain specified population with distribution function $F(x)$. The range of possible values for the variate is divided into k non-overlapping intervals $(-\infty, a_1), (a_1, a_2), ..., (a_{k-1}, \infty)$. Under the null hypothesis, the probability that a particular observation will fall in the ith interval is $p_i = F(a_i) - F(a_{i-1})$ (section 8.2). The chi-square test is performed in the usual manner (example 12.7.3). Another method is given in section 12.12.

Example 12.7.1. A six-sided die is tossed 50 times with the following results: 12 sixes, 9 fives, 9 fours, 6 threes, 9 twos, and 5 ones. Is there any evidence that the die is biased?

We need to test $H_0: p_1 = p_2 = p_3 = p_4 = p_5 = p_6 = \frac{1}{6}$;

H_1: null hypothesis is not true.

The critical region is the upper 5 per cent region of the χ_5^2-distribution; that is, greater than 11.070. The sample is reasonable large, so we can use the test statistic (12.7.1):

$$\chi_5^2 = (12 - 50 \times \tfrac{1}{6})^2/(50 \times \tfrac{1}{6}) + ... + (5 - 50 \times \tfrac{1}{6})^2/(50 \times \tfrac{1}{6}) = 3.76.$$

This result is not significant. We have no evidence that the die is biased.

Example 12.7.2.[1] In their classic genetics paper, Bateson and Punnett, in Peters [78], record the colour and shape of 427 grains of pollen. The data are reproduced in table 12.9.1. Test the null hypothesis that the purple and long genes are dominant *and* there is no linkage between the colour and shape loci.

Under this null hypothesis, each grain has a chance of $\frac{3}{4}$ of being purple and a chance $\frac{1}{4}$ of being red, a chance $\frac{3}{4}$ of being long and a chance $\frac{1}{4}$ of being round. Lack of linkage implies the independence of the probabilities and so the null hypothesis we need to test is

$$H_0: \begin{cases} P(\text{a grain is purple and long}) & = \frac{9}{16}, \\ P(\text{a grain is purple and round}) & = \frac{3}{16}, \\ P(\text{a grain is red and long}) & = \frac{3}{16}, \\ P(\text{a grain is red and round}) & = \frac{1}{16}. \end{cases}$$

The 5 per cent point for the χ_3^2-distribution is 7.81. Under the null hypothesis, the expected number of purple/long grains is $\frac{9}{16} \times 427 = 240.19$, the expected number of purple/round grains is $\frac{3}{16} \times 427 = 80.06$, the expected number of red/long grains is $\frac{3}{16} \times 427 = 80.06$ and the expected number of red/round grains is $\frac{1}{16} \times 427 = 26.69$. The test statistic is

$$\begin{aligned} \chi^2 &= (296 - 240.19)^2/240.19 + (19 - 80.06)^2/80.06 \\ &\quad + (27 - 80.06)^2/80.06 + (85 - 26.69)^2/26.69 \\ &= 222.1. \end{aligned}$$

This value is (highly) significant; so we have clear evidence to reject the null hypothesis.

Example 12.7.3. The following twenty observations come from an unknown population:

$$0.33, \quad -0.52, \quad -2.41, \quad -1.93, \quad 0.46, \quad -0.44, \quad -0.97,$$
$$-0.38, \quad 0.48, \quad 1.29, \quad -1.82, \quad -1.23, \quad -0.21, \quad 2.66,$$
$$-1.22, \quad -0.41, \quad -0.95, \quad 1.47, \quad -0.83, \quad -0.43.$$

Test the null hypothesis that the underlying distribution is normal with zero mean and unit variance.

To perform the chi-square test, let us divide the range of the unit normal distribution into six convenient intervals: $(-\infty, -1.5), (-1.5, -0.5),$ $(-0.5, 0), (0, 0.5), (0.5, 1.5)$ and $(1.5, \infty)$. The expected number of observations in the interval $(0.5, 1.5)$, for example, is

$$20(\Phi(1.5) - \Phi(0.5)) = 4.8346.$$

The expected numbers in the other intervals are obtained in a similar manner

---

[1] This example should be compared with example 12.9.1.

Table 12.7.1. *Testing the null hypothesis that an unknown distribution is normal with zero mean and unit variance by the chi-square method*

Interval	Number of observations in given interval		Contribution to chi-square
	Observed	Expected	
$-\infty$ to -1.5	3	1.3362	2.0717
-1.5 to -0.5	6	4.8346	0.2809
-0.5 to 0.0	5	3.8292	0.3580
0.0 to 0.5	3	3.8292	0.1796
0.5 to 1.5	2	4.8346	1.6620
1.5 to ∞	1	1.3362	0.0846
Total	20	20.0000	4.6368

(table 12.7.1). The 5 per cent critical value for the chi-square distribution with five degrees of freedom is 11.07. The chi-square value we obtain (4.64) is well below this value; so we accept the null hypothesis and conclude that the underlying distribution is normal with zero mean and unit variance.

The reader's attention is drawn to example 12.12.1.

Further reading: Brownlee [7] 140–4; Chakravarti *et al.* [9] 305; Cochran [13]; Remington and Schork [86] 232–3.

12.8   Equality of multinomial (binomial) probabilities in two or more experiments; the difference between two binomial probabilities

Usual form of data

	Number of trials resulting in an outcome of type				Total number of trials
	1	2	...	k	
Experiment 1	n_{11}	n_{12}	...	n_{1k}	$n_{1.}$
Experiment 2	n_{21}	n_{22}	...	n_{2k}	$n_{2.}$
$\vdots$	$\vdots$	$\vdots$		$\vdots$	$\vdots$
Experiment r	n_{r1}	n_{r2}	...	n_{rk}	$n_{r.}$
All experiments	$n_{.1}$	$n_{.2}$	...	$n_{.k}$	$n_{..}$

Note: $n_{1.} = n_{11} + n_{12} + ... + n_{1k},$
$n_{.1} = n_{11} + n_{21} + ... + n_{r1},$
$n_{..} = n_{.1} + ... + n_{.k} = n_{1.} + ... + n_{r.}.$

Statistical model

The $n_{..}$ trials are independent. In experiment 1, the probability of outcome 1 is p_{11}, the probability of outcome 2 is p_{12}, ...; in experiment 2, the probability of outcome 1 is p_{21}, the probability of outcome 2 is p_{22}, ...; etc. The distribution of outcomes for each experiment is therefore multinomial.

Hypotheses

$$H_0: \begin{cases} p_{11} = p_{21} = ... = p_{r1}, \\ p_{12} = p_{22} = ... = p_{r2}, \\ \cdots \cdots \cdots \cdots \\ p_{1k} = p_{2k} = ... = p_{rk}; \end{cases}$$

H_1: null hypothesis not true.

Critical region

Greater than the upper 5 per cent point of the chi-square distribution with $(r-1)(k-1)$ degrees of freedom.

Computation of test statistic

If the null hypothesis were true, the best estimate $\hat{p}_1$ the common value of $p_{11}, p_{21}, ..., p_{r1}$ would be given by $n_{.1}/n_{..}$. Experiment 1 involves $n_{1.}$ trials; so we would expect $n_{1.}\hat{p}_1 = n_{1.} \times n_{.1}/n_{..}$ of these trials to result in type 1 outcomes. We repeat this procedure for each of the rk cells in the table; that is, we multiply the two marginal totals and divide by the grand total to obtain the expected number in the cell. We than compute

■
$$\chi^2 = \sum_{\substack{\text{all } rk \\ \text{cells}}} \frac{(\text{observed number} - \text{expected number})^2}{\text{expected number}} \qquad (12.8.1)$$

Comments

1. In the binomial situation $k = 2$.

2. The test makes use of the chi-square approximation to the multinomial (section 10.5) and is only valid for reasonably large samples; the expected numbers in the various cells should not be too small. (See comments 1 and 2 of section 12.7.) If a cell is small and needs to be combined with another, combine the cell with another from the same experiment.

3. This statistical model is similar to, though not identical with that of contingency tables as given in section 12.9.

4. In the special case of two binomial experiments, the chi-square distribution has one degree of freedom; the test statistic is the square of a unit normal random variable. To test

$H_0: p_{11} - p_{21} = \delta$ (a specified constant);
$H_1: p_{11} - p_{21} \neq \delta$;

use the upper and lower $2\frac{1}{2}$ per cent regions of the unit normal distribution and the test statistic

$$T = (\hat{p}_{11} - \hat{p}_{21} - \delta)/(\hat{p}_{11}\hat{q}_{11}/n_1. + \hat{p}_{21}\hat{q}_{21}/n_2.)^{\frac{1}{2}} \qquad (12.8.2)$$

where $\hat{p}_{11} = n_{11}/n_1.$, $\hat{q}_{11} = 1 - \hat{p}_{11}$;
$\hat{p}_{21} = n_{21}/n_2.$, $\hat{q}_{21} = 1 - \hat{p}_{21}$.

To test $H_0: p_{11} - p_{21} \leqslant \delta$;
$H_1: p_{11} - p_{21} > \delta$;

use the upper 5 per cent region of the unit normal distribution and the test statistic (12.8.2).

Example 12.8.1. A study is being made of the length of stay in hospital of patients following a particular operation. Fifty patients are selected at random in each of three Australian states: New South Wales, Queensland and Tasmania. Each patient is recorded as having stayed under one week, between one and two weeks, between two and three weeks, between three and four weeks, or over four weeks. The results are given in table 12.8.1. Do patients stay in hospital the same length of time in all three states?

Table 12.8.1. *Length of stay in hospital*

State	Length of stay (weeks)					Total
	<1	1–2	2–3	3–4	>4	
	Number of patients					
New South Wales	8	16	16	6	4	50
Queensland	6	13	14	7	10	50
Tasmania	5	12	10	10	13	50
Total	19	41	40	23	27	150

We need to test H_0: for each duration, the proportion is the
same in each state;
H_1: null hypothesis is not true.
If the null hypothesis were true, the expected number of New South Wales patients staying less than one week would be $19 \times 50/150 = 6.\overset{.}{3}$, and the expected number of Tasmanian patients staying more than four weeks would be $27 \times 50/150 = 9.0$. Similar results hold for the remaining 13 cells. The 5 per cent critical region is the upper 5 per cent region of the χ_8^2 distri-

bution; that is, greater than 15.51. The test statistic is

$$\chi^2 = (8-6.\dot{3})^2/6.\dot{3}+ \dots +(13-9)^2/9 = 8.57.$$

This value is not significant; so we have no reason to reject the null hypothesis.

The reader should note that the data are ordered (patients are subdivided according to length of stay). The above test does not take this into account and it therefore wastes information.

> *Further reading*: Balaam [3] 115; Chakravarti *et al.* [9] 306–8, 311–12; Cochran [13]; Hoel [45] 178–80; Zar [105] 296–7.

12.9 Testing association in a contingency table

Usual form of data

Frequency data are often collected simultaneously for two categorical variables; for example, the hair colour and eye colour of an individual. In a sample of size $n_{..}, n_{ij}$ individuals are of hair category i and eye category j. The data are summarised in a *contingency table*:

Hair-colour category	Eye-colour category				
	1	2	...	s	Total
1	n_{11}	n_{12}	...	n_{1s}	$n_{1.}$
2	n_{21}	n_{22}	...	n_{2s}	$n_{2.}$
$\vdots$	$\vdots$	$\vdots$		$\vdots$	$\vdots$
r	n_{r1}	n_{r2}	...	n_{rs}	$n_{r.}$
Total	$n_{.1}$	$n_{.2}$	...	$n_{.s}$	$n_{..}$

Statistical model

A total of $n_{..}$ random individuals are observed, and the hair colour and eye colour of each individual are noted. The number of individuals with hair colour 2 and eye colour 1 is denoted by n_{21}. Similarly for the other cells. Note that each individual may fall into any one of the rs cells; so the distribution over all rs cells is multinomial.

Hypotheses

H_0: no association between the hair colour and eye colour of an individual

H_1: null hypothesis not true.

Critical region

Greater than the upper 5 per cent point of the chi-square distribution with $(r-1)(s-1)$ degrees of freedom.

Computation of test statistic

Calculate the expected number in each cell by multiplying together

6

the two marginal totals and dividing the product by the grand total, as in section 12.8. Then compute

■
$$\chi^2 = \sum_{\substack{\text{all } rs \\ \text{cells}}} \frac{(\text{observed number} - \text{expected number})^2}{\text{expected number}}.$$
(12.9.1)

Comments

1. This test is approximate, and it is only valid for large values of $n_{..}$ ($n_{..} > 20$, say). The expected numbers in the various cells should not be too small. For contingency tables with more than one degree of freedom, Cochran [13] recommends a minimum expectation of one, provided relatively few expectations are less than five (one cell out of five, or two cells out of ten or more).

2. Fisher's exact test is available for 2×2 tables (section 12.10). This exact test should be used in preference to (12.9.1) whenever $n_{..} < 20$, or $20 \leqslant n_{..} < 40$ and the smallest cell expectation is less than five.

3. The Yates continuity correction should be used when (12.9.1) is applied to a 2×2 table. Calculate the expected numbers in the usual way, but reduce the absolute value of each deviation between observed and expected by $\frac{1}{2}$ before substitution in (12.9.1).

4. Special computational formulae are sometimes used when the table is 2×2, $2 \times s$ or $r \times 2$.

5. Although the computation of the test statistic is exactly the same as in section 12.8, the underlying statistical model is different. We noted above that the distribution of the numbers in the various cells was multinomial. The actual test of association is based on the conditional distribution of the numbers in the various cells given the marginal totals, and in the case of a 2×2 table this conditional distribution turns out to be a hypergeometric distribution (section 10.13).

Example 12.9.1. In their classic genetics paper, Bateson and Punnett, in Peters [78], record the colour and shape of 427 grains of pollen. The data are reproduced in table 12.9.1. Do these data suggest that there is an association between colour and shape?

We need to test H_0: no association;

H_1: there is association.

The 5 per cent critical region is the upper 5 per cent region of the χ_1^2-distribution; that is, greater than 3.84. If there were no association, the expected number of purple/long grains would be $315 \times 323/427 = 238.28$, the expected number of purple/round grains would be $315 \times 104/427 = 76.72$, the expected number of red/long grains would be $112 \times 323/427 = 84.72$ and the expected number of

Table 12.9.1. *The colour and shape of 427 grains of pollen*

Shape	Colour		Total
	Purple	Red	
	Number of grains		
Long	296	27	323
Round	19	85	104
Total	315	112	427

red/round grains would be $112 \times 104/427 = 27.28$. The test statistic is

$$\chi^2 = (296 - 238.28 - 0.5)^2/238.28 + (19 - 76.72 + 0.5)^2/76.72$$
$$+ (27 - 84.72 + 0.5)^2/84.72 + (85 - 27.28 - 0.5)^2/27.28$$
$$= 215.1.$$

This value is (highly) significant; so we reject the null hypothesis and accept the alternative: there is evidence of an association between colour and shape.

This example should be compared with example 12.7.2 in which the expected segregation at each locus was taken into account.

Further reading: Bailey [2] 52–66; Balaam [3] 224; Brownlee [7] 211–17; Chakravarti *et al.* [9] 312–15; Hoel [45] 241–4; Mood and Graybill [70] 311–18; Remington and Schork [86] 235; Snedecor [91] 217–31; Zar [105] 60–9.

12.10   Fisher's exact test for a 2×2 contingency table

Usual form of data

The general $r \times s$ contingency table is described in section 12.9 in terms of two variables: hair colour and eye colour. The 2×2 table takes the form:

Hair-colour category	Eye-colour category		Total
	1	2	
1	n_{11}	n_{12}	$n_{1.}$
2	n_{21}	n_{22}	$n_{2.}$
Total	$n_{.1}$	$n_{.2}$	$n_{..}$

Statistical model

The distribution of the numbers falling in the four cells is multinomial. The conditional distribution of the number in any particular cell, given the marginal totals, is hypergeometric (comment 5 of section 12.9).

Hypotheses

H_0: no association between the hair colour and eye colour of an individual;

H_1: null hypothesis not true.

Critical region

P less than 0.05.

Computation of test statistic

A 2×2 contingency table with the marginal totals $n_{1.}, n_{2.}, n_{.1}$ and $n_{.2}$ and grand total $n_{..}$ is fully determined once we know the entry in the top left-hand cell (the table has one degree of freedom). It is possible therefore to find a range of permissible values for the top left-hand entry f_{11}. The maximum and minimum f_{11} values will be denoted by M and m respectively. Once a particular value for f_{11} is chosen, the entries f_{12}, f_{21} and f_{22} are known. The probability of obtaining a contingency table with top left-hand entry f_{11}, given the above marginal totals is

■
$$p(f_{11}) = \left(\frac{n_{1.}! \, n_{2.}! \, n_{.1}! \, n_{.2}!}{n_{..}!}\right) \frac{1}{f_{11}! \, f_{12}! \, f_{21}! \, f_{22}!}. \qquad (12.10.1)$$

Let us imagine that $n_{1.} < n_{2.}$ and $n_{.1} < n_{.2}$. (The rows and columns can always be rearranged so that this is true.) The following calculation procedure is followed when n_{11} is less than $n_{1.} n_{.1}/n_{..}$:

1. Calculate the left-hand tail probabilities $p(n_{11}), p(n_{11}-1), ...,$ $p(m)$.

2. Calculate the right-hand tail probabilities $p(M), p(M-1), ...,$ $p(M^*)$. M^* is the integer such that $p(M^*) < p(n_{11}) < p(M^*-1)$.

3. The test statistic P is the sum of the above left-and right-hand tail probabilities.

When $n_{11} > n_{1.} n_{.1}/n_{..}$, the following calculation process is followed:

1. Calculate the right-hand tail probabilities $p(n_{11}), p(n_{11}+1), ...,$ $p(M)$.

2. Calculate the left-hand tail probabilities $p(m), p(m+1), ..., p(m^*)$. m^* is the integer such that $p(m^*) < p(n_{11}) < p(m^*+1)$.

3. The test statistic P is the sum of the left- and right-hand tail probabilities.

Comments

1. Whenever possible, the more convenient chi-square test should be used (section 12.9). The conditions under which the chi-square test

is inappropriate and Fisher's exact test must be used are fully explained in comment 2 of section 12.9.

2. The calculation of individual probabilities can be avoided by the use of tables. Zar [105] 518–42 gives such a table.

Example 12.10.1. Is there any evidence of association in the following contingency table?

4	2	6
1	8	9
5	10	15

The chi-square test cannot be used for this table; so we are forced to use Fisher's exact method. The maximum and minimum values permitted in the top left-hand cell for a table with these marginal totals are $M = 5$ and $m = 0$ respectively. We also note that the actual cell entry (4) is greater than the expected value $(5 \times 6)/15 = 2$. We begin by calculating

$$\frac{6! \, 9! \, 5! \, 10!}{15!} = 87\,004.2.$$

Then $p(4) = 87\,004.2 \, \dfrac{1}{4! \, 2! \, 1! \, 8!} = 0.044\,955;$

$p(5) = 87\,004.2 \, \dfrac{1}{5! \, 1! \, 0! \, 9!} = 0.001\,998;$

$p(0) = 87\,004.2 \, \dfrac{1}{0! \, 6! \, 5! \, 4!} = 0.041\,958;$

$p(1) = 87\,004.2 \, \dfrac{1}{1! \, 5! \, 4! \, 5!} = 0.251\,748.$

The test criterion is $P = p(4) + p(5) + p(0) = 0.0889$. This value is not significant; so we have no evidence of association.

Further reading: Brownlee [7] 163–6; Zar [105] 291–5.

12.11 The chi-square test of goodness-of-fit

A distribution with γ parameters has been fitted to some data and we wish to test the goodness-of-fit. The chi-square method outlined in comment 4 of section 12.7 and demonstrated in example 12.7.3 cannot be used for this purpose because the parameters of the distribution are actually estimated from the data instead of being pre-specified. It turns out, however, that the chi-square test requires only a minor modification: a reduction in the number of degrees of freedom.

Usual form of data

A distribution with γ parameters has been fitted to some data. The range of possible values for the variable is divided into k non-overlapping intervals or cells. Each of the original n observations must lie in one of these cells. The expected number in each cell is calculated on the basis of the fitted distribution.

	Number of observations in cell				
	1	2	...	k	Total
Observed number	o_1	o_2	...	o_k	n
Expected number	e_1	e_2	...	e_k	n

Statistical model

The n observations are independent, and each observation has probability p_1 of falling in cell 1, p_2 of falling in cell 2, ..., p_k of falling in cell k. The distribution over the k cells is therefore multinomial.

Hypotheses

H_0: $p_1 = e_1/n, p_2 = e_2/n, ..., p_k = e_k/n$;
H_1: null hypothesis not true.

Critical region

Greater than the upper 5 per cent point of the chi-square distribution with $k - \gamma - 1$ degrees of freedom.

Computation of test statistic

$$\chi^2 = \frac{(o_1 - e_1)^2}{e_1} + ... + \frac{(o_k - e_k)^2}{e_k}.$$

(12.11.1)

Comments

1. The expected number in each cell must not be too small. In the case of a unimodal distribution the expectations will only be small at one or both tails; Cochran [13] recommends grouping so that the minimum expectation at each tail is at least one.

2. If the expected number in a particular cell is too small, combine it with a neighbour by adding the respective observed and expected values, and then reduce the number of degrees of freedom by one.

3. In the case of a continuous distribution, each cell will correspond to a range of possible values for the continuous variate (example 12.7.3).

4. The Kolmogorov–Smirnov method of section 12.12 is sometimes

used for testing goodness-of-fit; this test has the advantage of taking the order of the cells into account.

5. Note that the distribution of the numbers falling in the various cells is multinomial even though the distribution fitted might be normal or Poisson or some other.

Example 12.11.1. In example 10.11.1, a negative binomial distribution with mean 6.73 and parameter $k = 0.6$ is fitted to some data on the distribution of eggs of parasitic nematodes in sheep. The observed and fitted values are exhibited in table 10.11.1. Test the goodness-of-fit by the chi-square method.

The expected frequencies of 6, 7, 8 and 9 eggs are small. Let us therefore combine cells 6 and 7 into a single cell and cells 8 and 9 into another single cell. We are then left with eleven distinct cells.

When the negative binomial distribution was fitted, two parameters k and p were calculated from the data. The 5 per cent critical region for the chi-square goodness-of-fit test is therefore the upper 5 per cent region of the χ_8^2 distribution; that is, greater than 15.51. The test statistic is

$$\chi^2 = (20-20.0)^2/20.0 + (12-11.0)^2/11.0 + ... + (11-7.8)^2/7.8 = 12.1.$$

This value is not significant; so we conclude that the fit is adequate.

Further reading: Bailey [2] 67–77; Chakravarti *et al.* [9] 390–1; Cochran [13]; Colquhoun [16] 132–4; Mood and Graybill [70] 308–11; Zar [105] 41, 45–50.

12.12 The Kolmogorov–Smirnov one-sample test

The chi-square method for testing whether a given sample came from a certain specified population was outlined in comment 4 of section 12.7 and demonstrated in example 12.7.3. The Kolmogorov–Smirnov test is another non-parametric test which can be used for this purpose, and it has the advantage over the chi-square method that it takes the order of the observations into account. The test is designed to test the goodness-of-fit of an empirical to a theoretical distribution function.

The statistical model underlying the test assumes a continuous distribution so that the sample observations have zero probability of being equal. In practice, however, the test is often applied to grouped data and data from discrete distributions, and in both these situations equal observations are encountered. The significance level of the test is then lower than the nominal rate and the chance of a Type II error is increased.

The chi-square test of example 12.7.3 and the Kolmogorov–Smirnov test both require the null-hypothesis distribution to be fully specified in advance (for example, normal with zero mean and unit variance). In the goodness-of-fit context, the form of distribution is specified, but the parameters of the distribution are estimated from the sample. We have seen in section 12.11 that the chi-square test is easily modified by reducing the number of degrees of freedom, but it is not known what adjustment to the Kolmogorov–Smirnov test may be necessary. Nevertheless, it is sometimes used for testing goodness-of-fit. The reader is warned that the true significance level will then be slightly lower than the nominal level and the chance of Type II error will be increased. The effect is not likely to be great if the number of estimated parameters is small relative to the sample size.

Usual form of data

$x_1, x_2, ..., x_n$ (a sample of n independent observations ordered such that $x_1 < x_2 < ... < x_n$).

Statistical model

The n observations are independent and all come from the same population which is continuous.

Hypotheses

H_0: the distribution function of the population is $F(x)$;
H_1: the null hypothesis is not true.

Critical region

Greater than the upper 5 per cent point of the Kolmogorov–Smirnov distribution (table 12.12.1).

Calculation of test statistic

1. Calculate the cumulative differences
$D_1 = 1 - nF(x_1), D_2 = 2 - nF(x_2)$, etc.
2. Find $|D_i|_{max}$, the largest absolute cumulative difference.
3. Compute the test statistic $D = (|D_i|_{max})/n$.

Comment

Avoid using this test for goodness-of-fit purposes. The test should also be avoided with discrete data and grouped continuous data, because of the uncertainty surrounding the true significance level. The alternative chi-square test, however, will not always detect the accumulation of a large number of small deviations all the same sign. If the detection of such discrepancies is important, use the Kolmogorov–Smirnov test, but with care. Reject the null hypothesis when the test statistic exceeds the nominal critical value, and accept the null hypothesis when the test statistic is well below the nominal critical value. Be careful about making any decision when the test statistic is only slightly smaller than the nominal critical value.

Example 12.12.1. Twenty independent observations from an unknown distribution are listed in example 12.7.3. Test the null hypothesis that the unknown distribution is normal with zero mean and unit variance by the Kolmogorov–Smirnov method.

Table 12.12.1. *The upper 100α per cent points of the Kolmogorov-Smirnov distribution**

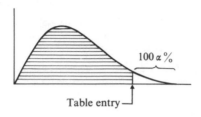

Table entry

n†	$\alpha = 0.05$	0.01	0.001
5	0.563	0.669	0.781
6	0.519	0.617	0.725
7	0.483	0.576	0.679
8	0.454	0.542	0.641
9	0.430	0.513	0.608
10	0.409	0.489	0.580
11	0.391	0.468	0.556
12	0.375	0.449	0.534
13	0.361	0.432	0.515
14	0.349	0.418	0.498
15	0.338	0.404	0.482
16	0.327	0.392	0.467
17	0.318	0.381	0.454
18	0.309	0.371	0.442
19	0.301	0.361	0.431
20	0.294	0.352	0.421
25	0.264	0.317	0.378
30	0.242	0.290	0.347
35	0.224	0.269	0.322
40	0.210	0.252	0.302
45	0.198	0.238	0.285
50	0.188	0.226	0.271

* *Source*: Jerrold H. Zar, *Biostatistical Analysis*, ($\bar{C}$) 1974, pp. 471–3. Reprinted by permission of Prentice-Hall, Inc., Englewood Cliffs, New Jersey, USA.
† For $n > 50$, critical values of D can be approximated by
$$\left\{ \frac{1}{2n} \ln \left(\frac{2}{\alpha} \right) \right\}^{\frac{1}{2}}.$$

Table 12.12.2. *Testing the null hypothesis that an unknown distri-*
bution is normal with zero mean and unit variance by the
Kolmogorov–Smirnov method

Observation x_i	Number of observations less than or equal to x_i		|Difference| $\|D_i\|$
	Observed	Expected	
-2.41	1	0.16	0.84
-1.93	2	0.54	1.46
-1.82	3	0.69	2.31
-1.23	4	2.19	1.81
-1.22	5	2.22	2.78
-0.97	6	3.32	2.68
-0.95	7	3.42	3.58
-0.83	8	4.07	3.93
-0.52	9	6.03	2.97
-0.44	10	6.60	3.40
-0.43	11	6.67	4.33
-0.41	12	6.82	5.18
-0.38	13	7.04	5.96
-0.21	14	8.34	5.66
0.33	15	12.59	2.41
0.46	16	13.54	2.46
0.48	17	13.69	3.31
1.29	18	18.03	0.03
1.47	19	18.58	0.42
2.66	20	19.92	0.08

We begin by listing the observations in ascending order (table 12.12.2), and then calculate the observed and expected numbers less than or equal to each observation. The observed number of observations less than or equal to -0.97 is six and the expected number is $20\Phi(-0.97) = 3.32$. The absolute value of the cumulative difference at $x = -0.97$ is therefore 2.68. The absolute cumulative differences at the other points are calculated similarly.

From table 12.12.2, we see that the maximum absolute difference is 5.96, and the test statistic is

$$D = 5.96/20 = 0.298.$$

The 5 per cent critical value is 0.294; so we reject the null hypothesis and conclude that the unknown distribution is not normal with zero mean and unit variance. This conclusion is contrary to the one arrived at via the chi-square test, which does not take into account the large *cumulative* difference between the observed distribution and the unit normal distribution.

The reader's attention is drawn to the warning in section 12.2.

Example 12.12.2. In example 10.11.1 a negative binomial distribution with mean 6.73 and parameter $k = 0.6$ was fitted to some data on the distribution of eggs of parasitic nematodes in sheep. The observed and fitted values are exhibited in table 10.11.1. Test the goodness-of-fit by the Kolmogorov-Smirnov method.

We begin by forming table 12.12.3. According to the formula at the foot of table 12.12.1, the 5 per cent critical value is 0.143. The test statistic

$$D = 7.4/90 = 0.082$$

is well below the nominal critical value. In fact, the formula at the foot of table 12.12.1 tells us that the nominal probability associated with $D = 0.082$ is about 0.6. So we accept the fit as adequate.

Further reading: Balaam [3] 225; Chakravarti *et al.* [9] 392–3; Massey [64]; Zar [105] 54–8.

Table 12.12.3. *The cumulative distribution of eggs of parasitic nematodes in sheep in Scotland, together with the cumulative expected distribution calculated on the basis of the negative binomial distribution (example 10.11.1)*

Number of eggs	Cumulative frequency		Cumulative difference
	Observed	Fitted	
0	20	20.0	0.0
1	32	31.0	1.0
2	46	39.1	6.9
3	53	45.6	7.4
4	56	50.9	5.1
5	62	55.4	6.6
6	65	59.3	5.7
7	68	62.6	5.4
8	70	65.5	4.5
9	73	68.1	4.9
10–14	77	77.1	−0.1
15–19	79	82.2	−3.2
20+	90	90.0	0.0

12.13 Tests involving a single mean

Usual form of data

$x_1, x_2, x_3, ..., x_n$ (a sample of size n).

Statistical model

The observations are independent and come from a normal population with mean μ and variance σ^2.

Hypotheses

(a) Equality	(b) Inequality	(c) Inequality
$H_0: \mu = \mu_0;$	$H_0: \mu \leqslant \mu_0;$	$H_0: \mu \geqslant \mu_0;$
$H_1: \mu \neq \mu_0.$	$H_1: \mu > \mu_0.$	$H_1: \mu < \mu_0.$

Critical region

(a) *Equality*: greater than upper $2\frac{1}{2}$ per cent point or less than the lower $2\frac{1}{2}$ per cent point of the t_{n-1}-distribution.

(b) *Inequality*: greater than upper 5 per cent point of the t_{n-1}-distribution.

(c) *Inequality*: less than lower 5 per cent point of the t_{n-1}-distribution.

Computation of test statistic

$\bar{x}$ = sample mean (formula (8.4.1)).

s^2 = sample variance (formula (8.4.3)).

Set $\mu = \mu_0$ in the statistic

■ $$t = n^{\frac{1}{2}}(\bar{x} - \mu)/s. \tag{12.13.1}$$

Comments

1. μ_0 is a specified number; for example, 103.6.

2. Sometimes the population variance σ^2 is known. The unit normal distribution should then be used instead of the t-distribution, and there is no need to compute s^2. The population standard deviation σ will replace s in formula (12.13.1).

3. The test is robust for moderate departures from normality. Nevertheless, the reader's attention is drawn to the non-parametric test of section 12.14.

4. Because of the symmetry of the t-distribution, it is sufficient when testing equality to compare $|t|$ with the upper $2\frac{1}{2}$ per cent region of the t_{n-1}-distribution.

Example 12.13.1. A statistician has procured eleven hand-grenades. He wishes to test the manufacturer's claim that the mean fuse-time[2] is 4.01 seconds, and obtains the following results with the sample of eleven:

4.21, 4.03, 3.99, 4.05, 3.89, 3.98, 4.01, 3.92, 4.23, 3.85 and 4.20.

What conclusion can he come to?

[2] Variability in fuse-time is perhaps even more important. This problem is dealt with in example 12.28.1.

The hand-grenade will not be suitable for military use if the mean fuse-time is too short (for obvious reasons!) or if the mean fuse time is too long (because an enemy can catch it and throw it back). We need to test

H_0: $\mu = 4.01$;
H_1: $\mu \neq 4.01$.

From the t_{10}-tables, the critical region is $|t| > 2.228$. The sample mean $\bar{x}$ is found to be 4.033 and the sample standard deviation s is 0.130. The test statistic is

$$t = (11)^{\frac{1}{2}}(4.033 - 4.01)/0.130 = 0.59.$$

This value is not significant; so we have no reason to doubt the manufacturer's claim.

Further reading: Balaam [3] 98, 123; Brownlee [7] 105–9, 113–18, 258–60, 295–7; Chakravarti *et al.* [9] 323; Guenther [39] 14–15; Hoel [45] 170–4; Mood and Graybill [70] 301–2; Remington and Schork [86] 200–9; Zar [105] 86–91.

12.14 Non-parametric tests involving a single median

Usual form of data
$x_1, x_2, x_3, \ldots, x_n$ (a sample of size n).
Statistical model
The observations are independent and all come from the same distribution.
Hypotheses

(a) Equality	(b) Inequality	(c) Inequality
H_0: median $= m$;	H_0: median $\leq m$;	H_0: median $\geq m$;
H_1: median $\neq m$.	H_1: median $> m$.	H_1: median $< m$.

Critical region
(a) *Equality*: greater than the upper $2\frac{1}{2}$ per cent point or less than the lower $2\frac{1}{2}$ per cent point of the unit normal distribution.
(b) *Inequality*: greater than the upper 5 per cent point of the unit normal distribution.
(c) *Inequality*: less than the lower 5 per cent point of the unit normal distribution.
Computation of test statistic
For each observation, record $+$ if the observation is greater than m and $-$ if the observation is less than m. Count the total number of $+$ signs and call this N. The test statistic is

■ $$T = (2N - n)/\sqrt{n}.$$ (12.14.1)

Comments

1. The population median is defined in section 8.5. In the case of a symmetric distribution the mean and median coincide, and (12.14.1) can be used to test hypotheses concerning the mean.

2. The test statistic (12.14.1) should only be used when n is large (>25, say). For smaller values of n, use N as the test statistic and the following critical regions:

(a) *Equality*: the upper $2\frac{1}{2}$ per cent region or the lower $2\frac{1}{2}$ per cent region of the binomial $(n, \frac{1}{2})$ distribution.

(b) *Inequality*: the upper 5 per cent point of the binomial $(n, \frac{1}{2})$ distribution.

(c) *Inequality*: the lower 5 per cent region of the binomial $(n, \frac{1}{2})$ distribution.

The upper and lower percentage points of the binomial $(n, \frac{1}{2})$ distribution are given in table 12.23.1.

3. When n is small, the power of the test is not very great.

Example 12.14.1. A population is known to have the normal distribution with variance one. The mean is unknown. A sample of size 35 is drawn from the population: twenty-five of the observations are greater than 10 and ten are less than 10. Does a claim that the population mean is 10 seem reasonable?

The normal distribution is symmetrical; so the median test can be used to test the null hypothesis that the mean is 10. For a 5 per cent level of significance, we use the upper and lower $2\frac{1}{2}$ per cent regions of the unit normal distribution; that is, $|T| > 1.96$. The test statistic is

$$T = (2 \times 25 - 35)/\sqrt{(35)} = 2.54.$$

This value is significant. So we reject the null hypothesis and conclude that the population mean is not 10.

Further reading: Balaam [3] 222–3.

12.15 Equality (inequality) of two means – equal variance case

Usual form of data

Sample 1 (size n_1)	Sample 2 (size n_2)
x_{11}	x_{12}
x_{21}	x_{22}
$\vdots$	$\cdot$
	$\cdot$
x_{n_11}	$\cdot$
	x_{n_22}

Totals	$T_{.1}$	$T_{.2}$

Statistical model

Both samples come from normal populations with variances σ^2. The first sample comes from a population with mean μ_1 and the second sample comes from a population with mean μ_2. All the observations are independent.

Hypotheses

(a) Equality	(b) Inequality
$H_0: \mu_1 = \mu_2$;	$H_0: \mu_1 \leqslant \mu_2$;
$H_1: \mu_1 \neq \mu_2$.	$H_1: \mu_1 > \mu_2$.

Critical region

(a) *Equality*: greater than the upper $2\frac{1}{2}$ per cent point or less than the lower $2\frac{1}{2}$ per cent point of the t-distribution with $n_1 + n_2 - 2$ degrees of freedom.

(b) *Inequality*: greater than the upper 5 per cent point of the t-distribution with $n_1 + n_2 - 2$ degrees of freedom.

Computation of test statistic

$\bar{x}_{.1} = T_{.1}/n_1$.

$\bar{x}_{.2} = T_{.2}/n_2$.

$s_1^2 =$ sample variance for first sample (formula (8.4.3)).

$s_2^2 =$ sample variance for second sample.

Set $\mu_1 = \mu_2$ in the statistic ⎤ *goes to 0*

$$t = \frac{\{(\bar{x}_{.1} - \bar{x}_{.2}) - (\mu_1 - \mu_2)\}(n_1 + n_2 - 2)^{\frac{1}{2}}}{\{(1/n_1) + (1/n_2)\}^{\frac{1}{2}} \{(n_1 - 1)s_1^2 + (n_2 - 1)s_2^2\}^{\frac{1}{2}}} .$$ (12.15.1)

Comments

1. The equality test (*a*) is equivalent to the one-way analysis of variance test for two samples (section 12.18).

2. The test is robust for moderate departures from normality. Note, however, the non-parametric test of section 12.17.

3. The test is robust for moderate departures from the equality of variance assumption when n_1 and n_2 are approximately equal. The approximate test of section 12.16 does not assume equal variances. A test of equality of variance is given in section 12.29.

4. Other hypotheses can be tested; for example:

(*c*) $H_0: \mu_1 - \mu_2 = \delta$ (a specified constant);

$\quad H_1: \mu_1 - \mu_2 \neq \delta$.

Use method (*a*) and set $\mu_1 - \mu_2$ equal to δ in formula (12.15.1).

(*d*) $H_0: \mu_1 - \mu_2 \leqslant \delta$;

$\quad H_1: \mu_1 - \mu_2 > \delta$.

Use method (*b*) and set $\mu_1 - \mu_2$ equal to δ in formula (12.15.1).

5. For values of $n_1 + n_2 - 2$ greater than 30, the normal distribution can be used instead of the *t*-distribution (section 9.3).

6. When the observations in the two samples are paired, the method of section 12.22 must be used.

7. It sometimes happens that the common population variance σ^2 is known. In this situation use σ^2 instead of s_1^2 and s_2^2 and the normal table instead of *t*.

8. Because of the symmetry of the *t*-distribution, it is sufficient when testing equality to compare $|t|$ with the upper $2\frac{1}{2}$ per cent of the *t*-distribution.

Example 12.15.1. Two different types of wheat are being compared for yield. Type *A* is a common variety and type *B* is a new hybrid variety. Twenty-five acres of each kind are planted and exposed to fairly uniform growing conditions. The average yield with variety *A* is 32.0 bushels per acre with variance 5.9. The average yield with variety *B* is 36.2 bushels with variance 11.2. Is the yield from variety *B* significantly higher than the yield from variety *A*?

We need to test $H_0: \mu_A \geqslant \mu_B$;

$\quad H_1: \mu_A < \mu_B$.

Let us use a 5 per cent significance level. The critical region is the lower 5 per cent region of the t_{48}-distribution: that is, less than -1.645. The test statistic is

$$t = \frac{(32.0 - 36.2)(48)^{\frac{1}{2}}}{(\frac{1}{25} + \frac{1}{25})^{\frac{1}{2}}(24 \times 5.9 + 24 \times 11.2)^{\frac{1}{2}}} = -5.08.$$

This value is (very) highly significant, and we conclude that variety *B* gives a higher yield.

Table 12.15.1. *Serum cholesterol levels in turtles* (mg/100 ml)

Male	Female
226.5	221.5
224.1	230.2
218.6	223.4
220.1	224.3
228.8	230.8
229.6	223.8
222.5	

Example 12.15.2. The data in table 12.15.1 give the serum cholesterol levels in seven male and six female turtles. Do male and female turtles have the same mean concentration?

We need to test

$$H_0: \mu_M = \mu_F;$$

$$H_1: \mu_M \neq \mu_F.$$

Let us use a 5 per cent significance level. The critical region is made up of the lower and upper $2\frac{1}{2}$ per cent regions of the t_{11} distribution (that is, $|t| > 2.201$).

From the data we compute $\bar{x}_M = 224.314$, $\bar{x}_F = 225.667$, $s_M^2 = 17.765$ and $s_F^2 = 14.951$. The test statistic is

$$|t| = \left| \frac{(224.314 - 225.667)(11)^{\frac{1}{2}}}{(\frac{1}{7} + \frac{1}{6})^{\frac{1}{2}}(6 \times 17.765 + 5 \times 14.951)^{\frac{1}{2}}} \right| = 0.599.$$

This value is not significant; so we accept the null hypothesis that male and female turtles have the same mean cholesterol concentrations.

Further reading: Balaam [3] 126–9; Brownlee [7] 119, 297–9; Chakravarti *et al.* [9] 324; Colquhoun [16] 148–51; Guenther [39] 22–3; Mood and Graybill [70] 303–6; Remington and Schork [86] 210–15; Snedecor [91] 85–95; Zar [105] 105–6.

12.16 Equality (inequality) of two means – unequal variance case

Usual form of data

See section 12.15.

Statistical model

The first sample comes from a normal population with mean μ_1 and variance σ_1^2. The second sample comes from a normal population with mean μ_2 and variance σ_2^2. All the observations are independent.

Hypotheses

(a) Equality

$H_0: \mu_1 = \mu_2$;

$H_1: \mu_1 \neq \mu_2$.

(b) Inequality

$H_0: \mu_1 \leqslant \mu_2$;

$H_1: \mu_1 > \mu_2$.

Critical region

(a) *Equality*: greater than the upper $2\frac{1}{2}$ per cent point or less than the lower $2\frac{1}{2}$ per cent point of the t_ν-distribution, where

$$\blacksquare \qquad \nu = \frac{\left(\dfrac{s_1^2}{n_1} + \dfrac{s_2^2}{n_2}\right)^2}{\left(\dfrac{s_1^2}{n_1}\right)^2 \left(\dfrac{1}{n_1-1}\right) + \left(\dfrac{s_2^2}{n_2}\right)^2 \left(\dfrac{1}{n_2-1}\right)} - 2. \qquad (12.16.1)$$

(b) *Inequality*: greater than the upper 5 per cent point of the t_ν-distribution.

Computation of test statistic

$\bar{x}_{.1} = T_{.1}/n_1$.

$\bar{x}_{.2} = T_{.2}/n_2$.

s_1^2 = sample variance for first sample (formula (8.4.3)).

s_2^2 = sample variance for second sample.

Set $\mu_1 = \mu_2$ in the statistic

$$\blacksquare \qquad t = \frac{(\bar{x}_{.1} - \bar{x}_{.2}) - (\mu_1 - \mu_2)}{\left(\dfrac{s_1^2}{n_1} + \dfrac{s_2^2}{n_2}\right)^{\frac{1}{2}}}. \qquad (12.16.2)$$

Comments

1. This is an approximate test. Use the exact test of section 12.15 if there is no evidence that the variances are unequal. A test for equality of variance is given in section 12.29.

2. The test is robust for moderate departures from normality. Note however the non-parametric test of section 12.17.

3. Avoid having to calculate ν. Always evaluate the test statistic before computing ν; in many cases it will be obvious whether the statistic is significant or not without calculating ν. If $n_1 + n_2$ is greater than 30, use the normal distribution instead of the t_ν-distribution (section 9.3).

4. This method must not be used when the observations are paired. See sections 12.22–12.24.

5. Other hypotheses can be tested; for example:

(c) $H_0: \mu_1 - \mu_2 = \delta$;

$\quad H_1: \mu_1 - \mu_2 \neq \delta$.

Use method (a) and set $\mu_1 - \mu_2$ equal to δ in formula (12.16.2).

(d) H_0: $\mu_1 - \mu_2 \leqslant \delta$;
   H_1: $\mu_1 - \mu_2 > \delta$.

Use method (b) and set $\mu_1 - \mu_2$ equal to δ in formula (12.16.2).

6. It sometimes happens that the population variances σ_1^2 and σ_2^2 are known. In this situation use the population variances instead of the sample variances and the normal table instead of the t-table; the test is exact.

7. Because of the symmetry of the t-distribution, it is sufficient when testing equality to compare $|t|$ with the upper $2\frac{1}{2}$ per cent region of the t_ν-distribution

Example 12.16.1. Two different types of wheat are being compared for yield; the experimental results are described in example 12.15.1. Is the yield from variety B significantly higher than the yield from variety A?

The yield with variety B seems to be somewhat more variable than the yield with variety A, and we may be uneasy about using the test of section 12.15. The approximate test of the present section can be used. We need to test

H_0: $\mu_A \geqslant \mu_B$;
H_1: $\mu_A < \mu_B$.

Let us again use a 5 per cent significance level, and because $n_1 + n_2$ is large the critical region is the lower 5 per cent region of the normal distribution; that is, less than -1.645. The test statistic is

$$t = (32.0 - 36.2)/(5.9/25 + 11.2/25)^{\frac{1}{2}} = -5.08.$$

This value is (highly) significant; so we reject the null hypothesis and conclude that variety B must give a higher yield.

This is the second test we have applied to these data, and we do it for demonstration purposes. The reader's attention is drawn to the warning in section 12.2.

Further reading: Brownlee [7] 299–304; Guenther [39] 22–3; Hoel [45] 171–5; Snedecor [91] 97–100.

12.17 Two independent samples – the Mann–Whitney test

This non-parametric test is often used to test differences between two means when the assumptions underlying the usual t-test are not valid (for example, the underlying populations are not normal). It is also used when the data are actually given as ranks (for example, the ranked heights of $n = n_1 + n_2$ students, n_1 of whom are boys and n_2 of whom are girls).

The reader should note that the null hypothesis actually tested is that the two populations from which the samples were obtained are identical.

A significant value of the test statistic leads us to conclude that the populations differ, but not necessarily in respect of their means; to infer that the samples come from populations with different means or medians, we must assume that they are identical in all other respects, for example, that they have the same variance. In practice, moderate differences in variability are acceptable, because the test statistic is insensitive to them.

Usual form of data

Sample 1 (size n_1)		Sample 2 (size n_2)	
Observation	Rank	Observation	Rank
x_{11}	r_{11}	x_{12}	r_{12}
x_{21}	r_{21}	x_{22}	r_{22}
.	.	.	.
.	.	.	.
.	.	.	.
$x_{n_1 1}$	$r_{n_1 1}$		
		$x_{n_2 2}$	$r_{n_2 2}$
Totals —	R_1	—	R_2

The $\{x_{ij}\}$ are the actual observations. Define $n = n_1 + n_2$. The $\{r_{ij}\}$ are the ranks of these observations and range from 1 to n.

Statistical model

All the observations are independent, and the observations within the same sample come from the same population.

Hypotheses

H_0: the two populations are identical;
H_1: the null hypothesis is not true.

Critical region

(i) *Small samples*: the upper $2\frac{1}{2}$ per cent region of the Mann-Whitney distribution with parameters n_1 and n_2 (table 12.17.1).
(ii) *Large samples*: the upper $2\frac{1}{2}$ per cent region of the unit normal distribution.

Computation of test statistic

(i) *Small samples*

$U_1 = n_1 n_2 + \frac{1}{2}n_1(n_1+1) - R_1$,
$U_2 = n_1 n_2 + \frac{1}{2}n_2(n_2+1) - R_2$,
$U = \max(U_1, U_2)$.

Compare U with small sample critical region.

(ii) *Large samples*

Compare the following statistic with the large sample critical region:

$$T = (U - \tfrac{1}{2}n_1 n_2)/\{n_1 n_2(n+1)/12\}^{\frac{1}{2}}. \tag{12.17.1}$$

Table 12.17.1. *The upper*† *5 per cent (ordinary type) and* $2\frac{1}{2}$ *per cent (heavy type) points of the Mann–Whitney distribution.* * *The critical region includes the tabled value*

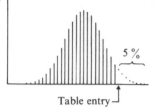

5 %

Table entry ⌐

n_2	n_1									
	2	3	4	5	7	10	12	15	17	20
4	–	12	15							
	--	–	**16**							
5	10	14	18	21						
	–	**15**	**19**	**23**						
6	12	16	21	25						
	–	**17**	**22**	**27**						
7	14	19	24	29	38					
	–	**20**	**25**	**30**	**41**					
8	15	21	27	32	43					
	16	**22**	**28**	**34**	**46**					
9	17	23	30	36	48					
	18	**25**	**32**	**38**	**51**					
10	19	26	33	39	53	73				
	20	**27**	**35**	**42**	**56**	**77**				
12	22	31	39	47	63	86	102			
	23	**32**	**41**	**49**	**66**	**91**	**107**			
14	25	35	45	54	72	99	117	144		
	27	**37**	**47**	**57**	**76**	**104**	**123**	**151**		
16	29	40	50	61	82	112	132	163	183	
	31	**42**	**53**	**65**	**86**	118	**139**	**170**	**191**	
18	32	45	56	68	91	125	148	182	204	
	34	**47**	**60**	**72**	**96**	**132**	**155**	**190**	**213**	
20	36	49	62	75	101	138	163	200	225	262
	38	**52**	**66**	**80**	**106**	**145**	**171**	**210**	**235**	**273**
25	44	61	77	93	125	171	202	247	278	323
	47	**65**	**82**	**98**	**131**	**179**	**211**	**258**	**290**	**337**
30	53	73	92	111	149	204	240	294	330	384
	55	**77**	**97**	**117**	**156**	**213**	**251**	**307**	**344**	**400**
35	61	84	107	129	172	236	279	341	383	445
	64	**89**	**113**	**136**	**181**	**247**	**291**	**356**	**399**	**463**
40	69	96	121	147	196	269	317	388	435	506
	73	**102**	**129**	**155**	**206**	**281**	**331**	**404**	**453**	**526**

* *Source*: Milton (1964, *J. American Statistical Association*, 59, 925–34). Derived with permission of the publisher.

† If required, the lower $100\alpha\%$ points of the Mann–Whitney distribution are obtained by subtracting the upper $100\alpha\%$ points from $n_1 n_2$

Comments

1. It does not matter whether the observations are ranked from smallest to largest or vice versa.

2. When two or more observations have exactly the same value, they are said to be *tied*. The rank assigned to each of them is the mean of the ranks which would have been assigned to them had they not been tied. A small adjustment is sometimes made to take account of ties (Zar [105] 112–13).

3. Note that $U_1 + U_2 = n_1 n_2$. If U_1 lies in the lower $2\frac{1}{2}$ per cent region, U_2 will lie in the upper $2\frac{1}{2}$ per cent region and so will U. If U_2 lies in the lower $2\frac{1}{2}$ per cent region, U_1 will lie in the upper $2\frac{1}{2}$ per cent region and so will U. It follows that the test described is a two-tailed 5 per cent test even though it does not appear so at first sight.

4. Other hypotheses can be tested. Let us assume that the samples come from populations which are identical except possibly in respect to their means. We may wish to test:

(*a*) $H_0: \mu_1 - \mu_2 = \delta$ (a specified constant);

$\qquad H_1: \mu_1 - \mu_2 \neq \delta$.

Subtract δ from each observation in the first sample, determine the ranks and use the above method.

(*b*) $H_0: \mu_1 - \mu_2 \leq \delta$;

$\qquad H_1: \mu_1 - \mu_2 > \delta$.

Subtract δ from each observation in the first sample and then rank in the usual way. If the values are ranked so that larger values are assigned larger (smaller) ranks, and R_1 turns out to be smaller (larger) than R_2, we automatically accept the null hypothesis. Otherwise, we compare U (or T) with the upper 5 per cent critical region.

5. Interpolation may be necessary in table 12.17.1. More extensive tables are given by Zar [105] 475–87.

6. Table 12.17.1 assumes that $n_1 \leq n_2$. This need cause no difficulty, since the critical points corresponding to parameters n_1 and n_2 are the same as those for parameters n_2 and n_1.

7. It is incorrect to use this test for paired observations. See section 12.24.

8. If the statistical model of section 12.15 is applicable to the data, the reader may be advised to use the more powerful *t*-test.

Example 12.17.1. The data in table 12.15.1 give the serum cholesterol levels in seven male and six female turtles. Do male and female turtles have the same mean concentrations? Let us assume that with the possible exception of the mean the two distributions are the same. We can then use the Mann-Whitney method to test

$$H_0: \mu_M = \mu_F;$$
$$H_1: \mu_M \neq \mu_F.$$

For a 5 per cent significance level, the critical region is $U \geqslant 36$ (interpolation in table 12.17.1). Ranking from largest to smallest, the sum of the male ranks is 54 and the sum of the female ranks is 37. It follows that $U_1 = 16$, $U_2 = 26$ and $U = 26$. This value is not significant; so we accept the null hypothesis that the means are the same.

This is the second test we have applied to these data, and we do it for demonstration purposes. The reader's attention is drawn to the warning in section 12.2.

> *Further reading*: Balaam [3] 227–8; Colquhoun [16] 143–7; Zar [105] 109–14.

12.18   Equality of several means – the one-way analysis of variance

In this section (and again in section 12.25) we consider an example of the very general technique known as the *analysis of variance,* which (in spite of its name) has as its object the provision of statistics for comparing population means. The general procedure is to determine how much of the variation in our experimental results is caused by population differences and how much is due to random variability. By comparing the contributions to variance from these two sources, we are able to determine the importance of the population differences.

Usual form of data

	Sample 1 (size n_1)	Sample 2 (size n_2)	...	Sample k (size n_k)
	x_{11}	x_{12}		x_{1k}
	x_{21}	x_{22}		x_{2k}
	.	.		.
	.	.		.
	.	.		.
	$x_{n_1 1}$			$x_{n_k k}$
		$x_{n_2 2}$		
Totals	$T_{.1}$	$T_{.2}$	...	$T_{.k}$

$$T_{..} = T_{.1} + T_{.2} + \ldots + T_{.k};$$
$$n = n_1 + n_2 + \ldots + n_k.$$

Statistical model

The samples all come from normal populations with variances σ^2. The first sample comes from a population with mean μ_1, the second sample comes from a population with mean μ_2, ..., and the kth sample comes from a population with mean μ_k. All the observations are independent.

Hypotheses

$H_0: \mu_1 = \mu_2 = \ldots = \mu_k$;

H_1: the means are not all equal.

Critical region

Greater than the upper 5 per cent point of the $F_{k-1, n-k}$ distribution.

Computation of test statistic

Square and add all the observations to obtain $\Sigma\Sigma x_{ij}^2$.

Compute the total sum of squares $\Sigma\Sigma x_{ij}^2 - T_{..}^2/n$.

Compute the between samples sum of squares

$$(T_{.1}^2/n_1 + \ldots + T_{.k}^2/n_k) - T_{..}^2/n.$$

Complete the analysis of variance table 12.18.1.[3] The test statistic is

$$\blacksquare \quad F = \frac{\text{between samples mean square}}{\text{residual mean square}}. \tag{12.18.1}$$

Comments

1. In the case of two samples, the test is equivalent to test (*a*) of section 12.15.

2. The test is robust for moderate departures from normality provided the sample sizes are reasonably large.

3. The test is robust for moderate departures from the constant variance assumption when the sample sizes are approximately equal. Equality of variance tests are given in sections 12.29–12.30.

4. If doubtful about the applicability of the model to the data under consideration try the non-parametric test of section 12.19.

5. Other hypotheses can be tested; for example:

$H_0: \mu_1 - \mu_2 = \delta, \mu_2 = \mu_3 = \ldots = \mu_k$;

H_1: null hypothesis not true.

Subtract δ from each observation in sample 1 and then use above test procedure.

6. If the test yields a significant result, we conclude that the means are not all equal. To determine where the actual differences lie, use

[3] In this table (and others) s.s. denotes sum of squares, d.f. denotes degrees of freedom, and m.s. denotes mean square.

a multiple-comparison method (section 12.21). It is *not* correct to compare the columns in pairs using a *t*-test or the above analysis of variance.

7. (A computational hint.) Each item of data may be multiplied or divided by the same constant, or have the same constant added to it or subtracted from it. The final *F* value will be unchanged.

8. In the computation of the between samples sum of squares certain totals are squared. The square is always divided by the number of numbers in the total.

Table 12.18.1. *The one-way analysis of variance*

Source (1)	s.s. (2)	d.f. (3)	m.s. $(4) = (2)/(3)$
Between samples	$(\Sigma T^2_{.j}/n_j) - T^2_{..}/n$	$k-1$	$\left(\begin{array}{c}\text{By}\\\text{division}\end{array}\right)$
Residual	By difference	$n-k$	
Total	$\Sigma\Sigma x^2_{ij} - T^2_{..}/n$	$n-1$	—

Example 12.18.1. A biologist has classified worms into three groups by a structural characteristic. A random sample is taken from each group (table 12.18.2). Is the mean length of a worm the same for each group?

Let us use a 5 per cent significance level. The critical region is the upper 5 per cent region of the $F_{2,15}$-distribution, that is, greater than 3.68.

Table 12.18.2. *The length of worms in three different groups* (cm)

	Group 1	Group 2	Group 3
	10.2	12.2	9.2
	8.2	10.6	10.5
	8.9	9.9	9.2
	8.0	13.0	8.7
	8.3	8.1	9.0
	8.0	10.8	
		11.5	
Totals	51.6	76.1	46.6

From table 12.18.2, we note that

$$T_{.1} = 51.6 \qquad n_1 = 6$$
$$T_{.2} = 76.1 \qquad n_2 = 7$$
$$T_{.3} = 46.6 \qquad n_3 = 5$$

$$T_{..} = 174.3 \qquad n = 18$$

We need to calculate

$$\Sigma\Sigma x_{ij}^2 = (10.2)^2 + (8.2)^2 + ... + (8.7)^2 + (9.0)^2 = 1726.31.$$

The total sum of squares is therefore

$$1726.31 - (174.3)^2/18 = 38.505.$$

The between samples sum of squares is

$$(51.6)^2/6 + (76.1)^2/7 + (46.6)^2/5 - (174.3)^2/18 = 17.583.$$

The analysis of variance table 12.18.3 can now be formed and the test statistic is

$$F = 8.792/1.395 = 6.3.$$

This F value is significant. So we reject the null hypothesis; the mean length of a worm is not the same in all three groups. Although we have concluded that the means are not all the same, we do not know where the differences lie. A multiple-comparison method is necessary for this purpose (section 12.21).

Table 12.18.3. *One-way analysis of variance example*

Source	s.s.	d.f.	m.s.
Between samples	17.583	2	8.792
Residual	20.922	15	1.395
Total	38.505	17	—

Further reading: Balaam [3] 161–5; Brownlee [7] 309–33; Chakravarti *et al.* [9] 342; Colquhoun [16] 171–90; Guenther [39] 26–64; Hoel [45] 250–6; Mood and Graybill [70] 372–3; Remington and Schork [86] 282–7; Scheffé [88] 331–68; Snedecor [91] 237–53; Zar [105] 133–9.

12.19 Several independent samples – the Kruskal–Wallis non-parametric one-way analysis of variance

This non-parametric test is often used to test the equality of several means when the assumptions underlying the usual one-way analysis of variance (section 12.18) are not valid (for example, the underlying populations are not normal). It is also used when the data are given as ranks (for example, tallest child, second tallest child, etc).

The reader should note that the null hypothesis actually tested is that the populations from which the samples were obtained are identical. Rigorously interpreted, all we can conclude from a significant value of the test statistic is that the populations differ, not necessarily that the means differ; to infer that the samples come from populations with different means or medians, we must assume that they are identical in all other respects, for example, that they have the same variance. In practice, moderate differences in variability are acceptable, because the test statistic is fairly insensitive to them.

Usual form of data
See section 12.18 (k samples; a total of n observations).

Statistical model
All the observations are independent, and the observations within the same sample come from the same population (that is, there are k populations).

Hypotheses
H_0: the k populations are identical;
H_1: the null hypothesis is not true.

Critical region
(i) *Small samples*: the upper 5 per cent region of the Kruskal-Wallis distribution (table 12.19.1).
(ii) *Larger samples*: the upper 5 per cent region of the χ^2_{k-1}-distribution.

Computation of test statistic
The n observations are ranked in ascending order from 1 to n. The rank sums $R_1, R_2, ..., R_k$ are computed for the k groups. The following test statistic is computed:

$$\blacksquare \quad H = \frac{12}{n(n+1)} (R_1^2/n_1 + ... + R_k^2/n_k) - 3(n+1). \qquad (12.19.1)$$

Comments
1. The test statistic H is used for both small and large samples.
2. When two or more observations have exactly the same value, they are said to be *tied*. The rank assigned to each of them is the mean of the ranks which would have been assigned to them had they not been tied. A small adjustment is sometimes made to H to take account of ties (Zar [105] 142).
3. When we reject H_0, we do not know which groups are different. It is not correct to do a Mann-Whitney test on all possible pairs. A multiple-comparison method must be used (Zar [105] 156-7).

4. Other hypotheses can be tested. For example.

$H_1: \mu_1 - \delta = \mu_2 = \ldots = \mu_k;$
$H_1: H_0$ not true.
(δ is a specified constant.)

Subtract δ from eaoh observation in the first sample, rank, and proceed as above.

5. It is incorrect to use this test for related observations. See section 12.26.

6. If the statistical model of section 12.18 is applicable to the data, the reader may be advised to use the more powerful F-test.

Table 12.19.1. *The upper 100α per cent points of the Kruskal–Wallis distribution.* The critical region includes the tabled value*

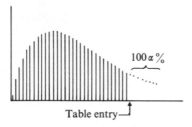

n_1	n_2	n_3	$\alpha = 0.10$	0.05	0.01	n_1	n_2	n_3	$\alpha = 0.10$	0.05	0.01
2	2	2	4.57	–	–	5	3	1	4.01	4.96	–
3	2	2	4.50	4.71	–	5	3	2	4.65	5.25	6.82
3	3	2	4.55	5.36	–	5	3	3	4.53	5.34	6.98
3	3	3	4.62	5.60	7.20	5	4	1	3.98	4.98	6.95
4	2	2	4.37	5.33	–	5	4	2	4.54	5.27	7.11
4	3	2	4.51	5.44	6.44	5	4	3	4.54	5.63	7.44
4	3	3	4.70	5.72	6.74	5	4	4	4.61	5.61	7.76
4	4	1	4.16	4.96	6.66	5	5	1	4.10	5.12	7.30
4	4	2	4.55	5.45	7.03	5	5	2	4.50	5.33	7.33
4	4	3	4.54	5.59	7.14	5	5	3	4.54	5.70	7.57
4	4	4	4.65	5.69	7.65	5	5	4	4.52	5.66	7.82
5	2	2	4.37	5.16	6.53	5	5	5	4.56	5.78	7.98

* *Source*: Kruskal and Wallis (1952, *J. American Statistical Association*, **47**, 583–621). Reprinted with permission of the publisher.

Example 12.19.1. A biologist has classified worms into three groups by a structural characteristic. A random sample is taken from each group and the length of each worm recorded (table 12.18.2). Is the mean length of a worm the same in each group?

This null hypothesis was tested in section 12.18 using the one-way analysis of variance. Let us now demonstrate the use of the Kruskal–Wallis test on these data. The values of n_1, n_2 and n_3 exceed those shown in table 12.19.1, and so as our 5 per cent critical region, we use the upper 5 per cent region of the χ^2_2-distribution; that is, greater than 5.99.

When the 18 observations are ranked in ascending order, from 1 to 18, the sums of the ranks for the three groups are $R_1 = 31$, $R_2 = 94$ and $R_3 = 46$. The test statistic is

$$ H = \frac{12}{18 \times 19} \left(\frac{31^2}{6} + \frac{94^2}{7} + \frac{46^2}{5} \right) - 3 \times 19 = 7.76. $$

This value is significant. So we conclude that the three samples do not come from identical populations. If we *assume* that the three distributions are the same (apart from the mean), we can conclude that the three population means are not all equal.

This is the second test we have applied to these data, and we do it for demonstration purposes. The reader's attention is drawn to the warning given in section 12.2.

Further reading: Colquhoun [16] 191–5; Kruskal and Wallis [56]; Zar [105] 139–42.

12.20 Several independent samples – the median test

Usual form of data
> See section 12.18.

Statistical model
> All the observations are independent, and the observations within the same sample come from the same population (that is, there are k populations).

Hypotheses
> H_0: the k populations are identical;
> H_1: the null hypothesis is not true.

Critical region
> The upper 5 per cent region of the χ^2_{k-1}-distribution.

Computation of test statistic
> 1. Combine all the samples to form one large sample.
> 2. Find the *median* of this combined sample (section 8.5).
> 3. For each sample, count the number of values exceeding the overall median. Record these numbers in a table like 12.20.1.
> 4. Calculate the expected number in each cell by multiplying together the two marginal totals which include that cell and dividing the product by the grand total n.

5. Compute the test statistic

$$\chi^2 = \sum_{\substack{all \\ 2k \\ cells}} \frac{(observed\ number - expected\ number)^2}{expected\ number}. \qquad (12.20.1)$$

Comments

1. Once the data have been arranged in the form of table 12.20.1, the actual test is really a test of equality of binomial probabilities in several samples. Refer to section 12.8.

2. Sometimes two or more of the individual sample values are equal to the median. Divide them evenly between the 'above median' and 'below median' groups. If the number of such cases is odd, they should be divided evenly and the last case ignored.

3. Note also the tests in sections 12.18 and 12.19.

4. If H_0 is rejected and we wish to conclude that the population means differ, we must *assume* that the underlying distributions are the same apart from the means.

Table 12.20.1

	Sample 1	Sample 2 ... Sample k	Total
Number of observations larger than median	L_1	L_2 $\quad$ ... L_k	$\sum_j L_j$
Number of observations less than median	$n_1 - L_1$	$n_2 - L_2$ $\quad$ ... $n_k - L_k$	$n - \sum_j L_{.j}$
Total	n_1	n_2 $\quad$ ... n_k	n

Example 12.20.1. Let us apply the method of this section to the worm data in table 12.18.2 and test the null hypothesis that the three distributions of worm length are the same.

For a 5 per cent level of significance, we use the upper 5 per cent region of the χ_2^2-distribution; that is, greater than 5.99. The median is 9.2, and we see that only one out of six observations in group 1 is greater than the median, only one out of seven observations in group 2 is less than the median, and in group 3 two observations out of five are greater than the median. These results are summarised in table 12.20.2, together with the expected numbers calculated using the marginal totals and the grand total (for example, $3 = (6 \times 9)/18$).

Table 12.20.2. *Number of worms greater than and less than the median. Expected numbers are shown in parentheses*

	Sample 1	Sample 2	Sample 3	Total
Number larger than median	1 (3)	6 (3.5)	2 (2.5)	9
Number less than median	5 (3)	1 (3.5)	3 (2.5)	9
Total	6	7	5	18

The expected numbers in the individual cells are rather small. Nevertheless, let us calculate chi-square in the usual manner:

$$\chi^2 = (1-3)^2/3 + (5-3)^2/3 + (6-3.5)^2/3.5 + \ldots + (3-2.5)^2/2.5 = 6.4.$$

This value is significant at the 5 per cent level; so we reject the null hypothesis and conclude the three distributions are not identical. If we can *assume* that the distributions are identical apart from the means, we can conclude that the means are not all the same.

This is the third test we have applied to these data, and we do it for demonstration purposes. The reader's attention is drawn to the warning given in section 12.2.

Further reading: Brownlee [7] 246–8.

12.21 Several independent samples – Scheffé's multiple comparisons

The analysis of variance of section 12.18 may tell us that the populations from which the samples were drawn do not have the same mean, but the test does not tell us which population means actually differ. We might be tempted to make a series of pair-wise comparisons of the populations using the *t*-method of section 12.15, but this approach is wrong. The chance of rejecting a true hypothesis is, say, 5 per cent on each test, and when several tests are made, the chance of rejecting at least one true hypothesis becomes considerably larger than 5 per cent. Scheffé's multiple-comparison method overcomes this problem. In fact, using Scheffé's method, and making all possible comparisons, the probability of obtaining one or more false conclusions when all the population means are actually equal is exactly 5 per cent.

Usual form of data
See section 12.18.

Statistical model
See section 12.18.

Hypotheses

$H_0: c_1\mu_1 + c_2\mu_2 + \ldots + c_k\mu_k = 0$, where the $\{c_i\}$ are certain specified constants adding to zero;

H_1: null hypothesis not true.

Critical region

Greater than the upper 5 per cent point of the $F_{k-1,n-k}$ distribution.

Computation of test statistic

1. Compute the mean for each sample: $\bar{x}_{.i} = T_{.i}/n_i$.
2. Denote the residual mean square in table 12.18.1 by s^2.
3. Compute the test statistic

■
$$S = \frac{(c_1\bar{x}_{.1} + c_2\bar{x}_{.2} + \ldots + c_k\bar{x}_{.k})^2}{(k-1)s^2(c_1^2/n_1 + \ldots + c_k^2/n_k)}.$$ (12.21.1)

Comments

1. This method is normally used to make a series of comparisons like

$H_0: \mu_1 - \mu_2 = 0$;
$H_1: \mu_1 - \mu_2 \neq 0$.

2. Other multiple-comparison methods are available; for example, Tukey's method and the Student–Newman–Keuls method (Zar [105] 151–61).

3. To avoid logical inconsistencies with pair-wise comparisons, compare the group having the largest sample mean with the group having the smallest sample mean, then with the group having the next smallest sample mean and so on. As soon as a non-significant comparison occurs, or no group with a smaller sample mean remains, replace the group having the largest sample mean by the group with the second-largest sample mean and start the process again.

Example 12.21.1. The tests of sections 12.18, 12.19 and 12.20 all indicate that the mean worm lengths are not the same for the three groups in table 12.18.2. Let us use Scheffé's method to detect where the differences lie.

The critical region is the upper 5 per cent region of the $F_{2,15}$-distribution, that is, greater than 3.68. The three sample means are 8.60, 10.87 and 9.32 and from table 12.18.3, the residual mean square s^2 is known to be 1.395.

According to comment 3, the first comparison should be between sample 2 and sample 1. We set $c_1 = -1$, $c_2 = 1$ and $c_3 = 0$, and compute the test statistic

$$S = \frac{(10.87 - 8.60)^2}{2 \times 1.395 \times (\frac{1}{6} + \frac{1}{7})} = 5.97.$$

$(k \cdot 1)$

always
$\frac{1}{n_x} + \frac{1}{n_y}$

This value lies in the critical region; so we reject the null hypothesis that $\mu_1 = \mu_2$ and conclude that $\mu_1 \neq \mu_2$.

According to comment 3, the next comparison we make is between sample 2 and sample 3. We use $c_1 = 0, c_2 = 1$ and $c_3 = -1$:

$$S = \frac{(10.87 - 9.32)^2}{2 \times 1.395 \times (\frac{1}{7} + \frac{1}{5})} = 2.51.$$

This value is not significant; so we accept the null hypothesis $\mu_2 = \mu_3$. We conclude therefore that $\mu_1 \neq \mu_2 = \mu_3$.

> *Further reading*: Balaam [3] 176–8; Colquhoun [16] 207–12; Guenther [39] 54–9; Scheffé [88] 68–82; Snedecor [91] 251–3; Zar [105] 151–61.

12.22 Paired observations – paired *t*-tests

Usual form of data

It often happens that data values are paired; for example:

	Weight before diet	Weight after diet
Man 1	x_{11}	x_{12}
Man 2	x_{21}	x_{22}
.	.	.
.	.	.
.	.	.
Man n	x_{n1}	x_{n2}

Statistical model

A man's weight after the diet is equal to his weight before the diet plus the diet effect. The diet effects are independent normal random variables with mean μ and variance σ^2

Hypotheses

(a) Equality	(b) Inequality	(c) Inequality
$H_0: \mu = \mu_0$;	$H_0: \mu \leqslant \mu_0$;	$H_0: \mu \geqslant \mu_0$;
$H_1: \mu \neq \mu_0$.	$H_1: \mu > \mu_0$.	$H_1: \mu < \mu_0$.

Critical region

(a) *Equality*: the upper $2\frac{1}{2}$ per cent region and the lower $2\frac{1}{2}$ per cent region of the t-distribution with $n-1$ degrees of freedom.

(b) *Inequality*: the upper 5 per cent region of the t-distribution with $n-1$ degrees of freedom.

(c) *Inequality*: the lower 5 per cent region of the t-distribution with $n-1$ degrees of freedom.

Computation of test statistic

Compute the n differences $d_1 = x_{12} - x_{11}, d_2 = x_{22} - x_{21}, \ldots$.
Compute the mean difference $d = (d_1 + \ldots + d_n)/n$.

$s_d^2 =$ sample variance of the differences $\{d_i\}$ (formula (8.4.3)).
Set $\mu = \mu_0$ in the test statistic

■ $t = n^{\frac{1}{2}} (\bar{d} - \mu)/s_d.$ (12.22.1)

Comments

1. To test the null hypothesis that diet has no effect, we use method (*a*) with $\mu_0 = 0$.

2. When the observations are not paired, use the *t*-test of section 12.15.

3. The non-parametric methods of sections 12.23 and 12.24 may be useful when the model underlying this *t*-test is not appropriate to the data. The test is, however, robust for moderate departures from normality.

4. When $\mu_0 = 0$, the equality test (*a*) is equivalent to the two-way analysis of variance test for two columns of figures (section 12.25).

5. Because of the symmetry of the *t*-distribution, it is sufficient when testing equality to compare $|t|$ with the upper $2\frac{1}{2}$ per cent region of the *t*-distribution.

Example 12.22.1. Two different types of wheat are being compared for yield. Type A is a common variety and type B is a new hybrid variety. Ten two-acre blocks are chosen and each block is divided into two one-acre plots. The growing conditions are uniform within each particular block. One plot is chosen at random from each block and planted with wheat of variety A; the remaining ten plots are planted with variety B. The yield results are shown in table 12.22.1. Do we have any evidence that the yield with type B is greater than the yield with type A?

We need to test

$H_0 : \mu \leqslant 0;$
$H_1 : \mu > 0.$

The 5 per cent critical region is the upper 5 per cent region of the t_9 distribution; that is, greater than 1.833. The ten differences are shown in table 12.22.1. The mean difference $\bar{d}$ is 0.57 and the sample variance of the differences s_d^2 is 0.4312. The test criterion is

$$t = (10)^{\frac{1}{2}} (0.57 - 0)/(0.4312)^{\frac{1}{2}} = 2.7.$$

This value is significant; so we reject the null hypothesis and conclude that the yield from type B is greater than that of type A.

Further reading: Balaam [3] 134–45; Colquhoun [16] 167–9; Guenther [39] 24–6; Snedecor [91] 52–3, 77–80; Zar [105] 121–4.

Table 12.22.1. *Yields of wheat (bushels)*

Block	Type A	Type B	Difference $(B-A)$
1	36.9	36.8	-0.1
2	35.2	37.1	1.9
3	31.2	31.4	0.2
4	34.1	34.1	0.0
5	36.1	35.9	-0.2
6	34.1	35.2	1.1
7	37.2	37.9	0.7
8	36.8	37.2	0.4
9	29.6	30.2	0.6
10	35.4	36.5	1.1

12.23 Two related samples – the sign test

Usual form of data

See section 12.22. We define differences

$$d_1 = x_{12} - x_{11}, d_2 = x_{22} - x_{21}, ..., d_n = x_{n2} - x_{n1}.$$

Statistical model

The difference d_i is a random observation from a population P_i.
The populations $\{P_i\}$ $(i = 1, ..., n)$ all have the same median.

Hypotheses

H_0: the common median is zero;
H_1: the common median is not zero.

Critical region

(i) *Small samples* $(n \leqslant 20)$: The upper $2\frac{1}{2}$ per cent region and the lower $2\frac{1}{2}$ per cent region of the binomial distribution with parameters n and $p = \frac{1}{2}$ (table 12.23.1).
(ii) *Larger samples*: the upper $2\frac{1}{2}$ per cent region and the lower $2\frac{1}{2}$ per cent region of the unit normal distribution.

Computation of test statistic

(i) *Small samples*:
For each pair, record the sign of the difference. Count the number of positive signs, and use this total N as the test statistic.
(ii) *Larger samples*:
Compute N as in (i) and then use

■ $T = (2N - n)/\sqrt{n}.$ (12.23.1)

Table 12.23.1. *The upper* 100α per cent points of the binomial distribution with parameters n and* $\frac{1}{2}$. *The critical region includes the tabled value*

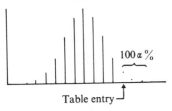

100 α %

Table entry

n	α = 0.005	0.01	0.025	0.05
5	—	—	—	5
6	—	—	6	6
7	—	7	7	7
8	8	8	8	7
9	9	9	8	8
10	10	10	9	9
11	11	10	10	9
12	11	11	10	10
13	12	12	11	10
14	13	12	12	11
15	13	13	12	12
16	14	14	13	12
17	15	14	13	13
18	15	15	14	13
19	16	15	15	14
20	17	16	15	15

* The lower 100α per cent point is obtained by subtracting the table entry from the binomial parameter *n*.

Comments

1. When the two numbers of a pair are equal, and the difference is zero, include $\frac{1}{2}$ in *N*.

2. Other hypotheses may be tested by this method. For example:

(*a*) H_0: common median $= \delta$ (a specified constant);

  H_1: common median $\neq \delta$.

Subtract δ from each difference and proceed as above.

(*b*) H_0: common median $\leqslant \delta$;

  H_1: common median $> \delta$.

Subtract δ from each difference and use the upper 5 per cent region of the binomial or normal distribution.

3. If the statistical model of section 12.22 is applicable to the data, the reader may be advised to use the more powerful t-test. The non-parametric Wilcoxon test of section 12.24 is also more powerful, but it makes more assumptions.

Example 12.23.1. Let us apply test (b) outlined in comment 2 to the wheat example 12.22.1 and use a 5 per cent significance level.

The sample size is small ($n = 10$); so the critical region is the upper 5 per cent region of the binomial distribution with $n = 10$ and $p = \frac{1}{2}$. According to table 12.23.1, the upper 5 per cent region of this distribution includes $N = 9$ and $N = 10$. The observed value $N = 7.5$ does not lie in the critical region. We therefore accept the null hypothesis that type B variety produces a yield no greater than type A.

This conclusion is contrary to the one arrived at via the t-test, and it serves to emphasise the fact that the sign test has less power than the t-test.

This is the second test we have applied to these data, and we do it for demonstration purposes. The reader's attention is drawn to the warning given in section 12.2.

Further reading: Balaam [3] 229; Brownlee [7] 242–6; Colquhoun [16] 153–6; Pollard [81] 119–31; Remington and Schork [86] 316; Zar [105] 290–1.

12.24   The Wilcoxon paired-sample test

This non-parametric test is often used instead of the paired t-test of section 12.22 when the assumptions underlying that test are not valid (for example, the underlying populations are not normal).

Our description of the test is in terms of men's weights before and after a certain diet. The reader should note that the null hypothesis actually tested is that the common distribution of the differences is symmetrical about zero. If the null hypothesis is rejected and we wish to infer from this that the distributions of weights before and after the diet have different means, we must assume either that the distributions both have the same shape or that they are both symmetric. The test is not very sensitive to these symmetry requirements, however, and modest departures from symmetry can be ignored in practice.

Usual form of data

See section 12.22. The number of pairs is denoted by n. We use d_i to denote the difference between the ith pair.

Statistical model

The differences are independent and identically distributed.

Hypotheses

H_0: the distribution of the differences is symmetrical about zero;
H_1: null hypothesis not true.

Table 12.24.1. *The upper†* 100α *per cent points of the Wilcoxon, paired-sample statistic.* * *The critical region includes the tabled value*

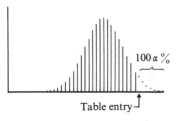

n	$\alpha = 0.05$	0.025	0.01	0.005
7	25	26	28	—
8	31	33	35	36
9	37	40	42	44
10	45	47	50	52
11	53	56	59	61
12	61	65	69	71
13	70	74	79	82
14	80	84	90	93
15	90	95	101	105
16	101	107	113	117
17	112	119	126	130
18	124	131	139	144
19	137	144	153	158
20	150	158	167	173

* *Source*: McCornack (1965, *J. American Statistical Association,* **60,** 864–71). Derived with permission of the publisher.
† The lower 100α per cent point is obtained by subtracting the upper 100α per cent point from $\frac{1}{2}n\,(n+1)$.

Critical region

(i) *Small samples*: the upper $2\frac{1}{2}$ per cent region and the lower $2\frac{1}{2}$ per cent region of the Wilcoxon signed rank distribution (table 12.24.1).

(ii) *Large samples*: the upper $2\frac{1}{2}$ per cent region and the lower $2\frac{1}{2}$ per cent region of the unit normal distribution.

Computation of test statistic

(i) *Small samples*: Rank all the differences regardless of sign in ascending order from 1 to n.

Attach the sign of the difference to the rank.

Calculate N, the sum of the positive ranks.

Compare N with the critical region.

(ii) *Large samples*: Compute N as in (i) and then calculate the statistic

■ $T = \{N - n(n+1)/4\} / \{n(n+1)(2n+1)/24\}^{\frac{1}{2}}.$ (12.24.1)

Comments

1. Zero differences are omitted altogether (and n is reduced accordingly).

2. Differences equal in absolute value are allotted ranks equal to the mean of the ranks which would have been assigned to them had they not been tied. A small adjustment is sometimes made to take account of ties (Zar [105] 126).

3. The absolute value of the sum of the negative ranks can be used instead of the sum of the positive ranks. The same results are obtained. In fact, one can compute both and just compare the larger with the upper $2\frac{1}{2}$ per cent critical value.

4. Other hypotheses can be tested. Let us assume that the pairs come from populations which are either symmetrical or identical except possibly in respect of their means. We may wish to test

(a) H_0: mean increase in weight $= \delta$ (a specified constant);

   H_1: mean increase in weight $\neq \delta$.

Subtract δ from each difference and proceed as above.

(b) H_0: mean increase in weight $\leqslant \delta$;

   H_1: mean increase in weight $> \delta$.

Subtract δ from each difference; compute the test criteria as above, but use the upper 5 per cent critical region.

5. If the statistical model of section 12.22 is applicable to the data, the reader may be advised to use the more powerful t-test.

Example 12.24.1. Let us apply test (b) of comment 4 to the wheat example 12.22.1, and use a 5 per cent significance level. We set $\delta = 0$. The sample size is small ($n = 10$) and one difference is zero. We therefore use table 12.24.1 with $n = 9$; the upper 5 per cent region includes values of N greater than or equal to 37.

The signed ranks are shown in table 12.24.2, and it is soon apparent that $N = 41.5$. This value lies in the critical region; so we reject the null hypothesis and conclude that wheat variety B produces a higher average yield than wheat variety A.

This is the third test we have applied to these data, and we do it for demonstration purposes. The reader's attention is drawn to the warning given in section 12.2.

Further reading: Brownlee [7] 258–60; Colquhoun [16] 160–6; Remington and Schork [86] 318.

Table 12.24.2. *The signed ranks corresponding to table 12.22.1*

Block	Difference	Signed rank
1	−0.1	−1
2	1.9	9
3	0.2	2.5
4	0.0	−
5	−0.2	−2.5
6	1.1	7.5
7	0.7	6
8	0.4	4
9	0.6	5
10	1.1	7.5

$N = 41.5$

12.25 Related (paired) samples – the two-way analysis of variance

In this section we consider an example of the general technique known as the *analysis of variance*. A brief description of the general procedure is given in the introductory paragraph of section 12.18.

Usual form of data

	Columns				
	1	2	...	c	Totals
Row 1	x_{11}	x_{12}	...	x_{1c}	$T_{1.}$
Row 2	x_{21}	x_{22}	...	x_{2c}	$T_{2.}$
.	.	.		.	.
.	.	.		.	.
.	.	.		.	.
Row r	x_{r1}	x_{r2}	...	x_{rc}	$T_{r.}$
Totals	$T_{.1}$	$T_{.2}$	...	$T_{.c}$	$T_{..}$

Statistical model

The observation in row i and column j is equal to an overall mean plus the sum of the row i effect and the column j effect plus a random component which has the normal distribution with zero mean and variance σ^2. Thus,

$$x_{ij} = \mu + \alpha_i + \beta_j + e_{ij},$$

where μ is the overall mean, α_i is the row i effect, β_j is the column j effect, and e_{ij} is the random component.
The $\{e_{ij}\}$ are independent.

Hypotheses

(a) *Equality of row effects*	(b) *Equality of column effects*
$H_0: \alpha_1 = \alpha_2 = ... = \alpha_r;$	$H_0: \beta_1 = \beta_2 = ... = \beta_c;$
$H_1:$ not all equal.	$H_1:$ not all equal.

Critical region

(a) *Row effects*: greater than the upper 5 per cent point of the F-distribution with $r-1$ and $(r-1)(c-1)$ degrees of freedom in the numerator and denominator respectively.

(b) *Column effects*: greater than the upper 5 per cent point of the F-distribution with $c-1$ and $(r-1)(c-1)$ degrees of freedom in the numerator and denominator respectively.

Computation of test statistic

Square and add all observations to obtain $\Sigma\Sigma x_{ij}^2$.

Compute the total sum of squares $\Sigma\Sigma x_{ij}^2 - T_{..}^2/rc$.

Compute the between rows sum of squares

$$(T_{1.}^2/c + ... + T_{r.}^2/c) - T_{..}^2/rc.$$

Compute the between columns sum of squares

$$(T_{.1}^2/r + ... + T_{.c}^2/r) - T_{..}^2/rc.$$

Compute the analysis of variance table[4] 12.25.1.

(a) *Row effects*. The test statistic is

■
$$F = \frac{\text{between rows mean square}}{\text{residual mean square}}. \tag{12.25.1}$$

(b) *Column effects*. The test statistic is

■
$$F = \frac{\text{between columns mean square}}{\text{residual mean square}}. \tag{12.25.2}$$

Comments

1. The paired t-test for equality in section 12.22 is a special case of the two-way analysis of variance column effect test.

2. The test seems to be reasonably robust for moderate departures from normality and moderate departures from the uniformity of variance assumption. Gross departures from these assumptions may be detected as follows. Add to each observation x_{ij} the grand mean $x_{..} = T_{..}/rc$ and subtract the respective row and column means $\bar{x}_{i.} = T_{i.}/c$ and $\bar{x}_{.j} = T_{.j}/r$. Check whether the resulting numbers could have come from a normal population with zero means and variance equal to the residual mean square.

3. The non-parametric method of section 12.26 may be useful when the model underlying the two-way analysis of variance is not appropriate to the data.

[4] See footnote 3.

Table 12.25.1. *The two-way analysis of variance*

Source (1)	s.s. (2)	d.f. (3)	m.s. (4) = (2)/(3)
Between rows	$(\Sigma T_{i.}^2/c) - T_{..}^2/rc$	$r-1$	
Between columns	$(\Sigma T_{.j}^2/r) - T_{..}^2/rc$	$c-1$	$\left(\dfrac{\text{By}}{\text{division}}\right)$
Residual	By difference	$(r-1)(c-1)$	
Total	$\Sigma\Sigma x_{ij}^2 - T_{..}^2/rc$	$rc-1$	—

4. Other hypotheses can be tested; for example:
(c) $H_0: \alpha_1 - \alpha_2 = 6$,
$\qquad \alpha_2 = \alpha_3 = ... = \alpha_r$;
$\quad H_1$: the null hypothesis is not true.
Use method (a) after subtracting 6 from each entry in row 1.
(d) $H_0: \beta_1 - \beta_2 = 5$,
$\qquad \beta_2 = \beta_3 = ... = \beta_c$;
$\quad H_1$: the null hypothesis is not true.
Use method (b) after subtracting 5 from each entry in column 1.
5. In the computation of the row and column sums of squares certain totals are squared. The square is always divided by the number of numbers in that total.
6. If the test yields a significant row (column) effect, we conclude that the row (column) effects are not all equal. A multiple comparison method must be used to determine where the differences lie (section 12.27). It is *not* correct to compare the rows (columns) in pairs using the paired t-test or the above analysis of variance.
7. *A single missing value.* Even in the best laboratories, missing values occur (test tubes are dropped and animals die). The following method can be used to overcome the problem of a missing value, but it must not be used when the missing value is a direct consequence of the experiment (for example, animal poisoned by the diet). If a single value is missing and the missing value belongs in the ith row and the jth column, estimate the missing value X as follows:

$$X = (rT'_{i.} + cT'_{.j} - T'_{..})/\{(r-1)(c-1)\}. \qquad (12.25.3)$$

Here $T'_{i.}$, $T'_{.j}$ and $T'_{..}$ are respectively the totals for the ith row, the jth column and the whole table, *excluding* the missing value. An approximate two-way analysis of variance can now be performed with the residual degrees of freedom reduced by one. An exact procedure is available, however. If we wish to test whether there is any difference between rows, the exact procedure is as follows.

Table 12.25.2. *The two-way analysis of variance with p missing values*

Source (1)	s.s (2)	d.f. (3)	m.s. $(4) = (2)/(3)$
Between columns		$c-1$	–
Additional reduction due to rows	⎧ See ⎫ ⎨ comment ⎬	$r-1$	⎛ By ⎞
Residual	⎩ 7 ⎭	$(r-1)(c-1)-p$	⎝ division ⎠
Total	(Comment 7)	$rc-1-p$	–

(i) Determine the between columns and total sums of squares using the one-way analysis of variance computation method (section 12.18) and excluding the missing value(s).

(ii) Determine the residual sum of squares by the two-way computation method *including* the estimate(s) of the missing value(s).

(iii) Determine the additional reduction in sum of squares due to rows by subtraction.

The analysis of variance table is 12.25.2. The critical region comes from the F-distribution with $r-1$ and $(r-1)(c-1)-1$ degrees of freedom. The whole process must be repeated with rows and columns interchanged to test whether there is any difference between columns.

8. *Two missing values.* Put a trial value in the first position and estimate the second missing value. Use this estimated value to obtain a better estimate for the first missing value. Repeat this process iteratively until a solution is obtained. Apply the method in comment 7 with the usual residual degrees of freedom reduced by two.

Example 12.25.1. The phosphorus content (mg/100 g) of each of four different organs was determined in each of three animal species. The results are shown in table 12.25.3. Test the hypothesis that there is no difference in phosphorus content among the three species.

Table 12.25.3. *The phosphorus content (mg/100 g) of animal organs*

	Heart	Lung	Liver	Kidney	Total
Species 1	86.7	102.7	204.6	184.6	578.6
Species 2	88.4	108.1	213.2	183.4	593.1
Species 3	81.2	99.8	201.1	179.0	561.1
Total	256.3	310.6	618.9	547.0	1732.8

Table 12.25.4. *Phosphorus content – analysis of variance*

Source	s.s.	d.f.	m.s.
Rows	128.375	2	64.188
Columns	31 253.100	3	10 417.700
Residual	30.365	6	5.061
Total	31 411.840	11	—

The critical region corresponding to a 5 per cent level of significance is the upper 5 per cent region of the $F_{2,6}$-distribution; that is, greater than 5.14. The sum of the squares of the twelve observations is 281 628.16. It follows that the total sum of squares is

$$281\ 628.16 - (1732.8)^2/12 = 31\ 411.840.$$

The row sum of squares is

$$(578.6)^2/4 + (593.1)^2/4 + (561.1)^2/4 - (1732.8)^2/12 = 128.375.$$

The column sum of squares is

$$(256.3)^2/3 + (310.6)^2/3 + (618.9)^2/3 + (547.0)^2/3 - (1732.8)^2/12$$
$$= 31\ 253.100.$$

We now complete the analysis of variance table 12.25.4. The test criterion is

$$F = 64.188/5.061 = 12.7.$$

This value is significant; so we reject the null hypothesis and conclude that there is a difference in phosphorus content among the three species.

Further reading: Balaam [3] 165–73; Brownlee [7] 467–501; Chakravarti *et al.* [9] 344; Colquhoun [16] 195–9; Guenther [39] 65–84; Hoel [45] 256–60; Mood and Graybill [70] 374–8; Remington and Schork [86] 288–300; Scheffé [88] 331–68; Snedecor [91] 291–327; Zar [105] 163–74.

12.26 Related samples – Friedman's non-parametric two-way analysis of variance

This non-parametric test is often used to test for column effects when the assumptions underlying the usual two-way analysis of variance of section 12.25 are not valid (for example, the underlying populations are not normal). It is also used when the data are actually given as scores or ranks.

Usual form of data

See section 12.25 (r rows, c columns).

Statistical model

The observations are independent. The observation in row i and column j comes from a population with distribution function $F_{ij}(x)$ and mean μ_{ij}. Within each row i, the distribution functions $\{F_{ij}(x)\}$ are identical except possibly in respect of their means which may depend upon the column. The column effects are such that if the c means across each row i are ranked smallest to largest from 1 to c, the rankings are the same in all rows. (Thus, for example, if $\mu_{13} > \mu_{14}$, then $\mu_{53} > \mu_{54}$.)

Hypotheses

H_0: there are no column effects;
H_1: the null hypothesis is not true.

Critical region

(i) *Small samples*: the upper 5 per cent region of the Friedman distribution (table 12.26.1).
(ii) *Larger samples*: the upper 5 per cent region of the χ^2_{c-1}-distribution.

Computation of test statistic

Rank the entries in each row from 1 to c; denote the sum of the ranks in column j by R_j $(j = 1, ..., c)$. Compute

$S = (R_1^2 + ... + R_c^2) - (R_1 + ... + R_c)^2/c$.

The test statistic is

$$\chi^2 = 12S/\{rc(c+1)\} . \tag{12.26.1}$$

Comments

1. When two or more observations have exactly the same value they are said to be tied. The rank assigned to each of them is the mean of ranks which would have been assigned to them had they not been tied. A small adjustment is sometimes made to take account of ties (Zar [105] 176).

2. When we reject H_0, we do not know which columns differ. It is not correct to perform this test (or those of sections 12.23 and 12.24) on all possible pairs of columns. A multiple comparison method must be used (Zar [105] 176–7).

3. The analysis of variance model of section 12.25 required the column effects to be the same size in all rows. The Friedman model does not; it merely assumes that the column effects are in the same direction in all rows.

4. The Friedman model assumes that the distributions within each row are identical except, perhaps, in respect of their means. In practice, modest departures from this assumption can be ignored.

5. To test row effects, interchange rows and columns.

6. If the statistical model of section 12.25 is applicable, the reader might be advised to use the more powerful two-way analysis of variance instead of the Friedman test.

Table 12.26.1. *The upper 100α per cent points of the Friedman distribution.* *The critical region includes the table entry*

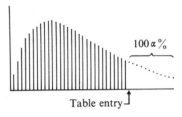

c	r	$\alpha = 0.05$	0.02	0.01	0.005	0.001
3	3	6.000	–	–	–	–
3	4	6.500	8.000	8.000	8.000	–
3	5	6.400	6.400	8.400	10.000	10.000
3	6	7.000	8.333	9.000	10.333	12.000
3	7	7.143	8.000	8.857	10.286	12.286
3	8	6.250	7.750	9.000	9.750	12.250
3	9	6.222	8.000	8.667	10.667	12.667
4	2	6.000	–	–	–	–
4	3	7.400	8.200	9.000	9.000	–
4	4	7.800	8.400	9.600	10.200	11.100

* *Source*: Friedman (1937, *J. American Statistical Association*, **32**, 675–701). Reprinted with permission of the publisher.

Example 12.26.1. Let us apply Friedman's method to the phosphorus data of example 12.25.1, and use a 5 per cent significance level. We wish to examine species effects; so we must interchange rows and columns. According to table 12.26.1, the critical region is the region in which S is greater than or equal to 6.500.

Table 12.26.2. *A ranking of the animal phosphorus data of section 12.25*

	Species 1	Species 2	Species 3
Heart	2	3	1
Lung	2	3	1
Liver	2	3	1
Kidney	3	2	1
Total	9	11	4

The ranked values are shown in table 12.26.2, and we calculate

$$S = (9^2+11^2+4^2)-(9+11+4)^2/3 = 26.0.$$

The test statistic is

$$\chi^2 = (12 \times 26.0)/(4 \times 3 \times 4) = 6.5.$$

This value is just significant. So we reject the null hypothesis and conclude that the three species do not all have the same mean phosphorus content.

This is the second test we have applied to these data, and we do it for demonstration purposes. The reader's attention is drawn to the warning given in section 12.2.

> *Further reading*: Balaam [3] 233; Brownlee [7] 260–2; Colquhoun [16] 200–3; Zar [105] 175–6.

12.27 Related (paired) samples – Scheffé's multiple comparisons

The analysis of variance of section 12.25 may tell us that there are non-zero column effects in a two-way table, but the test does not tell us where the differences lie. We might be tempted to make a series of pair-wise column comparisons using the t-method of section 12.22, but this approach is wrong. The introductory paragraph of section 12.21 explains why. Pair-wise comparisons can, however, be made using Scheffé's multiple-comparison method.

Usual form of data
See section 12.25 (r rows and c columns).
Statistical model
See section 12.25.
Hypotheses
$H_0: c_1\alpha_1+c_2\alpha_2+ \cdots +c_r\alpha_r = 0$, where the $\{c_i\}$ are certain specified constants which add to zero;
H_1: null hypothesis not true.
Critical region
Greater than the upper 5 per cent point of the F-distribution with $r-1$ and $(r-1)(c-1)$ degrees of freedom.
Computation of test statistic
1. Compute the mean for each row: $\bar{x}_{i.} = T_{i.}/c$.
2. Denote the residual mean square in table 12.25.1 by s^2.
3. Compute the test statistic

$$\blacksquare \quad S = \frac{(c_1\bar{x}_{1.}+c_2\bar{x}_{2.}+\ldots+c_k x_{k.})^2}{(r-1)s^2(c_1^2/c+\ldots+c_r^2/c)}. \tag{12.27.1}$$

Comments

1. This method is normally used to make a series of comparison like
$H_0: \alpha_1 - \alpha_2 = 0$;
$H_1: \alpha_1 - \alpha_2 \neq 0$.
2. For multiple comparison between columns, interchange r and c and use the column means $\{\bar{x}_{.j}\}$.
3. Other multiple-comparison methods are available; for example, Tukey's method and the Student-Newman-Keuls method (Guenther [39]; Zar [105] 174).
4. Read comment 3 of section 12.21.

Example 12.27.1. The tests of sections 12.25 and 12.26 indicate significant differences in phosphorus content among the three species in table 12.25.3. Let us now use Scheffé's method to detect where the differences lie.

The critical region is the upper 5 per cent region of the $F_{2,6}$-distribution, that is, greater than 5.14. The row means are 144.65, 148.275 and 140.275, and the residual mean square is $s^2 = 5.061$.

According to comment 4, the first comparison should be between row 2 and row 3. We set $c_1 = 0, c_2 = 1$ and $c_3 = -1$ and compute the test statistic

$$S = \frac{(148.275 - 140.275)^2}{2 \times 5.061 \times (\frac{1}{4} + \frac{1}{4})} = 12.6.$$

This value lies in the critical region, so we conclude that $\alpha_2 \neq \alpha_3$.

The next comparison is between row 2 and row 1. We set $c_1 = -1, c_2 = 1$, $c_3 = 0$, and compute

$$S = \frac{(148.275 - 144.65)^2}{2 \times 5.061 \times (\frac{1}{4} + \frac{1}{4})} = 2.6.$$

This value is not significant. We conclude that $\alpha_1 = \alpha_2 \neq \alpha_3$.

Further reading: Balaam [3] 176-8; Colquhoun [16] 207-12; Guenther [39] 79; Scheffé [88] 68-82; Zar [105] 151-61.

12.28 Tests for the value of a single variance

Usual form of data

$x_1, x_2, x_3, \ldots, x_n$ (sample size n).

Statistical model

The observations are independent and come from a normal population with mean μ and variance σ^2.

Hypotheses

(a) Equality	(b) Inequality	(c) Inequality
$H_0: \sigma^2 = \sigma_0^2;$	$H_0: \sigma^2 \leqslant \sigma_0^2;$	$H_0: \sigma^2 \geqslant \sigma_0^2;$
$H_1: \sigma^2 \neq \sigma_0^2.$	$H_1: \sigma^2 > \sigma_0^2.$	$H_1: \sigma^2 < \sigma_0^2.$

Critical region

(a) *Equality*: greater than upper $2\frac{1}{2}$ per cent point of χ^2_{n-1}-distribution or less than lower $2\frac{1}{2}$ per cent point of χ^2_{n-1}-distribution.

(b) *Inequality*: greater than upper 5 per cent point of χ^2_{n-1}-distribution.

(c) *Inequality*: less than lower 5 per cent point of χ^2_{n-1}-distribution.

Computation of test statistic

$s^2 = $ sample variance (formula (8.4.3)).

Set σ^2 equal to σ_0^2 in the statistic

■ $$\chi^2 = (n-1)\, s^2/\sigma^2. \tag{12.28.1}$$

Comments

1. σ_0^2 is a specified number (for example, 3.67).

2. The test is not robust for departures from normality. Tests for normality are outlined in section 12.35.

Example 12.28.1. A statistician has procured 11 hand-grenades.

He wishes to test the manufacturer's claim that the mean fuse-time is 4.01 seconds and the standard deviation 0.07 seconds. He obtains the following results with the sample of eleven: 4.21, 4.03, 3.99, 4.05, 3.89, 3.98, 4.01, 3.92, 4.23, 3.85 and 4.20. What conclusion can he come to?

The mean fuse-time was examined in example 12.13.1, where we calculated $\bar{x} = 4.033$ and $s^2 = 0.0170$. We now wish to test the hypothesis

$$H_0: \sigma^2 \leqslant 0.0049;$$
$$H_1: \sigma^2 > 0.0049.$$

The 5 per cent critical region is the upper 5 per cent region of the χ^2_{10}-distribution, that is, greater than 18.31. The test statistic is

$$\chi^2 = \frac{10 \times 0.0170}{0.0049} = 34.7.$$

This value is (highly) significant. So we reject the manufacturer's claim.

This is the second test we have applied to these data. The reader's attention is drawn to the warning given in section 12.2.

Further reading: Balaam [3] 108; Chakravarti *et al.* [9] 326; Guenther [39] 16–17; Mood and Graybill [70] 307; Remington and Schork [86] 200–9; Snedecor [91] 59–62; Zar [105] 98.

12.29   Equality (inequality) of two variances

Usual form of data

Two samples as in section 12.15. The first contains n_1 observations and has sample variance s_1^2. The second contains n_2 observations and has sample variance s_2^2.

Statistical model

The first sample comes from a normal population with mean μ_1 and variance σ_1^2. The second sample comes from a population with mean μ_2 and variance σ_2^2. All the observations are independent.

Hypotheses

(a) Equality	(b) Inequality
$H_0: \sigma_1^2 = \sigma_2^2;$	$H_0: \sigma_1^2 \leqslant \sigma_2^2;$
$H_1: \sigma_1^2 \neq \sigma_2^2.$	$H_1: \sigma_1^2 > \sigma_2^2.$

Critical region

(a) *Equality*: greater than upper $2\frac{1}{2}$ per cent point of the $F_{p,q}$-distribution (p and q are defined below).

(b) *Inequality*: greater than upper 5 per cent point of the F_{n_1-1, n_2-1}-distribution.

Computation of test statistic

Set $\sigma_1^2 = \sigma_2^2$ in the statistic

■
$$F = \frac{s_1^2}{\sigma_1^2} \Big/ \frac{s_2^2}{\sigma_2^2}. \tag{12.29.1}$$

Case (a) only. If the test statistic is greater than one, use $p = n_1 - 1$ and $q = n_2 - 1$; if the statistic is less than one, use its reciprocal and set $p = n_2 - 1; q = n_1 - 1$.

Comments

1. The reciprocal result for the equality test follows as a consequence of the definition of an F-distribution (section 9.4).

2. Other hypotheses can be tested; for example:

(c) $H_0: \sigma_1^2 = \kappa\sigma_2^2$, where κ is a specified constant;
   $H_1: \sigma_1^2 \neq \kappa\sigma_2^2.$

Use method (a) and set σ_1^2 equal to $\kappa\sigma_2^2$ in formula (12.29.1).

(d) $H_0: \sigma_1^2 \leqslant \kappa\sigma_2^2;$
   $H_1: \sigma_1^2 > \kappa\sigma_2^2.$

Use method (b) and set σ_1^2 equal to $\kappa\sigma_2^2$ in formula (12.29.1).

Example 12.29.1. The yields of two different varieties of wheat are compared in sections 12.15 and 12.16. Is the variance of the yield from type B the same as the variance of type A?

We need to test $H_0: \sigma_A^2 = \sigma_B^2;$
$$H_1: \sigma_A^2 \neq \sigma_B^2.$$

From section 12.15, we know that $n_A = n_B = 25$; $s_A^2 = 5.9$ and $s_B^2 = 11.2$. The 5 per cent critical region is the upper $2\frac{1}{2}$ per cent region of the $F_{24, 24}$-distribution, that is, F greater than 1.98. The test statistic is

$$F = \frac{11.2}{5.9} = 1.90.$$

This value is not significant; we conclude that the two variances are equal.

This is the third test we have applied to these data. The reader's attention is drawn to the warning in section 12.2.

> *Further reading*: Balaam [3] 145–50; Brownlee [7] 282–8; Chakravarti *et al.* [9] 327; Guenther [39] 18–19; Remington and Schork [86] 210–15; Snedecor [91] 96; Zar [105] 101–2.

12.30 Equality of several variances – Bartlett's test

Usual form of data

k samples as in section 12.18. The ith sample contains n_i observations and has sample variance s_i^2. We define $n = n_1 + ... + n_k$.

Statistical model

The samples all come from normal populations. The mean and variance of the first population are μ_1 and σ_1^2 respectively, the mean and variance of the second population are μ_2 and σ_2^2 respectively, ... All the observations are independent.

Hypotheses

H_0: $\sigma_1^2 = \sigma_2^2 = ... = \sigma_k^2$;
H_1: the variances are not all equal.

Critical region

The upper 5 per cent region of the χ_{k-1}^2-distribution.

Computation of test statistic

$$C = 1 + \{1/(n_1 - 1) + ... + 1/(n_k - 1) - 1/(n - k)\} / \{3(k - 1)\}.$$

$$S^2 = \{(n_1 - 1)s_1^2 + ... + (n_k - 1)s_k^2\} / (n - k).$$

■ $$\chi^2 = [(n - k)\ln S^2 - \{(n_1 - 1)\ln s_1^2 + ... + (n_k - 1)\ln s_k^2\}]/C.$$

$$(12.30.1)$$

Comments

1. This is an approximate test.
2. If $k = 2$, use the exact test of section 12.29.
3. The more the sample variances differ from one another, the larger (12.30.1) becomes. The statistic is small when they are all approximately equal.
4. Other hypotheses can be tested; for example:

H_0: $\sigma_1^2 = 2\sigma_2^2$, $\sigma_2^2 = \sigma_3^2 = ... = \sigma_k^2$;
H_1: null hypothesis not true.

Use the above method with $s_1^2/2$ instead of s_1^2.

5. If natural logarithms are not readily available, use logarithms to base 10 in (12.30.1) and then multiply the result by 2.3026.
6. Bartlett's test is not robust for departures from normality of the data (Box [5]). Tests for normality are outlined in section 12.35.

Example 12.30.1. Use Bartlett's method and a 5 per cent significance level to test the data in table 12.18.2 for homogeneity of variance.[5]
The critical region is the upper 5 per cent region of the χ_2^2-distribution. We find that $s_1^2 = 0.7240$, $s_2^2 = 2.5657$ and $s_3^2 = 0.4770$.

$$C = 1 + (\tfrac{1}{5} + \tfrac{1}{6} + \tfrac{1}{4} - \tfrac{1}{15})/(3 \times 2) = 1.091\,67.$$
$$S^2 = (5 \times 0.7240 + 6 \times 2.5657 + 4 \times 0.4770)/15 = 1.3948.$$

$$\chi^2 = \{15 \ln (1.3948) - 5 \ln (0.7240) - 6 \ln (2.5657) \\ - 4 \ln (0.4770)\} / 1.091\,67 = 3.585.$$

This value is not significant; there is no evidence of heterogeneity of variance.
This is the fourth test we have applied to these data. The reader's attention is drawn to the warning in section 12.2.

Further reading: Box [5]; Brownlee [7] 290–5; Chakravarti et al. [9] 328; Guenther [39] 20–1; Samiuddin and Atiqullah [87]; Snedecor [91] 285–9; Zar [105] 131–3.

12.31 Fisher's z-transformation for testing correlation hypotheses

Usual form of data
n points: $(x_1, y_1), (x_2, y_2), ..., (x_n, y_n)$.
Statistical model
Each point is a random point from the bivariate normal distribution[6] with correlation coefficient[7] ρ. The n points are independent.
Hypotheses

(a) Equality	(b) Inequality	(c) Inequality
$H_0: \rho = \rho_0;$	$\rho \leqslant \rho_0;$	$\rho \geqslant \rho_0;$
$H_1: \rho \neq \rho_0.$	$\rho > \rho_0.$	$\rho < \rho_0.$

Critical region
(a) *Equality*: greater than the upper $2\tfrac{1}{2}$ per cent point or less than the lower $2\tfrac{1}{2}$ per cent point of the unit normal distribution.
(b) *Inequality*: greater than the upper 5 per cent point of the unit normal distribution.
(c) *Inequality*: less than the lower 5 per cent point of the unit normal distribution.

[5] *Homoscedasticity* is sometimes used as a synonym for homogeneity of variance; *heteroscedasticity* refers to heterogeneity of variance.
[6] The bivariate normal distribution is described in section 9.6.
[7] The correlation coefficient is defined in section 8.8.

Table 12.31.1. *Fisher's z-transformation*:* $z = \frac{1}{2} \ln \{(1+r)/(1-r)\}$

r	z	r	z	r	z	r	z
0.00	0.000	0.25	0.255	0.50	0.549	0.75	0.973
0.01	0.010	0.26	0.266	0.51	0.563	0.76	0.996
0.02	0.020	0.27	0.277	0.52	0.576	0.77	1.020
0.03	0.030	0.28	0.288	0.53	0.590	0.78	1.045
0.04	0.040	0.29	0.299	0.54	0.604	0.79	1.071
0.05	0.050	0.30	0.310	0.55	0.618	0.80	1.099
0.06	0.060	0.31	0.321	0.56	0.633	0.81	1.127
0.07	0.070	0.32	0.332	0.57	0.648	0.82	1.157
0.08	0.080	0.33	0.343	0.58	0.662	0.83	1.188
0.09	0.090	0.34	0.354	0.59	0.678	0.84	1.221
0.10	0.100	0.35	0.365	0.60	0.693	0.85	1.256
0.11	0.110	0.36	0.377	0.61	0.709	0.86	1.293
0.12	0.121	0.37	0.388	0.62	0.725	0.87	1.333
0.13	0.131	0.38	0.400	0.63	0.741	0.88	1.376
0.14	0.141	0.39	0.412	0.64	0.758	0.89	1.422
0.15	0.151	0.40	0.424	0.65	0.775	0.90	1.472
0.16	0.161	0.41	0.436	0.66	0.793	0.91	1.528
0.17	0.172	0.42	0.448	0.67	0.811	0.92	1.589
0.18	0.182	0.43	0.460	0.68	0.829	0.93	1.658
0.19	0.192	0.44	0.472	0.69	0.848	0.94	1.738
0.20	0.203	0.45	0.485	0.70	0.867	0.95	1.832
0.21	0.213	0.46	0.497	0.71	0.887	0.96	1.946
0.22	0.224	0.47	0.510	0.72	0.908	0.97	2.092
0.23	0.234	0.48	0.523	0.73	0.929	0.98	2.298
0.24	0.245	0.49	0.536	0.74	0.950	0.99	2.647

* *Source*: P. G. Hoel, *Elementary statistics*, Wiley, New York, 1960. Reprinted with permission of the publisher.

Computation of test statistic

Compute the sample correlation coefficient r (section 8.8).

$z = \frac{1}{2} \ln \{(1+r)/(1-r)\}$.

$\zeta_0 = \frac{1}{2} \ln \{(1+\rho_0)/(1-\rho_0)\}$.

The test statistic is

■ $T = (n-3)^{\frac{1}{2}} (z - \zeta_0)$.   (12.31.1)

Comments

1. This is an approximate test. The exact distribution of r (which depends upon both ρ and n) is rather complicated (see, for example, Kendall and Stuart [53] 387).

2. An exact test is available for testing the null hypothesis $\rho = 0$ against the alternative $\rho \neq 0$. The critical region comprises the upper and lower $2\frac{1}{2}$ per cent regions of the t_{n-2}-distribution, and the test statistic is

■ $$t = (n-2)^{\frac{1}{2}} r/(1-r^2)^{\frac{1}{2}}. \tag{12.31.2}$$

3. The z-transformation remains asymptotically normally distributed when the underlying population is not bivariate normal, but the approach is less rapid (that is, larger n required). Skewness of the marginal distributions has relatively little effect on the test, but the effect of departures from the normal kurtosis value ($\beta_2 = 3$) can be considerable. Various data transformations may be tried in an attempt to produce a bivariate population with normal marginal distributions.

4. A non-parametric test of zero correlation is described in section 12.34.

5. Table 12.31.1 can be used for the z-transformation.

6. Statistical tests associated with regression analysis are described in Part III.

Example 12.31.1. The marks obtained by 15 students in two questions of a statistics examination are shown in table 12.31.2. Test the hypothesis that the correlation between marks in these two questions is 0.50.

The critical region is $|T| > 1.96$, and the sample correlation coefficient $r = 0.764$. We compute

$$z = \tfrac{1}{2}\ln(1.764/0.236) = 1.006,$$

$$\varsigma_0 = \tfrac{1}{2}\ln(1.500/0.500) = 0.549.$$

The test statistic is

$$T = (12)^{\frac{1}{2}}(1.006 - 0.549) = 1.583.$$

This value is not significant; so we accept the null hypothesis.

Further reading: Bailey [2] 78–90; Balaam [3] 196–206; Brownlee [7] 397–417; Chakravarti *et al.* [9] 330; Hoel [45] 187–94; Kendall and Stuart [54] 468–9; Snedecor [91] 173–80; Zar [105] 236–40.

Table 12.31.2. *The marks of fifteen students in two examination questions*

Student	Mark for question 1	Mark for question 2
1	27	15
2	18	5
3	15	10
4	10	9
5	3	2
6	18	10
7	6	8
8	15	9
9	15	10
10	12	7
11	13	8
12	17	14
13	19	13
14	17	11
15	9	6

12.32 Equality (inequality) of two correlation coefficients

Usual form of data

Two samples, each like the one in section 12.31. The first sample contains n_1 points and the second contains n_2 points. The correlation coefficient[8] for the first sample is r_1 and the correlation coefficient for the second sample is r_2.

Statistical model

The samples come from bivariate normal populations[9] with correlation coefficients ρ_1 and ρ_2 respectively. The observations are independent.[10]

Hypotheses

(a) Equality	(b) Inequality	(c) Inequality
$H_0: \rho_1 = \rho_2$;	$H_0: \rho_1 \geqslant \rho_2$;	$H_0: \rho_1 \leqslant \rho_2$;
$H_1: \rho_1 \neq \rho_2$.	$H_1: \rho_1 < \rho_2$.	$H_1: \rho_1 > \rho_2$.

[8] The population and sample correlation coefficients (ρ and r) are defined in section 8.8.

[9] The bivariate normal distribution is described in section 9.6.

[10] Each observation is composed of two values: an x-value and a y-value. The pairs are assumed to be independent, but the two numbers making up a pair are not assumed to be independent.

Critical region

(a) *Equality*: greater than the upper $2\frac{1}{2}$ per cent point or less than the lower $2\frac{1}{2}$ per cent point of the unit normal distribution

(b) *Inequality*: less than the lower 5 per cent point of the unit normal distribution.

(c) *Inequality*: greater than the upper 5 per cent point of the unit normal distribution.

Computation of test statistic

$$z_1 = \tfrac{1}{2} \ln \{(1+r_1)/(1-r_1)\}.$$

$$z_2 = \tfrac{1}{2} \ln \{(1+r_2)/(1-r_2)\}.$$

$$S^2 = (n_1-3)^{-1}+(n_2-3)^{-1}.$$

The test statistic is

■ $\qquad T = (z_1 - z_2)/S.$ (12.32.1)

Comments

1. The test is an approximate one and makes use of Fisher's z-transformation (table 12.31.1).

2. The robustness of the z-transformation is discussed in comment 3 of section 12.31.

Example 12.32.1. The heights and weights of ten males and eight females are given in table 12.32.1. Is the height/weight correlation coefficient the same for males and females?

Table 12.32.1. *The heights and weights of ten males and eight females chosen at random from a population*

Males		Females	
Height (cm)	Weight (kg)	Height (cm)	Weight (kg)
166	63	161	60
187	98	158	61
170	82	165	68
171	80	154	52
182	86	166	67
188	85	170	68
175	84	167	69
183	79	177	80
179	75		
178	76		

We need to test $H_0: \rho_M = \rho_F$;

$$H_1: \rho_M \neq \rho_F.$$

The 5 per cent critical region is composed of the upper and lower $2\frac{1}{2}$ per cent regions of the unit normal distribution, that is, $|T| > 1.96$. We find that

$$n_M = 10, \qquad r_M = 0.6744, \qquad z_M = 0.8188,$$
$$n_F = 8, \qquad r_F = 0.9665, \qquad z_F = 2.0362.$$

$S^2 = 0.342\,86$, and the test statistic is

$$T = (0.8188 - 2.0362)/(0.342\,86)^{\frac{1}{2}} = -2.08.$$

This value lies in the critical region. So we reject the null hypothesis and conclude that $\rho_M \neq \rho_F$.

Further reading: Chakravarti *et al.* [9] 331; Zar [105] 241.

12.33 Equality of several correlation coefficients

Usual form of data

k samples, each like the one in section 12.31. The first sample contains n_1 points, ..., and the kth sample contains n_k points. The correlation coefficient[11] for the first sample is r_1, ..., and the correlation coefficient for the kth sample is r_k.

Statistical model

The samples come from bivariate normal[12] populations with correlation coefficients $\rho_1, \rho_2, ..., \rho_k$. The observations are independent.[13]

Hypotheses

$$H_0: \rho_1 = \rho_2 = ... = \rho_k;$$
$$H_1: \text{null hypothesis not true.}$$

Critical region

The upper 5 per cent region of the χ^2_{k-1}-distribution.

Computation of test statistic

$$z_1 = \tfrac{1}{2}\ln\ \{(1+r_1)/(1-r_1)\}\ .$$
$$\vdots$$
$$z_k = \tfrac{1}{2}\ln\ \{(1+r_k)/(1-r_k)\}\ .$$
$$N = (n_1+n_2+...+n_k) - 3k.$$
$$\bar{z} = \{(n_1-3)z_1+...+(n_k-3)z_k\}\ /N.$$

The test statistic is

$$\chi^2 = \{(n_1-3)z_1^2+...+(n_k-3)z_k^2\} - N\bar{z}^2. \qquad (12.33.1)$$

[11] The population and sample correlation coefficients (ρ and r) are defined in section 8.8.

[12] The bivariate normal distribution is described in section 9.6.

[13] See footnote 10.

Comments
>1. The test is an approximate one and makes use of Fisher's
>z-transformation (table 12.31.1).
>2. The robustness of the z-transformation is discussed in comment 3
>of section 12.31.
>3. In the case $k = 2$, the test is equivalent to test (a) of section
>12.32; in fact (12.33.1) is the square of (12.32.1).

Example 12.33.1. The heights and weights of ten males and eight
females are given in table 12.32.1. Use the test statistic (12.33.1)
to test the null hypothesis that the male and female correlation coefficients
are equal.

The 5 per cent critical region is the upper 5 per cent region of the χ_1^2-
distribution, that is, greater than 3.84. We find that

$$n_M = 10, \qquad r_M = 0.6744, \qquad z_M = 0.8188,$$
$$n_F = 8, \qquad r_F = 0.9665, \qquad z_F = 2.0362.$$

Furthermore,

$$N = 10 + 8 - 3 \times 2 = 12,$$
$$\bar{z} = (7 \times 0.8188 + 5 \times 2.0362)/12 = 1.3261.$$

The test statistic is

$$\chi^2 = 7 \times (0.8188)^2 + 5 \times (2.0362)^2 - 12 \times (1.3261)^2 = 4.32.$$

This value lies in the critical region. So we reject the null hypothesis and
conclude that $\rho_M \neq \rho_F$.

Note that in this case $k = 2$, and the χ^2-test value (4.32) is the square of
the test value in example 12.32.1 (-2.08).

Further reading: Chakravarti *et al.* [9] 331; Zar [105] 242.

12.34 A non-parametric test of zero correlation – Spearman's rank correlation coefficient r_S

The correlation procedures of sections 12.31, 12.32 and 12.33
cannot be used when the bivariate population is markedly non-normal.
Various data transformations might be tried in an attempt to produce a
bivariate distribution more closely resembling the normal. Alternatively,
we can operate with the ranks of measurements for each variable and use
the Spearman rank correlation coefficient r_S. Such a method must of course
be used when the data are actually given as ranks.

The Spearman rank correlation coefficient r_S of a sample (x_1, y_1), ...,
(x_n, y_n) of size n is defined as the ordinary correlation coefficient of the
ranked variates. That is, we rank the x-values in ascending order from 1 to n,

and the y-values in ascending order from 1 to n, and we calculate the correlation between the rankings of each element of the pair (x_i, y_i) $(i = 1, ..., n)$. The calculation method of section 8.8 can be used, but the reader is advised to use the considerably simpler formula (12.34.2).

From an infinite sample, we would obtain the population rank correlation coefficient ρ_S. What is the connection between ρ_S and the ordinary correlation coefficient ρ of the two random variables X and Y? We recall from section 8.2 that the distribution function $F(x)$ of the random variable X gives the probability that X is less than or equal to x. Likewise, the distribution function $G(y)$ of the random variable Y gives the probability that Y is less than or equal to y. Let us use the functions $F(x)$ and $G(y)$ to define transformed random variables $U = F(X)$ and $V = G(Y)$. Then the population rank correlation coefficient of X and Y is equal to the ordinary correlation coefficient of U and V. Using this result, it is possible to prove that when X and Y have the bivariate normal distribution with correlation coefficient ρ,

■ $$\rho_S = \frac{6}{\pi} \arcsin \frac{\rho}{2} = \frac{3}{\pi} \left(\rho + \frac{\rho^3}{24} + \frac{3\rho^5}{640} + ... \right). \tag{12.34.1}$$

In fact, ρ_S and ρ are roughly equal in the case of the bivariate normal distribution.

We now summarise a test of zero correlation.

Usual form of data

    n points: $(x_1, y_1), (x_2, y_2), ..., (x_n, y_n)$.

Statistical model

    The sample comes from a bivariate population. The observations are independent.[14]

Hypotheses

    $H_0: \rho_S = 0$;
    $H_1: \rho_S \neq 0$.

Critical region

    (i) *Small samples*: the upper $2\frac{1}{2}$ per cent region and the lower $2\frac{1}{2}$ per cent region of the Spearman rank correlation distribution (table 12.34.1).
    (ii) *Larger samples*: the upper $2\frac{1}{2}$ per cent region and the lower $2\frac{1}{2}$ per cent region of the t_{n-2}-distribution.

Computation of test statistic

    (i) *Small samples*

Rank the x-values in ascending order from 1 to n.
Rank the y-values in ascending order from 1 to n.
Calculate the difference between the rank of each x-value and the rank of the corresponding y-value.

[14] See footnote 10.

Table 12.34.1. *The upper† 100α per cent points of the Spearman rank correlation distribution.* The critical region includes the table entry*

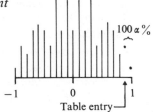

Sample size n	$\alpha = 0.05$	0.025	0.01
5	0.900	1.000	1.000
6	0.828	0.885	0.942
7	0.714	0.785	0.892
8	0.642	0.738	0.833
9	0.600	0.683	0.783
10	0.563	0.648	0.745
11	0.527	0.609	0.700
12	0.497	0.580	0.671
13	0.478	0.555	0.643
14	0.459	0.534	0.622
15	0.443	0.518	0.600
16	0.427	0.500	0.582
17	0.412	0.485	0.564
18	0.399	0.472	0.548
19	0.390	0.458	0.533
20	0.379	0.445	0.520
21	0.369	0.435	0.508
22	0.360	0.424	0.496
23	0.352	0.415	0.485
24	0.344	0.406	0.475
25	0.336	0.398	0.465
26	0.330	0.389	0.456
27	0.324	0.382	0.448
28	0.318	0.375	0.440
29	0.311	0.369	0.432
30	0.306	0.362	0.425

* The entries for $n \geqslant 11$ are reprinted, with permission of the publishers, from the table of Glasser and Winter (1961, *Biometrika,* **48**, 444–8). For $n < 11$, entries were calculated using the exact distribution of Σd_i^2 tabulated by Olds (1938, *Annals of Mathematical Statistics,* **9**, 133–48).

† The lower 100α per cent point is obtained by changing the sign of the upper 100α per cent point.

Denote these differences by $d_1, d_2, ..., d_n$.
The Spearman rank correlation coefficient r_S is used as the test statistic:

■ $r_S = 1 - 6(d_1^2 + ... + d_n^2)/(n^3 - n)$. (12.34.2)

(ii) *Larger samples*
Compute r_S as in (i). Then use the test statistic

■ $t = (n-2)^{\frac{1}{2}} r_S/(1 - r_S^2)^{\frac{1}{2}}$. (12.34.3)

Comment

When two or more values are tied, assign to each of them the average of the ranks which would have been assigned to them had they not been tied.

Example 12.34.1. The marks of fifteen students in two examination questions are given in table 12.31.2. Test the null hypothesis that the marks in the two questions are uncorrelated.
   According to table 12.34.1, the 5 per cent critical region is $|r_S| > 0.518$. To calculate the test critierion, we form the rank table 12.34.2. The rank correlation coefficient

$$r_S = 1 - 6 \{0^2 + (10.5)^2 + ... + 0^2\}/\{(15)^3 - 15\} = 0.704.$$

Table 12.34.2. *A ranking of students' marks in two examination questions*

Student	Question 1 rank	Question 2 rank	|Difference|
1	15	15	0
2	12.5	2	10.5
3	8	10	2
4	4	7.5	3.5
5	1	1	0
6	12.5	10	2.5
7	2	5.5	3.5
8	8	7.5	0.5
9	8	10	2
10	5	4	1
11	6	5.5	0.5
12	10.5	14	3.5
13	14	13	1
14	10.5	12	1.5
15	3	3	0

This value is significant at the 5 per cent level; so we reject the null hypothesis and conclude that $\rho_S \neq 0$.

This is the second test we have applied to these data. The warning in section 12.2 should be noted.

> *Further reading*: Kendall [52] 128-9; Pollard [81] 145-50; Snedecor [91] 190-9; Zar [105] 243.

12.35   Testing normality

It sometimes happens that we have a sample of n independent observations $x_1, ..., x_n$ from an unknown population, and we wish to test whether the underlying population is normal. The chi-square goodness-of-fit test of section 12.11 can be used for this purpose. We calculate the sample mean $\bar{x}$ and the sample variance s^2 and fit a normal curve with mean $\mu = \bar{x}$ and variance $\sigma^2 = s^2$. We then calculate the expected numbers in consecutive intervals according to the fitted normal curve and perform the usual chi-square goodness-of-fit test. The Kolmogorov–Smirnov test should only be used with caution (section 12.12).

Tests of skewness and kurtosis are described, for example, by Snedecor [91] 201-3, but the above tests of distribution are preferable.

> **Example 12.35.1.** The following twenty observations come from an unknown population:
>
> | 8.2, | 3.9, | 2.9, | 0.0, | 1.2, | 2.4, | 0.1, |
> | 4.2, | 7.7, | 9.8, | 3.5, | 8.5, | 1.3, | 7.7, |
> | 8.6, | 3.8, | 3.9, | 0.7, | 4.1, | 6.0. | |

Test the null hypothesis that the population is normal.

We begin by calculating the sample mean $\bar{x} = 4.425$ and the sample standard deviation $s = 3.093$. According to the fitted normal distribution with mean $\mu = 4.425$ and standard deviation $\sigma = 3.093$, the expected number of outcomes in the interval $\mu + 0.5\sigma$ to $\mu + 1.5\sigma$ is

$$20(\Phi(1.5) - \Phi(0.5)) = 4.8346.$$

The expectations for the other intervals in table 12.35.1 are calculated in a similar manner (the intervals have been chosen for convenience; alternative intervals would have done). The chi-square value turns out to be 8.44 which is greater than the upper 5 per cent point of the chi-square distribution with three degrees of freedom. We therefore reject the null hypothesis and conclude that the population is not normal.

> *Further reading*: Balaam [3] 225; Chakravarti *et al.* [9] 394; Massey [64]; Remington and Schork [86] 218-24, 233; Snedecor [91] 201-2; Zar [105] 79-85.

Table 12.35.1. *The chi-square method for testing normality*

Interval	Number of outcomes		Contribution to chi-square
	Observed	Expected	
$-\infty$ to $\mu-1.5\sigma$	0	1.3362	1.34
$\mu-1.5\sigma$ to $\mu-0.5\sigma$	6	4.8346	0.28
$\mu-0.5\sigma$ to μ	7	3.8292	2.63
μ to $\mu+0.5\sigma$	0	3.8292	3.83
$\mu+0.5\sigma$ to $\mu+1.5\sigma$	6	4.8346	0.28
$\mu+1.5\sigma$ to ∞	1	1.3362	0.08
Totals	20	20.0000	8.44

$$\frac{(obs - exp)^2}{exp.}$$

12.36 Testing a population to see whether it is Poisson

It sometimes happens that we have a sample of n independent observations $x_1, ..., x_n$ from an unknown distribution on the non-negative integers, and we wish to test whether the underlying population is Poisson. The chi-square goodness-of-fit test of section 12.11 can be used for this purpose. We calculate the sample mean $\bar{x}$ and fit a Poisson distribution with parameter $\lambda = \bar{x}$. We then compute expected numbers in consecutive intervals according to the fitted distribution and perform the usual chi-square goodness-of-fit test. The Kolmogorov–Smirnov test should only be used with caution (section 12.12).

The mean and variance of the Poisson distribution are equal. If the sample mean $\bar{x}$ and the sample variance s^2 are markedly different, we may doubt the appropriateness of the Poisson model. Rather than go through the whole process of fitting a Poisson distribution and testing goodness-of-fit, the following preliminary test of equality of mean and variance can be used.

Usual form of data

$x_1, ..., x_n$ (a sample of size n). The sample mean and variance are $\bar{x}$ and s^2 respectively. The $\{x_i\}$ are non-negative integers.

Statistical model

The $\{x_i\}$ are independent observations from a unimodal distribution on the non-negative integers. The distribution has mean μ and variance σ^2, and is close to normal.

Hypotheses

$H_0: \sigma^2 = \mu;$
$H_1: \sigma^2 \neq \mu.$

Critical region

Greater than the upper $2\frac{1}{2}$ per cent point or less than the lower $2\frac{1}{2}$ per cent point of the chi-square distribution with $n-2$ degrees of freedom.

Computation of test statistic

The test statistic is

■ $\chi^2 = (n-1)s^2/\bar{x}.$ (12.36.1)

Comments

1. This is an approximate test. Do not use it if $\bar{x}$ is less than about ten.

2. The test is based on (12.28.1). The unknown population variance σ^2 has been replaced by the sample mean $\bar{x}$ which provides an estimate of the Poisson variance.

3. The test assumes a unimodal distribution which is approximately normal; it tests dispersion rather than the form of distribution. The null hypothesis that the distribution is Poisson should be rejected if the test criterion is significant. Otherwise, the full goodness-of-fit test should be performed. The null hypothesis that the distribution is Poisson is rejected if this test criterion is significant.

Further reading: Cochran [13]; Massey [64]; Remington and Schork [86] 246–8; Zar [105] 302–5.

12.37   Testing other distributional forms

In sections 12.35 and 12.36, we described how the chi-square goodness-of-fit test can be used to test whether a population was normal or Poisson. The same procedure is followed with other distributional forms. We estimate the parameters of the distribution using the methods of chapter 13, and then calculate expected numbers in consecutive intervals. The chi-square test is then applied in the usual manner. The Kolmogorov-Smirnov test should only be used with caution (section 12.12).

12.38   Tests involving Poisson-distributed variables

The Poisson distribution has only one parameter: its mean λ. So hypotheses about Poisson distributions are always hypotheses about means. Normal-based tests can be used when the means are greater than about nine or ten and the sample sizes are not extremely small. Some of the normal-based tests require homogeneity of variance (for example, the one-way analysis of variance of section 12.18). The square root transformation of section 14.6 can be used to achieve this. The square root transformation also gives a distribution which is closer to normal.

12.39 Exercises

1. A manufacturer is considering buying one of two machines. The more expensive machine A is claimed to produce three fewer defectives per hundred output. Some trials are made. In these trials, machine A produces fifteen defectives from an output of 120, while machine B produces fourteen from an output of 130. Do these results contradict the claim made in respect of the two machines?

2. Is there any evidence of association in the following contingency table?

5	2	7
3	8	11
8	10	18

3. Test the goodness-of-fit of the logarithmic series distribution in example 10.12.2.

4. Assume that the population in example 12.12.1 is normal and test the null hypothesis that the mean is zero.

5. Assume that the population in example 12.12.1 is normal and test the null hypothesis that the variance is one.

6. The following two samples come from Poisson populations. Do the populations have the same mean?

14, 7, 13, 8, 5, 10, 5, 4, 7, 16, 11, 11, 6, 12, 10;
12, 7, 7, 7, 10, 8, 13, 7, 11, 14, 9, 7, 8, 6.

7. Apply the one-way analysis of variance procedure to the turtle cholesterol problem of example 12.15.2. Confirm that the F-value you obtain is the square of the t-value in example 12.15.2.

8. Imagine that the top left-hand entry in table 12.25.3 is missing. Test the null hypothesis that there is no difference in phosphorus content among the three species.

9. Test the null hypothesis that the marks in table 12.31.2 are uncorrelated.

13

Point and interval estimation

Summary The opening sections of this chapter summarise the general principles of point estimation, the method of maximum likelihood and confidence intervals. These methods are then applied to parameter estimation problems for the common discrete distributions (sections 13.5–13.11), means and linear combinations of means from normal populations (sections 13.12–13.18), variances and ratios of variances of normal populations (sections 13.19–13.24), parameter estimation problems for the log-normal distribution (section 13.25), and the correlation coefficient of a bivariate normal distribution (section 13.26). A method of obtaining confidence limits for a distribution function is described in section 13.27.

13.1   Point estimation

Consider a random sample of n observations $x_1, ..., x_n$. The distributional form of the population is known (for example, normal), but one or more of the parameters is unknown (for example, either μ or σ^2 or both unknown). The problem is to estimate the unknown parameter(s).

The function of the observations which we choose to estimate a parameter is known as an *estimator*, and the numerical value obtained from it using a particular set of data is called an *estimate.*

Often several different functions will suggest themselves as estimators (for example, the sample mean and the sample median can both be used to estimate the mean of a normal population) and we need to decide which to use. The following criteria are used by statisticians:

1. The estimator should be *unbiased,* so that its expectation is equal to the true value of the parameter. Thus, on average, the estimate obtained is equal to the underlying parameter. The estimator should not provide estimates which are on average too high or too low.

2. The estimator should be *consistent.* By this we mean that for any small quantity ϵ, the probability that the absolute value of the deviation of the estimator from the true parameter value is less than ϵ tends to 1 as n tends to infinity. Thus, for an estimate based on a large number of observations, there is a very small probability that its value will differ seriously from the true value of the parameter.

3. The estimator should be *efficient.* That is, the variance of the estimator should be minimal.

Further reading: Bailey [2] 21–6; Brownlee [7] 88–91; Fraser [26] 214–24; Hoel [45] 137–8; Hoel [46] 190–6; Pollard [81] 151–6; Reminton and Schork [86] 148–51; Spiegel [93] 156–66.

13.2 The principle of maximum likelihood

The method of *maximum likelihood* provides estimators which are usually quite satisfactory. They have the desirable properties of being consistent, asymptotically normal and asymptotically efficient for large samples under quite general conditions. They are often biased, but the bias is frequently removable by a simple adjustment (example 13.2.2). Other methods of obtaining estimators are also available, but the maximum likelihood method is the most frequently used.

Maximum likelihood estimators also have another desirable property: *invariance.* Let us denote the maximum likelihood estimator[1] of the parameter θ by $\hat{\theta}$. Then, if $f(\theta)$ is a single-valued function of θ, the maximum likelihood estimator of $f(\theta)$ is $f(\hat{\theta})$. Thus, for example, $\hat{\sigma} = (\widehat{\sigma^2})^{\frac{1}{2}}$.

The *principle of maximum likelihood* tells us that we should use as our estimate that value which maximises the likelihood[2] of the observed event. The following examples demonstrate the technique.

> **Example 13.2.1.** Derive the maximum likelihood estimator $\hat{p}$ of the binomial probability p for a coin-tossing experiment in which a coin is tossed n times and r heads are obtained.

From (10.1.1), we know that the likelihood of the observed event is

$$L = \binom{n}{r} p^r (1-p)^{n-r}.$$

According to the principle of maximum likelihood, we choose the value of p which maximises L. This value also maximises

$$\ln L = \ln \binom{n}{r} + r \ln p + (n-r) \ln (1-p).$$

At the maximum, the derivative with respect to p must be zero. That is,

$$\frac{\partial}{\partial p} \ln L = r/p - (n-r)/(1-p) = 0,$$

and

■ $$\hat{p} = r/n. \tag{13.2.1}$$

> **Example 13.2.2.** A sample of n observations $x_1, ..., x_n$ is drawn from a normal population. Derive the maximum likelihood estimators $\hat{\mu}$ and $\hat{\sigma}^2$ for the mean and variance of the population.

---

[1] This is the usual notation.
[2] The distinction between *likelihood* and *probability* is explained in a footnote on page 8 of Kendall and Stuart [54].

The likelihood of the observed event is

$$L = \left[\frac{1}{\sigma(2\pi)^{\frac{1}{2}}}\exp\left\{-\frac{1}{2}\left(\frac{x_1-\mu}{\sigma}\right)^2\right\}\right] \cdots \left[\frac{1}{\sigma(2\pi)^{\frac{1}{2}}}\exp\left\{-\frac{1}{2}\left(\frac{x_n-\mu}{\sigma}\right)^2\right\}\right].$$

According to the principle of maximum likelihood, we choose those values of μ and σ^2 which maximise L. The same values maximise $\ln L$. So we equate the partial derivatives of $\ln L$ with respect to $\hat{\mu}$ and $\hat{\sigma}^2$ to zero (section 1.6) and find that

■ $\hat{\mu} = \bar{x}$ (13.2.2)

and $\hat{\sigma}^2 = \frac{1}{n}\sum_{i=1}^{n}(x_i - \bar{x})^2.$

The latter estimator is slightly biased, but the bias can be removed by multiplying by $n/(n-1)$ and using the estimator

■ $\hat{\sigma}^2 = \frac{1}{n-1}\sum_{i=1}^{n}(x_i - \bar{x})^2.$ (13.2.3)

This is the same as sample variance defined in (8.4.3).

Further reading: Bailey [2] 21–6; Balaam [3] 104–6; Brownlee [7] 91–5; Fraser [26] 224–8; Hoel [45] 137–55; Mather [65] 203; Mather [66] 158–9.

13.3   Other methods

The method of maximum likelihood can sometimes lead to a large amount of computation, which may not be justified, or very occasional difficulties which are best avoided. Alternative procedures are available and of these the method of moments is perhaps the most common. The first few sample moments are calculated and these are equated to the population moments. The method is demonstrated in chapter 11 and examples also appear in sections 10.11 and 10.12.

The method of least squares is dealt with in Part III of this book.

Further reading: Fraser [26] 242–3; Mood and Graybill [70] 186–7; Sokolnikoff and Sokolnikoff [92] 544.

13.4   Confidence intervals

The methods of sections 13.2 and 13.3 can be used to obtain what are referred to as *point estimates* of a parameter, but we must remember that a point estimator is a random variable distributed in some way around the true value of the parameter. The true parameter value may be higher or lower than our estimate. It is often useful therefore to obtain an interval within which we are reasonably confident the true value will lie, and the generally accepted method is to construct what are known as *confidence limits*.

The following procedure will yield upper and lower 95 per cent confidence limits with the property that when we say that these limits include the true value of the parameter, 95 per cent of all such statements will be true and 5 per cent will be incorrect.

1. Choose a (test) statistic involving the unknown parameter and no other unknown parameter.
2. Place the appropriate sample values in the statistic.
3. Obtain an equation for the unknown parameter by equating the test statistic to the upper $2\frac{1}{2}$ per cent point of the relevant distribution.
4. The solution of the equation gives one limit.
5. Repeat the process with the lower $2\frac{1}{2}$ per cent point to obtain the other limit.

Warning. The comment in section 12.5 about the appropriateness of a particular statistical model is most important.

We can also construct 95 per cent confidence intervals using unequal tails (for example, using the upper 2 per cent point and the lower 3 per cent point). We usually want our confidence interval to be as short as possible, however, and with a symmetric distribution such as the normal or t, this is achieved using equal tails. The same procedure very nearly minimises the confidence interval with other non-symmetric distributions (for example, chi-square) and has the advantage of avoiding rather tedious computation.

When the appropriate statistic involves the square of the unknown parameter, *both* limits are obtained by equating the statistic to the upper 5 per cent point of the relevant distribution. Examples occur in sections 13.7, 13.15 and 13.18. The use of two tails in this situation would result in a pair of non-intersecting intervals.

When two or more parameters are involved, it is possible to construct a region within which we are reasonably confident the true parameter values will lie. Such regions are referred to as *confidence regions*. Thus, in the trinomial example of section 13.7, we obtain a region within which we are 95 per cent confident that p_1, p_2 and p_3 will lie *simultaneously*. The implied interval for p_1 does not form a 95 per cent confidence interval, however. Nor is it true that an 85.7375 per cent confidence region for p_1, p_2 and p_3 can be obtained by considering the intersection of the three separate 95 per cent confidence intervals, because the statistics used to obtain the individual confidence intervals are not independent. This problem is obvious with a multi-parameter distribution such as the multinomial, but it even occurs with the normal distribution because the statistic (12.13.1) which we use to obtain a confidence interval for the mean and the statistic (12.28.1) which we use to obtain a confidence interval for the variance are not independent. The problem is not likely to be of great concern unless a large number of parameters is involved.

Example 13.4.1. A sample of size nine is drawn from a normal population with unknown mean and variance. The sample mean is 4.2 and the sample variance 1.69. Obtain a 95 per cent confidence interval for the mean μ.

The appropriate statistic is (12.13.1). The confidence limits are obtained from the equations $\quad t_{2.5}\ n-1=8$

$$\sqrt{9}\,(4.2 - \mu)/(1.69)^{\frac{1}{2}} = 2.306;$$

$$\sqrt{9}\,(4.2 - \mu)/(1.69)^{\frac{1}{2}} = -2.306.$$

Rearrange to give μ
$$4.2 - \mu = 2.306 \left(\frac{1.69}{\sqrt{9}}\right)^{\frac{1}{2}}$$
$$= 1$$

$$(\bar{x} - \mu) = t\,\frac{s}{\sqrt{n}}$$

We find that the 95 per cent confidence interval is from 3.2 to 5.2.

Further reading: Bailey [2] 21–6; Brownlee [7] 121–7; Fraser [26] 276–89; Hoel [46] 200–2; Mood and Graybill [70] 248–70; Pollard [81] 151–6; Remington and Schork [86] 152–87; Spiegel [93] 156–66; Wilks [102] 195–215.

13.5 The binomial parameter p

Notation and statistical model

See section 12.6. In a sample of size n, r 'successes' are observed.

Point estimator

■ $\quad \hat{p} = r/n.$ (13.5.1)

95 per cent confidence limits

Equate (13.5.2) to the upper and lower $2\frac{1}{2}$ per cent points of the F-distribution with $2(r + 1)$ and $2(n - r)$ degrees of freedom:

■ $\quad F = \left(\dfrac{n - r}{r + 1}\right)\left(\dfrac{p}{1 - p}\right).$ (13.5.2)

Comments

1. The relationship between the binomial and F-distributions is explained in section 10.3. The method for obtaining the lower percentage points of the F-distribution is outlined in example 9.4.3.
2. When n is large (greater than 20, say) the normal approximation to the binomial can be used. Equate (12.6.1) to the upper and lower $2\frac{1}{2}$ per cent points of the normal distribution.
3. Binomial tables can also be used (section 10.1). Examine the tables corresponding to the given value of n but different values of p. Choose the value of p for which the upper $2\frac{1}{2}$ per cent point is equal to r. We then have one limit. The other limit is obtained by choosing the value of p for which the lower $2\frac{1}{2}$ per cent point is equal to r.
4. Charts from which confidence limits for p can be read off are given in the *Biometrika Tables* [76] 204.
5. The problem of estimating the parameter of a truncated binomial distribution is discussed by Johnson and Kotz [47] 73–6.

Example 13.5.1. In a binomial experiment eighteen trials were made and fourteen were successful. Estimate p (the probability of success) and determine a 95 per cent confidence interval for p.

According to (13.5.1) the point estimate is $\hat{p} = 14/18 = 0.78$. To obtain 95 per cent confidence limits, we equate $\frac{4}{15} p/(1-p)$ to the upper and lower $2\frac{1}{2}$ per cent points of the $F_{30,8}$-distribution:

$$\frac{4}{15} p/(1-p) = 3.89,$$
$$\frac{4}{15} p/(1-p) = 1/2.65.$$

We obtain the limits 0.586 and 0.936.

To obtain the confidence interval by the normal approximation, we need to solve the equations

$$(14 - 18p)/\{18p(1-p)\}^{\frac{1}{2}} = \pm 1.96.$$

We square both sides and multiply throughout by $18p(1-p)$ to obtain

$$(14 - 18p)^2 = (1.96)^2 \{18p(1-p)\},$$

which may be re-arranged as

$$393.1488p^2 - 573.1488p + 196 = 0.$$

The roots of this quadratic are 0.548 and 0.910, and these roots are the approximate 95 per cent confidence limits.

Further reading: Bailey [2] 21–6; Brownlee [7] 91–2, 129–30; Hoel [46] 196–200; Johnson and Kotz [47] 58–9; Mood and Graybill [70] 260–4; Spiegel [93] 156–66; Wilks [102] 195–200.

13.6   The difference between two binomial probabilities

Notation and statistical model

Two samples. The first sample is of size n_1 and r_1 'successes' are observed; the second sample is of size n_2 and r_2 'successes' are observed. The parameters are n_1 and p_1 for the first experiment, and n_2 and p_2 for the second; p_1 and p_2 are unknown. Define $q_1 = 1-p_1, q_2 = 1-p_2$ and $\delta = p_1 - p_2$.

Point estimator

■ $\hat{\delta} = r_1/n_1 - r_2/n_2.$ (13.6.1)

95 per cent confidence limits

Compute $\hat{p}_1 = r_1/n_1, \hat{q}_1 = 1 - \hat{p}_1,$
$\qquad \hat{p}_2 = r_2/n_2, \hat{q}_2 = 1 - \hat{p}_2.$
Equate (13.6.2) to the upper and lower $2\frac{1}{2}$ per cent points of the unit normal distribution

■ $T = (\hat{p}_1 - \hat{p}_2 - \delta)/(\hat{p}_1\hat{q}_1/n_1 + \hat{p}_2\hat{q}_2/n_2)^{\frac{1}{2}}.$ (13.6.2)

Comment

The confidence interval method relies on the normal approximation to the binomial. The sample sizes n_1 and n_2 should therefore be greater than about 20 (section 10.2).

Further reading: Zar [105] 296–7.

13.7 The multinomial parameters $\{p_i\}$

Notation and statistical model

See sections 10.4 and 12.7. In a sample of size n, n_i outcomes are of type i ($i = 1, 2, ..., k$).

Point estimators

■
$$\left. \begin{array}{l} \hat{p}_1 = n_1/n, \\ \cdots\cdots\cdots \\ \hat{p}_k = n_k/n. \end{array} \right\} \tag{13.7.1}$$
■

95 per cent confidence limits

The methods of section 13.5 can be used to obtain a confidence interval for an individual p-parameter. For several parameters, we can specify a *confidence region*. This region is obtained by equating (12.7.1) to the upper 5 per cent point of the χ^2_{k-1}-distribution and recalling the constraint (10.4.2).

Comments

1. It is not correct to apply the confidence interval method of section 13.5 to each individual p-parameter (see section 13.4). Procedures for obtaining simultaneous confidence intervals for $p_1, p_2, ..., p_k$ are outlined by Johnson and Kotz [47] 289.

2. Comment 1 of section 12.7 should be noted.

Example 13.7.1. In a trinomial experiment of 50 trials, ten of the outcomes are of type 1, twenty-four of the outcomes are of type 2, and sixteen of the outcomes are of type 3. Find the maximum likelihood estimates of p_1, p_2 and p_3, and calculate 95 per cent confidence limits for p_1. Obtain a 95 per cent confidence region for p_1, p_2 and p_3.

The maximum likelihood estimates of p_1, p_2 and p_3 are $\frac{10}{50} = 0.2$, $\frac{24}{50} = 0.48$ and $\frac{16}{50} = 0.32$ respectively. To obtain 95 per cent limits for p_1, we solve the equations

$$\tfrac{40}{11} p_1/(1 - p_1) = 1.87, \quad \longleftarrow \text{From F table}$$
$$\tfrac{40}{11} p_1/(1 - p_1) = 1/2.12. \longleftarrow$$

We obtain the limits 0.115 and 0.340.

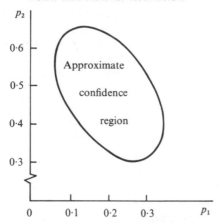

Fig. 13.7.1. An approximate 95 per cent confidence region for p_1 and p_2.

The boundaries of the 95 per cent confidence region for p_1, p_2 and p_3 are defined by the equations

$$(10-50p_1)^2/50p_1 + (24-50p_2)^2/50p_2 + (16-50p_3)^2/50p_3 = 5.99,$$
$$p_1 + p_2 + p_3 = 1.$$

The variable p_3 can be eliminated from the first equation by writing $1-p_1-p_2$ instead of p_3. The resulting confidence region for p_1 and p_2 is then found to have a shape not unlike an ellipse. An approximation to this confidence region can be obtained by replacing the expectations in the denominators by the observed numbers and plotting the ellipse

$$(10-50p_1)^2/10 + (24-50p_2)^2/24 + (50p_1+50p_2-34)^2/16 = 5.99.$$

This approximate confidence region is given in fig. 13.7.1.

Further reading: Johnson and Kotz [47] 288–90.

13.8 The Poisson parameter λ

Notation and statistical model

See section 10.6. The sample size is n; the mean is $\bar{x}$.

Point estimator

■ $\hat{\lambda} = \bar{x}.$ (13.8.1)

95 per cent confidence limits

    (i) $n\bar{x}$ *large* ($\geqslant 10$, say). Equate (13.8.2) to the upper and lower $2\frac{1}{2}$ per cent points of the unit normal distribution.

■ $T = n^{\frac{1}{2}}(\bar{x}-\lambda)/\lambda^{\frac{1}{2}}.$ (13.8.2)

    (ii) $n\bar{x}$ *small.* Equate $2n\lambda$ to the upper and lower $2\frac{1}{2}$ per cent points of the chi-square distribution with $2(n\bar{x}+1)$ degrees of freedom.

Comments

1. The connection between the Poisson and chi-square distributions is explained in section 10.9.

2. Formula (13.8.2) makes use of the normal approximation to the Poisson (section 10.8).

3. An alternative approach is to use Poisson tables (section 10.6). Examine the tables corresponding to different values of the Poisson parameter. Choose the table for which the upper $2\frac{1}{2}$ per cent point is equal to $n\bar{x}$. Divide the tabled Poisson parameter by n. This gives us one confidence limit. The other confidence limit is obtained by choosing the table for which the lower $2\frac{1}{2}$ per cent point is equal to $n\bar{x}$ and dividing the tabled Poisson parameter by n.

4. A table from which confidence limits for λ can be read off is given in the *Biometrika Tables* [76] 203.

5. The problem of estimating the parameter of a truncated Poisson distribution is discussed by Johnson and Kotz [47] 104-9. An approximate method is given in section 14.12.

Example 13.8.1. The south of London has been divided into 576 small areas of equal size. Clarke [12] reports that 229 of the areas experienced zero flying bomb hits during World War II, 211 experienced exactly one hit, 93 experienced two hits, 35 experienced three hits, 7 experienced four hits and one experienced five or more hits. Calculate the maximum likelihood estimate of the Poisson parameter λ (the mean number of hits in an area). Calculate also a 95 per cent confidence interval for λ.

The sample mean is 0.929, and this is the maximum likelihood estimate of λ. The 95 per cent confidence limits are obtained from the equations

$$\sqrt{576}(0.929 - \lambda)/\lambda^{\frac{1}{2}} = \pm 1.96.$$

These equations are solved by squaring and multiplying by λ. We then obtain the quadratic

$$576\lambda^2 - 1074.0496\lambda + 497.1116 = 0$$

with roots 0.854 and 1.011. These are the 95 per cent confidence limits for λ.

Further reading: Bailey [2] 21-6; Johnson and Kotz [47] 94-8, 104-9; Mather [66] 160-1.

13.9   The geometric distribution parameter p

Notation and statistical model

See section 10.10. The sample size is n; the mean is $\bar{x}$.

Point estimator

■ $\hat{p} = 1/(1 + \bar{x})$.   (13.9.1)

95 per cent confidence limits

(i) *Larger samples* ($n > 10$, say). Equate (13.9.2) to the upper and lower $2\frac{1}{2}$ per cent points of the unit normal distribution.

■ $T = (np\bar{x} - nq)/(nq)^{\frac{1}{2}}.$ (13.9.2)

(ii) *Smaller samples.* Use negative binomial tables (section 10.11). Examine the various tables in which the k-parameter is equal to n. Choose the table for which the upper $2\frac{1}{2}$ per cent point is equal to $n\bar{x}$. The p-parameter for that table gives one limit. The other confidence limit is obtained by choosing the table for which the lower $2\frac{1}{2}$ per cent point is equal to $n\bar{x}$.

Comments

1. The maximum likelihood estimator (13.9.1) is the same as the moment estimator obtained by equating (10.10.2) to $\bar{x}$.

2. Not all authors use the same parameters to describe the negative binomial distribution. Be careful.

Example 13.9.1. A sample of size fifteen is drawn from a geometric population. The mean of the sample is 4.3̇. Calculate the maximum likelihood estimate of the parameter p, and obtain a 95 per cent confidence interval for this parameter.

The maximum likelihood estimate is $\hat{p} = 1/5.3̇ = 0.1875$. To obtain a 95 per cent confidence interval for p, we need to solve the equations

$$(65p - 15q)/(15q)^{\frac{1}{2}} = \pm 1.96.$$

Squaring, multiplying by $15q$, writing $1 - p$ for q, and rearranging, we obtain the following quadratic:

$$6400p^2 - 2342.376p + 167.376 = 0$$

with roots 0.097 and 0.269. These are the 95 per cent confidence limits for p.

13.10   The parameters of the negative binomial distribution

Notation and statistical model

See section 10.11. The sample size is n; the mean is $\bar{x}$ and the sample variance is s^2.

Point estimator

■
■ $\left.\begin{array}{l} \hat{p} = \bar{x}/s^2, \\ \hat{k} = \bar{x}\hat{p}/(1 - \hat{p}). \end{array}\right\}$ (13.10.1)

Comments

1. No simple confidence interval method is available.

2. The estimators (13.10.1) are based on the method of moments.

The method of maximum likelihood can be used, but it is rather complicated (see, for example, Johnson and Kotz [47] 131-5).

3. The problem of estimating the parameters of a truncated negative binomial distribution is discussed by Johnson and Kotz [47] 136-7.

An example is given in section 10.11.

Further reading: Johnson and Kotz [47] 131-5.

13.11 The logarithmic series distribution parameter α

Notation and statistical model

See section 10.12. The sample is of size n and the mean is $\bar{x}$.

Point estimator

■ $\{\hat{\alpha}/(1-\hat{\alpha})\}/\{-\ln(1-\hat{\alpha})\} = \bar{x}.$ (13.11.1)

Comments

1. No simple confidence-interval method is available.

2. The point-estimator equation (13.11.1) needs to be solved. iteratively or by trial and error.

3. Equation (13.11.1) gives the maximum likelihood estimate. The same formula is obtained by the method of moments.

4. The problem of estimating the parameter of a truncated logarithmic series distribution is discussed by Johnson and Kotz [47] 178.

An example is given in section 10.12.

Further reading: Johnson and Kotz [47] 175-7.

13.12 The mean of a normal population

Notation and statistical model

See section 12.13. The sample size is n; the mean is $\bar{x}$ and the sample variance is s^2.

Point estimator

■ $\hat{\mu} = \bar{x}.$ (13.12.1)

95 per cent confidence limits

Equate (12.13.1) to the upper and lower $2\frac{1}{2}$ per cent points of the t_{n-1}-distribution.

Comments

1. Sometimes the population variance σ^2 is known. The unit normal distribution should then be used instead of the t-distribution, and there is no need to compute s^2. The population standard deviation σ will replace s in formula (12.13.1).

2. The method is robust for moderate departures from normality.

Example 13.12.1. A sample of size nine is obtained from a normal population with variance one but unknown mean. The sample mean is 4.11, and this is the point estimate of μ. Obtain a 95 per cent confidence interval for μ.

Noting the above comment, we equate (12.13.1) to the upper and lower $2\frac{1}{2}$ per cent points of the unit normal distribution and replace s by σ:

$$\sqrt{9}(4.11-\mu)/1 = \pm 1.96.$$

$1.96 = t \text{ at } 5\% \text{ for } \infty$

Solving for μ, we find that the 95 per cent confidence interval is 4.11 ± 0.65.

Further reading: Bailey [2] 43–51; Balaam [3] 102–4; Hoel [45] 138; Mather [66] 160–1; Mood and Graybill [70] 251–3; Pollard [81] 151–6; Spiegel [93] 156–66; Wilks [102] 203–9.

13.13 The difference between two means – equal variance case

Notation and statistical model

See section 12.15. Define $\delta = \mu_1 - \mu_2$.

Point estimator

■ $\hat{\delta} = \bar{x}_{.1} - \bar{x}_{.2}.$ $\hspace{3cm}$ (13.13.1)

95 per cent confidence limits

Equate (12.15.1) to the upper and lower $2\frac{1}{2}$ per cent points of the $t_{n_1 + n_2 - 2}$-distribution.

Comments

1. It sometimes happens that the common variance σ^2 is known. In this situation use σ^2 instead of s_1^2 and s_2^2 and the normal table instead of t.
2. The method is robust for moderate departures from normality, and robust for moderate departures from equality of variance provided n_1 and n_2 are approximately equal. See also section 13.14.
3. Do not use this method when the observations are paired. See section 13.17.

Example 13.13.1. Two different types of wheat are being compared for yield. Type A is a common variety and type B is a new hybrid variety. Twenty-five acres of each kind are planted and exposed to fairly uniform growing conditions. The average yield with variety A is 32.0 bushels per acre with variance 5.9. The average yield from variety B is 36.2 bushels with variance 11.2. Obtain the maximum likelihood estimate of the difference in mean yield for the two varieties and a 95 per cent confidence interval for the difference.

The maximum likelihood estimate is $\hat{\delta} = \bar{x}_B - \bar{x}_A = 4.2$. To obtain the 95 per cent confidence limits, we solve the equations

$$\frac{\{(36.2-32.0)-\delta\}\sqrt{48}}{(2/25)^{\frac{1}{2}}(24 \times 5.9 + 24 \times 11.2)^{\frac{1}{2}}} = \pm 1.96$$

and obtain the limits 2.58 and 5.82.

> *Further reading*: Balaam [3] 129–33; Hoel [46] 264; Wilks [102] 212–15.

13.14 The difference between two means – unequal variance case

Notation and statistical model
See section 12.16. Define $\delta = \mu_1 - \mu_2$.
Point estimator

■ $\hat{\delta} = \bar{x}_{.1} - \bar{x}_{.2}.$ (13.14.1)

95 per cent confidence limits
Equate (12.16.2) to the upper and lower $2\frac{1}{2}$ per cent points of the t_ν-distribution (ν is defined by (12.16.1)).
Comments
1. The confidence interval is approximate. The exact method of section 13.13 is to be preferred if there is no evidence that the variances are unequal.
2. The method is robust for moderate departures from normality.
3. Avoid having to calculate ν. If $n_1 + n_2$ is greater than 20, use the normal table instead of the t-distribution.
4. Do not use this method when the observations are paired. See section 13.17.

Example 13.14.1. Two different types of wheat are being compared for yield; the experimental results are described in example 13.13.1. Obtain the maximum likelihood estimate of the difference in mean yield for the two varieties and a 95 per cent confidence interval for the difference.
As in example 13.13.1, the maximum likelihood estimate is $\hat{\delta} = 4.2$. To obtain the 95 per cent confidence limits, we solve the equations

$$\frac{(36.2-32.0)-\delta}{(5.9/25 + 11.2/25)^{\frac{1}{2}}} = \pm 1.96$$

and obtain the limits 2.58 and 5.82.

> *Further reading*: Bailey [2] 43–51; Balaam [3] 129–33; Brownlee [7] 299–303; Hoel [46] 262–5.

13.15 A linear combination of the means of k populations with a common variance

Notation and statistical model

See section 12.18. The population means $\mu_1, ..., \mu_k$ are unknown. We wish to estimate $L = c_1\mu_1 + ... + c_k\mu_k$, where the $\{c_i\}$ are constants which may be positive, negative or zero. The mean and variance for sample i are $\bar{x}_{.i}$ and s_i^2 respectively; $n = n_1 + n_2 + ... + n_k$.

Point estimator

■ $\hat{L} = c_1\bar{x}_{.1} + ... + c_k\bar{x}_{.k}$. (13.15.1)

95 per cent confidence limits

(i) *When there is a single set of coefficients $\{c_i\}$ chosen prior to the experiment.*
Compute

■ $s^2 = \{(n_1-1)s_1^2 + ... + (n_k-1)s_k^2\}/(n-k)$. (13.15.2)

Then equate (13.15.3) to the upper 5 per cent point of the $F_{1, n-k}$-distribution.

■ $F = (\hat{L}-L)^2/\{(c_1^2/n_1 + ... + c_k^2/n_k)s^2\}$. (13.15.3)

(ii) *When several different sets of coefficients are used, or the coefficients are chosen after a study of the data.*
Formula (13.15.3) must not be used. The usual problem is to make comparisons between the means, in which case the sum of the $\{c_i\}$ is zero. Formula (13.15.4) can then be used. We calculate s^2 using (13.15.2), and equate (13.15.4) to the upper 5 per cent point of the $F_{k-1, n-k}$-distribution.

■ $S = \{1/(k-1)\}(\hat{L}-L)^2/\{(c_1^2/n_1 + ... + c_k^2/n_k)s^2\}$. (13.15.4)

Comments

1. We recall from example 9.4.2 that an $F_{1, n-k}$-variate is the square of a t_{n-k}-variable. The confidence limits obtained by procedure (i) can also be obtained by equating the square root of (13.15.3) to the upper and lower $2\frac{1}{2}$ per cent points of the t_{n-k}-distribution. Procedure (i) can therefore be called a t-method. The method of section 13.13 is a special case.

2. It is not correct to use many t-intervals all calculated on the same data and each supposedly having a 95 per cent confidence level; nor is it correct to use the t-method when the coefficients $\{c_i\}$ are chosen after a study of the data. The opening paragraph of section 12.21 explains why. Procedure (ii) is based on Scheffé's multiple comparison method (section 12.21).

3. The robustness of the method is dealt with in comments 2 and 3 of section 12.18.

4. The above methods can only be used when the samples come from populations with a common variance. When the variances differ, the approximate method of section 13.16 can be used instead of procedure (i).

5. The special case $c_1 = 1, c_2 = c_3 = ... = c_k = 0$ should be noted (Zar [105] 139).

6. Do not use these methods when the samples are related. See section 13.18.

7. If a one-way analysis of variance table has already been prepared using the data, s^2 is equal to the residual mean square. It is not necessary to compute $s_1^2, s_2^2, ..., s_k^2$.

Example 13.15.1. In the production of a certain machine three components are connected end-to-end. Two of the components are of the same type (type A) and the other component is of type B. An estimate of the average total length of the three components is required, and the following sample values have been obtained:

$$n_A = 10; \quad \bar{x}_A = 3.02; \quad s_A^2 = 0.0004; \quad n_B = 15; \quad \bar{x}_B = 10.04;$$
$$s_B^2 = 0.0006.$$

Find the maximum likelihood estimate of the mean total length and a 95 per cent confidence interval for the total length.

The maximum likelihood estimate is

$$\hat{L} = 2 \times 3.02 + 1 \times 10.04 = 16.08.$$

The variance estimate $s^2 = (9 \times 0.0004 + 14 \times 0.0006)/23 = 0.00052$. To obtain 95 per cent confidence limits for L we use procedure (i) and solve the equation

$$(16.08 - L)^2/\{(2^2/10 + 1^2/15)(0.00052)\} = 4.28,$$

which yields the limits 16.08 ± 0.03.

Further reading: Brownlee [7] 474; Hoel [46] 294; Mood and Graybill [70] 374; Scheffé [88] 30; Zar [105] 151–62.

13.16 A linear combination of the means of k populations with unequal variances

Notation and statistical model

See section 13.15. The *single* set of constants $\{c_i\}$ is defined *prior* to the experiment.

Point estimator

■ $\hat{L} = c_1 \bar{x}._1 + \ldots + c_k \bar{x}._k.$ (13.16.1)

95 per cent confidence limits

    Equate (13.16.2) to the upper 5 per cent point of the $F_{1,\nu}$-distribution.

■ $F = (\hat{L} - L)^2 / (c_1^2 s_1^2 / n_1 + \ldots + c_k^2 s_k^2 / n_k).$ (13.16.2)

    The parameter ν is given by

■
$$\nu = \frac{\left(\dfrac{c_1^2 s_1^2}{n_1} + \ldots + \dfrac{c_k^2 s_k^2}{n_k}\right)^2}{\left(\dfrac{c_1^2 s_1^2}{n_1}\right)^2 \dfrac{1}{n_1 - 1} + \ldots + \left(\dfrac{c_k^2 s_k^2}{n_k}\right)^2 \dfrac{1}{n_k - 1}} - 2. \qquad (13.16.3)$$

Comments

    1. This approximate method replaces procedure (i) of section 13.15 when the variances of the samples are not equal. If the variances are equal, procedure (i) of section 13.15 is to be preferred.

    2. The method is robust for moderate departures from normality.

    3. We recall from example 9.4.2 that an $F_{1,\nu}$-variate is the square of a t_ν-variable. The same 95 per cent confidence limits may be obtained by equating the square root of (13.16.2) to the upper and lower $2\frac{1}{2}$ per cent points of the t_ν-distribution. The method of section 13.14 is a special case.

    4. Avoid having to calculate ν. If n is greater than about 25, use the χ_1^2-distribution instead of the $F_{1,\nu}$-distribution (the normal instead of the t_ν-distribution).

    5. Do not use this method when the samples are related. See section 13.18.

    Example 13.16.1. In the production of a certain machine three components are connected end-to-end. Two of the components are of the same type (type A) and the other component is of type B. An estimate of the average total length of the three components is required, and the following sample values have been obtained:

$$n_A = 10, \quad \bar{x}_A = 3.04, \quad s_A^2 = 0.0004, \quad n_B = 15, \quad \bar{x}_B = 10.15,$$
$$s_B^2 = 0.0095.$$

Find the maximum likelihood estimate of the mean total length and a 95 per cent confidence interval for the total length.

    The maximum likelihood estimate is

$$\hat{L} = 2 \times 3.04 + 1 \times 10.15 = 16.23.$$

The total number of sample values n is 25; so we can use the χ_1^2-distribution instead of $F_{1,\nu}$. To obtain 95 per cent confidence limits for L we solve the equation

$$(16.23 - L)^2 \bigg/ \left(\frac{2^2 \times 0.0004}{10} + \frac{1^2 \times 0.0095}{15}\right) = 3.84.$$

The limits 16.17 and 16.29 are obtained.

Further reading: Brownlee [7] 299–303.

13.17 The mean difference between paired observations

Notation and statistical model
See section 12.22.

Point estimator

■ $\hat{\mu} = \bar{d}.$ (13.17.1)

95 per cent confidence limits
Equate (12.22.1) to the upper and lower $2\frac{1}{2}$ per cent points of the t_{n-1}-distribution.

Comments
1. The method is robust for moderate departures from normality.
2. It is incorrect to use this method when the observations are not paired.

Example 13.17.1. A wheat experiment is described in example 12.22.1.
The sample size is 10 and the average difference in yield between the two varieties is $\bar{d} = 0.57$. This is the maximum likelihood estimate of the mean difference in yield. The sample variance is $s_d^2 = 0.4312$. Calculate 95 per cent confidence limits for the mean difference in yield.
To obtain the confidence limits, we need to solve the equations

$$\sqrt{(10)}(0.57 - \mu)/(0.4312)^{\frac{1}{2}} = \pm 2.262.$$

We obtain the limits 0.57 ± 0.47.

Further reading: Balaam [3] 129–33; Zar [105] 121–4.

13.18 The means of several related (paired) populations

Notation and statistical model
See section 12.25. The row effects $\alpha_1, \alpha_2, ..., \alpha_r$ are unknown. We wish to estimate $L = c_1\alpha_1 + ... + c_r\alpha_r$, where the c_i are constants which sum to zero. The residual mean square in the analysis of variance table 12.25.1 will be denoted by s^2. The row means are $\bar{x}_1. = T_1./c, ..., \bar{x}_r. = T_r./c$.

Point estimator

∎ $\hat{L} = c_1 \bar{x}_1 . + ... + c_r \bar{x}_r . .$ (13.18.1)

95 per cent confidence limits

(i) *When there is a single set of coefficients chosen prior to the experiment.* Equate (13.18.2) to the upper 5 per cent point of the $F_{1,\,(r-1)\,(c-1)}$-distribution.

∎ $F = c(\hat{L} - L)^2 / \{(c_1^2 + ... + c_r^2)s^2\}.$ (13.18.2)

(ii) *When several different sets of coefficients are used or the coefficients are chosen after a study of the data.* Equate (13.18.3) to the upper 5 per cent point of the $F_{r-1,\,(r-1)\,(c-1)}$-distribution.

∎ $S = \{c/(r-1)\}\,(\hat{L} - L)^2 / \{(c_1^2 + ... + c_r^2)s^2\}.$ (13.18.3)

Comments

1. We recall from example 9.4.2 that an $F_{1,n}$-variate is the square of a t_n-variable. The confidence limits obtained by procedure (i) can also be obtained by equating the square root of (13.18.2) to the upper and lower $2\frac{1}{2}$ per cent points of the $t_{(r-1)(c-1)}$-distribution. Procedure (i) can therefore be called a t-method. The method of section 13.17 is a special case.

2. It is not correct to use many t-intervals all calculated on the same data and each supposedly having a 95 per cent confidence level; nor is it correct to use a t-method when the coefficients $\{c_i\}$ are chosen after a study of the data. The opening paragraph of section 12.21 explains why. Procedure (ii) is based on Scheffé's multiple-comparison method (section 12.27).

3. The robustness of the method is dealt with in comment 2 of section 12.25.

4. To apply the techniques to the column effects, interchange r and c, and use the column means $\{\bar{x}_{.j}\}$.

Example 13.18.1. The phosphorus contents (mg/100 g) of the heart, lung, liver and kidney for three animal species are shown in table 12.25.3. Obtain 95 per cent confidence limits for the difference between the sum of the heart and liver contents and the sum of the lung and kidney contents.

In this example, we apply the technique to the column effects, and we select $c_1 = c_3 = 1; c_2 = c_4 = -1$. The column means are 85.4, 103.5, 206.3 and 182.3 respectively. The maximum likelihood estimate is therefore

$$\hat{L} = 85.4 - 103.5 + 206.3 - 182.3 = 5.9.$$

These data were studied in section 12.25, and it would seem that the coefficients were chosen as a result of the earlier analysis. Procedure (ii) is therefore called for. According to the analysis of variance table 12.25.4 the residual mean square $s^2 = 5.061$. To obtain 95 per cent confidence limits for L, we need to solve the equation

$$\{3/(4-1)\}\,(5.9-L)^2/(4\times 5.061) = 4.76.$$

The limits are 5.9 ± 9.8.

Further reading: Guenther [39] 79; Scheffé [88] 30; Zar [105] 151–62.

13.19.   The variance of a normal population with unknown mean

Notation and statistical model

The underlying population is normal with unknown mean and unknown variance σ^2. The sample $x_1, ..., x_n$ is of size n, and the sample variance is s^2.

Point estimator

■ $\hat{\sigma}^2 = s^2.$ (13.19.1)

95 per cent confidence limits

Equate (12.28.1) to the upper and lower $2\frac{1}{2}$ per cent points of the χ^2_{n-1}-distribution.

Comments

1. Occasionally the mean of the population will be known. The method of section 13.20 should then be used.

2. The method is not very robust for departures from normality.

Example 13.19.1. A hand-grenade example is given in section 12.28. The sample size is eleven and the sample variance of the fuse-time is 0.0170. The unbiased point estimate of σ^2 is therefore 0.0170. The 95 per cent confidence limits are obtained by solving the equations

$$\frac{10\times 0.0170}{\sigma^2} = 20.48 \quad \text{and} \quad \frac{10\times 0.0170}{\sigma^2} = 3.25.$$

The limits are 0.0083 and 0.0523.

Further reading: Balaam [3] 104–6, 108–9; Hoel [46] 254–7; Mood and Graybill [70] 254–5; Zar [105] 97.

13.20   The variance of a normal population with known mean

Notation and statistical model

The underlying population is normal with known mean μ but unknown variance σ^2. The sample $x_1, ..., x_n$ is of size n, and the sample mean is $\bar{x}$.

Point estimator

■ $\quad \hat{\sigma}^2 = S^2 = \frac{1}{n} \sum_{i=1}^{n} x_i^2 - \mu(2\bar{x} - \mu).$ (13.20.1)

95 per cent confidence limits

Equate (13.20.2) to the upper and lower $2\frac{1}{2}$ per cent points of the χ_n^2-distribution.

■ $\quad \chi^2 = nS^2/\sigma^2.$ (13.20.2)

Comments

1. When the population mean μ is unknown, use the method of section 13.19.

2. The method is not very robust for departures from normality.

Example 13.20.1. A scientist wishes to test the skill of a technician with a micrometer screw gauge. He gives the technician a piece of brass known to be 6.764 cm in length and an accurate gauge, and asks the technician to take 10 readings. The latter returns with the following readings: 6.772, 6.763, 6.759, 6.768, 6.771, 6.764, 6.756, 6.762, 6.761 and 6.770. Obtain a 95 per cent confidence interval for the variance of the technician.

Using (13.20.1), we find that the maximum likelihood estimate is 268×10^{-7}. To obtain 95 per cent confidence limits we solve the equations

$$\frac{10 \times 268 \times 10^{-7}}{\sigma^2} = 20.48 \quad \text{and} \quad \frac{10 \times 268 \times 10^{-7}}{\sigma^2} = 3.25.$$

The limits are 131×10^{-7} and 825×10^{-7}.

Note: if the scientist were worried about possible bias in the technician's readings, it would probably be better for him to use the method of section 13.12 to estimate the technician's mean, and the method of section 13.19 to estimate his variance.

13.21 The common variance of several populations

Notation and statistical model

k independent normal populations with unknown means and unknown common variance σ^2. The sample from the ith population is of size n_i with sample variance s_i^2. We calculate s^2 using (13.15.2).

Point estimator

■ $\quad \hat{\sigma}^2 = s^2.$ (13.21.1)

95 per cent confidence limits

Equate (13.21.2) to the upper and lower $2\frac{1}{2}$ per cent points of the χ^2_{n-k}-distribution.

■　　$\chi^2 = \dfrac{(n-k)s^2}{\sigma^2}$.　　　　　　　　　　(13.21.2)

Comments

1. If a one-way analysis of variance table has already been prepared using the data, s^2 is equal to the residual mean square. It is not necessary to compute $s_1^2, s_2^2, ..., s_k^2$. See section 12.18.
2. The method is not very robust for departures from normality.

Example 13.21.1. In the production of a certain machine, two different types of components are used: type A and type B. The lengths of the components are of interest and it is known that they have the same variance. The following data have been collected.

Type A: sample size 10; mean 3.02; variance 0.0004.
Type B: sample size 15; mean 10.04; variance 0.0006.

Estimate the common variance of the two types of component.
A point estimate of the common variance is provided by

$$s^2 = (9 \times 0.0004 + 14 \times 0.0006)/23 = 0.00052.$$

Confidence limits are obtained from the equations

$$\frac{23 \times 0.00052}{\sigma^2} = 38.08 \quad \text{and} \quad \frac{23 \times 0.00052}{\sigma^2} = 11.69.$$

The limits are 0.00031 and 0.00102.

Further reading: Balaam [3] 111, 159.

13.22　　Paired observations – the variance of the difference

Notation and statistical model

See section 12.22.

Point estimator

■　　$\hat{\sigma}^2 = s_d^2$.　　　　　　　　　　(13.22.1)

95 per cent confidence limits

Equate (13.22.2) to the upper and lower $2\frac{1}{2}$ per cent points of the χ^2_{n-1}-distribution.

■　　$\chi^2 = (n-1)s_d^2/\sigma^2$.　　　　　　　　　　(13.22.2)

Comments

    1. The case of several related samples is dealt with in section 13.23.
    2. The method is not very robust for departures from normality.

Example 13.22.1. Estimate the variance of the difference in yield
between wheat varieties A and B in example 12.22.1.
    The point estimate is provided by $s_d^2 = 0.4312$. Confidence limits are
obtained by solving the equations

$$9 \times 0.4312/\sigma^2 = 19.02 \quad \text{and} \quad 9 \times 0.4312/\sigma^2 = 2.70.$$

We obtain the limits 0.204 and 1.437.

13.23   Related observations – the variance of the error term e_{ij} in the two-way analysis of variance

Notation and statistical model

    See section 12.25. The residual mean square in the analysis of
variance table (12.25.1) is denoted by s^2.

Point estimator

■ $\hat{\sigma}^2 = s^2.$ (13.23.1)

95 per cent confidence limits

    Equate (13.23.2) to the upper and lower $2\frac{1}{2}$ per cent points of the
chi-square distribution with $(r-1)(c-1)$ degrees of freedom.

■ $\chi^2 = (r-1)(c-1)s^2/\sigma^2.$ (13.23.2)

Comments

    1. In the case of two related samples, the method of section 13.22
can be used. The two approaches are mathematically equivalent.
    2. The method is not very robust for departures from normality.

Example 13.23.1. The phosphorus content (mg/100 g) of each of four
different organs was determined in each of three animal species. The
results are given in table 12.25.3. Estimate the variance of the error term.
    According to the analysis of variance table 12.25.4 the residual mean square
is 5.061 and this provides a point estimate of σ^2. To obtain 95 per cent confidence limits we solve the equations

$$6 \times 5.061/\sigma^2 = 14.45 \quad \text{and} \quad 6 \times 5.061/\sigma^2 = 1.24,$$

which yield the limits 2.1 and 24.5.

    Further reading: Balaam [3] 168.

13.24 The ratio of two variances

Notation and statistical model

Two independent normal populations with unknown means. The unknown variances are σ_1^2 and σ_2^2. We need to estimate the ratio $\gamma = \sigma_1^2/\sigma_2^2$. Samples of size n_1 and n_2 are taken from the respective populations, and sample variances s_1^2 and s_2^2 are obtained.

Point estimator

■
$$\hat{\gamma} = \frac{n_2 - 3}{n_2 - 1} \frac{s_1^2}{s_2^2}. \tag{13.24.1}$$

95 per cent confidence limits

Equate (12.29.1) to the upper and lower $2\frac{1}{2}$ per cent points of the F_{n_1-1, n_2-1}-distribution.

Comments

1. The method for obtaining the lower percentage points of the F-distribution is demonstrated in example 9.4.3.
2. The method is not very robust for departures from normality.

Example 13.24.1. The variances of two independent samples are 11.2 and 4.7. The sample sizes are respectively 17 and 51. Estimate the ratio of the first population variance to the second.

A point estimate is given by $\hat{\gamma} = (48/50)(11.2/4.7) = 2.29$. To obtain 95 per cent confidence limits for the ratio $\gamma = \sigma_1^2/\sigma_2^2$, we solve the equations

$$(11.2/\sigma_1^2)/(4.7/\sigma_2^2) = 2.10 \quad \text{and} \quad (4.7/\sigma_2^2)/(11.2/\sigma_1^2) = 2.48.$$

The limits are 1.13 and 5.91.

Further reading: Balaam [3] 145–50; Zar [105] 101–3.

13.25 The parameters of the log-normal distribution

The parameters of the log-normal distribution are θ, μ and σ^2 (section 9.5). If the value of the parameter θ is known (and it is often possible to take θ equal to zero), then the estimation of the parameters μ and σ^2 is straightforward. In fact, by using the values $\{z_i = \ln(x_i - \theta)\}$ instead of the actual observations $\{x_i\}$, the problem is reduced to that of estimating the mean and variance of a normal distribution (sections 13.12 and 13.19). The point estimators of μ and σ^2 are respectively the sample mean $\bar{z}$ and the sample variance s_z^2 of the $\{z_i\}$. Confidence intervals for μ and σ^2 can be found by applying the methods of sections 13.12 and 13.19 to the transformed values $\{z_i\}$.

When θ is unknown, the estimation problem is difficult. The parameter θ is a threshold value below which the distribution function is zero and above which it is positive. Estimation of θ is particularly inaccurate. However, estimation of the particular parameter values is usually not as important as an accurate estimate of the cumulative distribution function, and it turns out that the latter is fairly insensitive to the actual value of θ (Johnson and Kotz [47] 122). The most satisfactory approach would seem to be to choose a value of θ slightly less than the smallest x observation and then estimate μ and σ^2 as above. θ should be chosen very close to the smallest x observation when the sample size is large.

Further reading: Johnson and Kotz [47] 119–27.

13.26 The correlation coefficient ρ

Notation and statistical model
See section 12.31. The sample correlation coefficient is r and the sample size is n.

Point estimator

■ $\hat{\rho} = r.$ (13.26.1)

95 per cent confidence limits
Use Fisher's transformation to obtain z from r (section 12.31). Equate (13.26.2) to the upper and lower $2\frac{1}{2}$ per cent points of the unit normal distribution to obtain confidence limits for ζ (the transform of ρ). Then apply Fisher's transformation inversely to the ζ limits to obtain the confidence limits for ρ. Table 12.31.1 will be helpful.

■ $T = (n-3)^{\frac{1}{2}} (z - \zeta).$ (13.26.2)

Comments
1. This procedure provides approximate confidence limits.
2. The robustness of the method is dealt with in comment 3 of section 12.31.

Example 13.26.1. The marks obtained by fifteen students in two questions of a statistics examination are shown in table 12.31.2.

A point estimate of the correlation between marks in the two questions is provided by the sample correlation coefficient r which is 0.764. The corresponding z-value is 1.006. To obtain 95 per cent confidence limits for ζ, we solve the equations

$$(12)^{\frac{1}{2}} (1.006 - \zeta) = \pm 1.96.$$

The limits for ζ turn out to be 0.440 and 1.572, and we conclude that the limits for ρ are 0.413 and 0.917.

13.27   Confidence limits for distribution functions

Notation and statistical model

A population with continuous distribution function $F(x)$ which is unknown. A sample of n independent observations $x_1, x_2, ..., x_n$, ordered such that $x_1 < x_2 < ... < x_n$.

95 per cent confidence limits

To obtain 95 per cent confidence limits for $F(x)$ at the point x_j, equate (13.27.1) to the upper 5 per cent point of the Kolmogorov-Smirnov distribution (table 12.12.1).

■　　$Z = |F(x_j) - j/n|.$　　　　　　　　　　　　(13.27.1)

Comments

1. This procedure produces 95 per cent confidence limits for $F(x)$ at each of the n sample points.
2. Avoid using this method with data from discrete distributions, or grouped continuous data. The confidence interval obtained will be wider than the true 95 per cent interval. Further detail is given in section 12.12.

Example 13.27.1. The serum cholesterol levels of seven male turtles were given in table 12.15.1, and the data are reproduced in the first column of table 13.27.1. To obtain 95 per cent limits for the distribution function of male turtle serum cholesterol level at the point $x = 222.5$, we add 0.483 to $\frac{3}{7}$ and subtract 0.483 from $\frac{3}{7}$. The upper limit turns out to be 0.912. The calculated lower limit is negative. We know that the distribution

Table 13.27.1. *Confidence limits for the distribution function of serum cholesterol level in male turtles*

Serum cholesterol level (x_j)	Number of observations less than or equal to x_j (j)	95% confidence limits for $F(x_j)$ Lower	Upper
218.6	1	0.000	0.626
220.1	2	0.000	0.769
222.5	3	0.000	0.912
224.1	4	0.088	1.000
226.5	5	0.231	1.000
228.8	6	0.374	1.000
229.6	7	0.517	1.000

function $F(x)$ must lie on the interval $(0, 1)$, and so we choose the lower limit as zero. The confidence limits at the other six points are calculated similarly.

Further reading: Kendall and Stuart [54] 457-8.

13.28 Exercises

1. In a binomial experiment of thirty trials, nine successes are observed. In another such experiment involving forty trials, eighteen successes are observed. Obtain a 95 per cent confidence interval for the difference between the probabilities of success in the two experiments.

2. The mean of fifteen independent observations from a Poisson population is 0.60. Obtain 95 per cent confidence limits for the Poisson parameter λ.

3. Obtain 95 per cent confidence limits for the common variance of the three worm populations in example 12.18.1.

4. The following observations come from a three-parameter log-normal distribution. Estimate the parameters. 10.78, 6.79, 6.20, 3.74, 5.21, 5.93, 4.11, 6.97, 9.99, 18.43, 6.55, 11.36, 5.28, 9.99, 11.62, 6.72, 6.79, 4.84, 6.90, 8.24.

5. Use the results of section 12.32 to develop a method for obtaining confidence limits for the difference between two correlation coefficients.

14
Some special statistical techniques

Summary   The topics treated in this chapter fall naturally under three headings: random numbers (sections 14.1–14.3); data transformations (sections 14.4–14.7); censored and truncated distributions (sections 14.8–14.12). Topics treated under 'random numbers' include the generation and use of random observations on the unit interval, and transformations for obtaining random observations on other distributions. The 'data transformation' sections describe the arcsine, square-root, logarithmic and reciprocal transformations. Maximum likelihood methods for estimating the parameters of censored and truncated normal and Poisson distributions are described in the final portion of the chapter.

14.1   Random numbers

Problems often arise in statistical analysis which call for the use of random numbers. Survey samplers use such numbers to select random individuals from a population in order to estimate the parameters of the population. Experimental agriculturalists often find it necessary to assign different treatments (for example, fertilisers) to different blocks in a random manner in order to avoid biasing comparisons between treatments. Such an assignment relies on random numbers. Even the theoretician makes use of random numbers. He may wish to study the distribution of a complicated random variable which is defined in terms of other random variables with known distributions. Random observations on the random variables with the known distributions are generated using random numbers, and these random observations are then used to calculate random observations on the complicated random variable (example 14.1.1). Using these sample values, the theoretician can make inferences about the unknown distribution (shape, moments, etc.). W. S. Gosset ('Student') [94], [95] used this type of approach in 1908 to study the *t*-distribution and the distribution of the sample correlation coefficient. The term 'Monte Carlo methods' refers to that branch of mathematics concerned with the use of randomising methods.

Various tables of random numbers have been published (for example, Fisher and Yates [25] 134–9; Rand Corporation [85]; Zar [105] 577–80). These tables are made up of equally-likely decimal digits arranged in groups, and the groups usually contain five digits, although some tables use smaller groupings. Individual digits, or pairs of digits or groups of digits can be used for sampling purposes. To choose 30 men at random from a population of

926, for example, we may number all the men in the population from 1 to 926 and then choose 30 three-digit numbers from a table of random numbers. If a three-digit number we have chosen falls outside the range 1 to 926, we discard it and select three more digits.

A random fraction between 0 and 1 can be obtained by choosing a group of digits (for example, 5 or 10 digits) and imagining a decimal point in front of the group. We usually regard the distribution of such random fractions as continuous, although it is clear that the fractions can only take on values that are multiples of 10^{-n}, where n is the number of digits used in each fraction.

Tables of random numbers must be used carefully, and in such a way that the randomness of the digits used is preserved from one occasion to the next. One method is to start at the beginning of the table and keep a record of how much of the table has been used. Next time the table is used, start with the first unused number, and so work through the table gradually. Alternatively, open the table and, without looking, place a finger on a block of numbers. Use this block of numbers in a *predetermined* manner to choose the page, row, column and direction in which the required random numbers are to be read.

Random numbers are also generated by computers and programmable calculators. These numbers are usually fractions between 0 and 1 and distributed uniformly over the unit interval. Again, we normally regard the distribution as continuous, although it is clear that on a digital computer only multiples of some small number such as 2^{-30} are possible. The numbers are perhaps better referred to as *pseudo-random,* because they are generated by a completely specified rule. Once the first number has been chosen, the whole sequence is determined. The mathematical procedure for generating the sequence is so devised that no reasonable statistical test will detect lack of randomness. The great advantage of a specified rule is that the sequence can always be reproduced for checking purposes. The 'mid-square' method was the first procedure devised for generating pseudo-random numbers: the middle digits of the square of a random number are used to form the next random number. This method, however, was soon shown to be unsatisfactory. *Linear congruential generators* are now most commonly used. These generators operate with integers and produce random observations on the unit interval by division.

When a first-order linear congruential generator is used, the random integer at time $n-1$ (W_{n-1}) is multiplied by a constant integer k_1, and the product is divided by another constant integer p. The remainder is an integer, and this is used as the random integer at time n (W_n). The following equation represents the process:

$$W_n = k_1 W_{n-1} \pmod{p}. \tag{14.1.1}$$

Table 14.1.1. *Integer constants recommended by J. G. Skellam for use in linear congruential generators. In each case the full cycle length of $p^s - 1$ is attained (s is the order of the generator)*

Order of generator (s)	Recommended constants			
	p	k_1	k_2	k_3
1	999 563	470 001	—	—
2	998 917	366 528	508 531	—
2	999 563	254 754	529 562	—
3	997 783	360 137	519 815	616 087
3	997 783	286 588	434 446	388 251

The random integer W_n is divided by p to produce a random observation on the unit interval. The constants p and k_1 need to be chosen carefully. J. G. Skellam, in some unpublished notes, recommends the constants $p = 999\,563$ and $k_1 = 470\,001$. The full cycle length of $p - 1$ is attained.

Second- and third-order linear congruential generators are defined by the equations

$$W_n = k_1 W_{n-1} + k_2 W_{n-2} \pmod{p}; \tag{14.1.2}$$
$$W_n = k_1 W_{n-1} + k_2 W_{n-2} + k_3 W_{n-3} \pmod{p}. \tag{14.1.3}$$

In each case, the random integer W_n is divided by p to produce a random observation on the unit interval. The constants p, k_1, k_2 and k_3 in table 14.1.1 are recommended by Skellam.

Example 14.1.1. The number of successes in a binomial experiment with parameters $n = 20$ and $p = 0.4$ is a random variable X, and the distribution of X is well known (section 10.1). A random variable Y is obtained by sampling from a binomial population with parameters $n = (1+X)^2$ and $p = 1/(1+X)$, where X is the random variable defined above. The distribution of Y is rather complicated, but it can be obtained empirically as follows:

1. Generate a random number on the unit interval.
2. Record a 'success' if this random number is less than or equal to 0.4, and a 'failure' otherwise.
3. Perform steps 1 and 2 twenty times. The number of successes recorded in these 20 trials is a single observation on the random variable X. Calculate $(1+X)^2$ and $1/(1+X)$.
4. Generate a random number on the unit interval.
5. Record a 'success' if this random number is less than or equal to $1/(1+X)$, and a failure otherwise.

6. Perform steps 4 and 5 $(1 + X)^2$ times. The number of successes recorded in these $(1 + X)^2$ trials is a single observation on the random variable Y.

7. Perform the sequence 1 to 6 a large number of times N, and denote the observations on the Y-distribution by $Y_1, ..., Y_N$. An estimate of the distribution of Y can be obtained by plotting the relative frequency of the various outcomes. The above procedure is very straightforward on a digital computer or large programmable calculator.

Further reading: Fraser [26] 24–5; Hammersley and Handscomb [40] 1–9, 25–31; Kendall and Stuart [53] 213–26; Zar [105] 16–17, 577–80.

14.2 Random observations on the unit normal distribution

The normal distribution occupies a central position in statistical theory, and random observations from the unit normal distribution are often required. These can be generated using a table of random numbers and a table of the normal integral. A random fraction U between 0 and 1 is read from the table of random numbers, and the normal table is used inversely to find X such that $\Phi(X) = U$. X is a random observation from the unit normal distribution.

The same method can be used on a computer or programmable calculator, but it is usually inconvenient to store a table of the normal integral and it is not practical to compute the inverse of the normal integral each time it is required. Other methods are usually employed.

One method relies on the central limit theorem.[1] N random fractions on the unit interval are added together. The sum is a random variable with mean $N/2$ and variance $N/12$. We subtract $N/2$ from the total and divide by $\sqrt{(N/12)}$. The resulting random variable has zero mean and unit variance, and it is approximately normal for reasonably large N. When speed of calculation is important and the problem does not depend critically on the extreme tails of the normal distribution, there is some advantage in choosing N equal to 12.

An elegant method has been devised by Box and Müller [6]. Two independent random fractions on the unit interval U_1 and U_2 are chosen. We calculate

- $$X_1 = (-2 \ln U_1)^{\frac{1}{2}} \cos (2\pi U_2),$$ (14.2.1)
- $$X_2 = (-2 \ln U_1)^{\frac{1}{2}} \sin (2\pi U_2).$$ (14.2.2)

Then X_1 and X_2 are independent random observations from the unit normal distribution. The method is exact for a truly mathematical situation in which U_1 and U_2 are independent rectangular random variables (that is, absolutely continuous). In practice, the distribution of U_1 and U_2 is discrete. If the generator used to produce U_1 and U_2 has a small modulus p (less than 1 000 000 say) there are certain imperfections in the tails of the X_1 and X_2 distributions (Neave [72]), but these are not serious for most purposes. The Box–Müller method may not be as fast as the above central limit procedure.

[1] See section 9.1.

To generate a random observation from the normal distribution with mean μ and variance σ^2, obtain a random observation from the unit normal distribution. Multiply this observation by σ and add μ.

Example 14.2.1. A random observation from the normal distribution with mean 10 and variance 4 is required. We consult a table of random numbers and happen to obtain the random fraction 0.310 88. Interpolating in the normal integral table we find that $\Phi(-0.493) = 0.310\,88$. Our random observation from the normal distribution with mean 10 and variance 4 is therefore $10 + 2 \times (-0.493) = 9.014$.

$$(1 - 0.31088) = 0.68911 \quad \ln \Phi x \text{ table} = -0.493$$

Example 14.2.2. A method for generating random observations from the binomial distribution with parameters $n = 20$ and $p = 0.4$ was outlined in example 14.1.1. This method works well when n is not too large, but it may be slow when n is large. For large n, the normal approximation to the binomial may be used (section 10.2). We generate a random observation from the unit normal distribution, multiply this number by $(npq)^{\frac{1}{2}}$, add np, and round the result to the nearest integer.

Example 14.2.3. The normal approximation to the Poisson (section 10.8) can be used to generate random observations from the Poisson distribution whenever the parameter λ is reasonably large ($\geqslant 10$, say). We generate a random observation from the unit normal distribution, multiply this number by $\sqrt{\lambda}$, add λ, and round the result to the nearest integer.

Example 14.2.4. A random observation from the χ_n^2 distribution can be obtained by generating n independent observations from the unit normal distribution, squaring each and adding the n squares together (section 9.2).

Example 14.2.5. A random observation from the χ_m^2 distribution and an independent random observation from the χ_n^2 distribution can be obtained using $m + n$ independent observations from the unit normal distribution and the method of example 14.2.4. A random observation from the $F_{m, n}$ distribution can be obtained by dividing the first chi-square observation by the second and multiplying the answer by n/m (section 9.4).

Further reading: Hammersley and Handscomb [40] 39–40.

14.3 Random observations from a population with continuous distribution function $F(x)$

A method for generating random observations on the unit normal distribution is described in the first paragraph of section 14.2. The same approach can be used with any continuous distribution. Thus, to generate a random observation from the population with continuous distribution function $F(x)$, we obtain a random fraction U between 0 and 1, and compute $X = F^{-1}(U)$. X is a random observation from the required distribution.

This method is usually straightforward when $F(x)$ has an explicit mathematical form. When this is not true, a table of the $F(x)$ function is required, and another method might be more appropriate.

> **Example 14.3.1.** The exponential distribution over the range $(0, \infty)$ is defined by the probability-density function
>
> $$f(x) = \lambda e^{-\lambda x}, \tag{14.3.1}$$

or the cumulative distribution function

$$F(x) = 1 - e^{-\lambda x}. \tag{14.3.2}$$

The parameter λ is positive, and we note that $F^{-1}(y) = -\{\ln (1-y)\}/\lambda$.

To obtain a random observation X on the exponential distribution with parameter λ, we generate a random fraction U on the unit interval and then calculate $X = -\{\ln (1-U)\}/\lambda$.

> *Further reading*: Johnson and Kotz [48] 207–32.

14.4 Data transformations

Many statistical procedures assume that the variance of a random variable X is completely unrelated to the mean (for example, the analysis of variance of sections 12.18 and 12.25 and the regression analysis of Part III). Situations often arise, however, in which this assumption cannot be made (the Poisson mean and variance, for example, are always equal). It is then necessary to transform the random variable to another random variable with mean and variance which are unrelated. Suitable transformations for the binomial proportion and the Poisson distribution are described in sections 14.5 and 14.6. As well as producing a variance which is unrelated to the mean, these transformations tend to produce distributions which are closer to normal than the original distributions.

> *Further reading*: Brownlee [7] 144–6; Johnson and Leone [51] 54–6; Zar [105] 182–3.

14.5 The arcsine (or angular) transformation of a binomial proportion

In a binomial experiment of n trials, r successes are observed. The observed proportion of successes is $P = r/n$, and this random variable has expectation p. The variance of P is $p(1-p)/n$, which depends upon the mean p as well as the number of trials. The variance of

$$\blacksquare \qquad Z = \sin^{-1}\sqrt{P}, \tag{14.5.1}$$

on the other hand, is almost independent of p. In fact, if angles are measured in radians, the variance of Z is approximately $1/(4n)$; if angles are measured

9

in degrees, the variance of Z is approximately $821/n$. The expectation of Z is almost independent of n and is approximately equal to $\sin^{-1}\sqrt{p}$.

Another variant of the arcsine transformation sometimes used is the following:

■ $Y = 2\sqrt{n} \sin^{-1}\sqrt{P}.$ 　　　　　　　　　　　　　　　(14.5.2)

The variance of Y is almost independent of both p and n. If angles are measured in radians, the variance of Y is approximately one; if degrees are used the variance is approximately 3283. The expectation of Y depends upon both parameters.

The above transformations work well except at the extreme ends of the range of possible values (that is, near 0 per cent and 100 per cent). The transformation

■ $W = \sin^{-1}\{(r + \frac{3}{8})/(n + \frac{3}{4})\}^{\frac{1}{2}},$ 　　　　　　　　　　　(14.5.3)

on the other hand, is satisfactory over the whole range of proportions (Kendall and Stuart [55] 114).

As well as producing a variance which is almost independent of the mean, the arcsine transformation gives a distribution which is closer to normal than the original distribution.

All computers and most programmable and non-programmable calculators have square-root and arcsine facilities. So the transformation is straightforward. Tables are also published (for example, Zar [105] 505–9).

>   **Example 14.5.1.** In an experiment, approximately twenty plants are observed for a given time, and the number reaching a particular stage of development is recorded. This experiment is carried out in three different types of soil with two different fertilisers. The results are shown in table 14.5.1. Is there any evidence that the proportion developing in the given time is different for the two fertilisers?

The paired t-test of section 12.22 and the two-way analysis of variance of section 12.25 immediately come to mind. Both these tests require a constant variance. The variance of the binomial proportion P in the two-way table 14.5.1, however, depends upon the unknown mean for the particular cell. If the arcsine transformation (14.5.1) is applied to these binomial proportions, the variances should be approximately equal because the size of each sample is about 20. The transformed variables will also be approximately normal. We can then apply the two-way analysis of variance of section 12.25 (or equivalently the paired t-test of section 12.22) to the transformed data which are shown in table 14.5.2. We obtain an F-value on 1 and 2 degrees of freedom of 1.7, which is not significant, so that there is no evidence that the proportion developing in the given time is different for the two fertilisers.

Note: according to the above theory, the common variance of the transformed

Table 14.5.1. *The numbers of plants reaching a particular stage of development in a given time under different growing conditions*

	Fertiliser A			Fertiliser B		
	Number observed n	Number developing r	Proportion developing $P = r/n$	Number observed n	Number developing r	Proportion developing $P = r/n$
Soil 1	15	9	0.600	20	11	0.550
Soil 2	18	10	0.556	19	14	0.737
Soil 3	20	10	0.500	20	13	0.650

Table 14.5.2. *The results of applying the arcsine transformation (14.5.1) to the proportions in table 14.5.1. Angles are measured in radians*

	Fertiliser		
	A	B	Totals
Soil 1	0.8861	0.8355	1.7216
Soil 2	0.8411	1.0321	1.8732
Soil 3	0.7854	0.9377	1.7231
Totals	2.5126	2.8053	5.3179

[handwritten margin note: Take ·600 Convert to RAD Use 14·5·1 = 0·8861]

variables should be about $1/(4 \times 20) = 0.0125$. The residual mean square in the analysis of variance table turns out to be about 0.0084. These two values are quite compatible.

Further reading: Brownlee [7] 144–6; Johnson and Leone [51] 54–6; Kendall and Stuart [55] 114; Thöni [97]; Zar [105] 185–6.

14.6   The square-root transformation for the Poisson distribution

The variance of a Poisson random variable X is equal to the mean (section 10.6). To obtain a random variable with mean and variance which are unrelated, the *square-root transformation*

■ $$Y = \sqrt{(X + \tfrac{3}{8})} \qquad\qquad (14.6.1)$$

should be used (Anscombe [1]). The fraction $\tfrac{3}{8}$ can be omitted when the mean value of X is greater than about four. The transformation

■ $$Y = \sqrt{X} + \sqrt{(X + 1)} \qquad\qquad (14.6.2)$$

Table 14.6.1. *The application of the square-root transformation to Poisson data. The ith Poisson observation is denoted by X_i*

X_i	$\sqrt{(X_i + \frac{3}{8})}$	X_i	$\sqrt{(X_i + \frac{3}{8})}$	X_i	$\sqrt{(X_i + \frac{3}{8})}$
14	3.79	11	3.37	8	2.89
7	2.72	11	3.37	13	3.66
13	3.66	6	2.52	7	2.72
8	2.89	12	3.52	11	3.37
5	2.32	10	3.22	14	3.79
10	3.22	12	3.52	9	3.06
5	2.32	7	2.72	7	2.72
4	2.09	7	2.72	8	2.89
7	2.72	7	2.72	6	2.52
16	4.05	10	3.22		
				Totals 265	88.30

is to be preferred when the Poisson mean is less than three (Kendall and Stuart [55] 90).

The square-root transformation (14.6.1) should in fact be used to produce a random variable with mean and variance which are unrelated, whenever the variance of the observed random variable X is proportional to the mean (the Poisson is an important special case with proportionality constant $k = 1$). In the general situation, the variance of Y is approximately equal to $k/4$.

The square-root transformation also tends to produce a distribution which is closer to normal than the original distribution.

Example 14.6.1. Twenty-nine random observations from a Poisson distribution with mean 9 are given in table 14.6.1. The square-root transforms (with $\frac{3}{8}$ adjustment) are also given. The sample variance of the transforms is 0.251. It is worth noting that the sum of the squares of the transforms can be obtained by simply adding $29 \times (\frac{3}{8})$ to the sum of the original Poisson observations.

Further reading: Anscombe [1]; Johnson and Leone [51] 54–6; Kendall and Stuart [55] 89–90; Zar [105] 187–8.

14.7 The logarithmic and reciprocal transformations

We have seen in section 14.6 how the square-root transformation can be used to produce a random variable with mean and variance which are unrelated when the variance of the actual observed random variable is proportional to the mean. We now describe two other frequently-used transformations.

The *logarithmic transformation* (14.7.1) or (14.7.2) should be used to produce a random variable with mean and variance which are unrelated when the standard deviation of the observed random variable X is proportional to the mean.

- $Y = \ln X$;   (14.7.1)

- $Y = \ln (X+1)$.   (14.7.2)

If the constant of proportionality is k, the variance of Y will be approximately k^2. Formula (14.7.2) is to be preferred when counts are being transformed, to allow for cases where the count X may be zero. Logarithms to base 10 may be used equally well, in which case the variance becomes $0.1886 k^2$.

The logarithmic transformation is also used in the analysis of variance and in regression models when the factor effects are multiplicative rather than additive (the logarithms are additive).

The *reciprocal transformation* (14.7.3) or (14.7.4) should be used to produce a random variable with mean and variance which are unrelated when the standard deviation of the observed random variable X is proportional to the square of the mean.

- $Y = 1/X$,   (14.7.3)

- $Y = 1/(X+1)$.   (14.7.4)

If the constant of proportionality is k, the variance of Y will be approximately k^2. Formula (14.7.4) is to be preferred when counts are being transformed, to allow for cases where the count X may be zero.

As with the arcsine and square-root transformations, the logarithmic and reciprocal transformations tend to produce distributions which are closer to normal than the original random variable.

All computers and most programmable and non-programmable calculators have square-root and logarithm facilities. The transformations are therefore straightforward.

> *Further reading*: Johnson and Leone [51] 54–6; Thöni [97]; Zar [105] 182–9.

14.8   Censored and truncated populations

A random sample of size N is drawn from a population. Observations smaller than a given value A or greater than another value B $(>A)$ are counted, but the values are not actually recorded. The sample is said to be *censored* below at A and above at B.

Censored samples often occur. A device commonly used in Australia to measure the thickness of meat on a pig carcase, for example, has a limited range: some observations are therefore too small or too large and they can only be counted. We shall denote the number of observations less than A by N_1, the number in the recording range by N_2, and the number of observations greater than B by N_3 $(N = N_1 + N_2 + N_3)$.

Another method of censoring is to omit the N_1 smallest observations and the N_3 largest. The results of section 14.9 can be applied if we set A equal to the smallest recorded observation and B equal to the largest recorded observation. N_1 and N_3 must be pre-determined.

In some situations N, N_1 and N_3 are unknown. The data are then purely a set of N_2 observations from a *truncated distribution*, truncated below at A and above at B.

14.9 Estimating the parameters of a censored normal sample

A random sample of size N is drawn from a normal population with mean μ and variance σ^2. Both these parameters are unknown. The sample is censored below at A and above at B. Of the N observations, N_1 turn out to be less than A, N_2 fall in the recording range and N_3 turns out to be greater than B $(N = N_1 + N_2 + N_3)$. The N_2 recorded observations are denoted by $x_1, ...,$ x_{N_2}. The following two sample moments of the recorded observations are calculated:

- $$\bar{X} = (x_1 + ... + x_{N_2})/N_2,$$ (14.9.1)

- $$S^2 = (x_1^2 + ... + x_{N_2}^2)/N_2 - \bar{X}^2.$$ (14.9.2)

Maximum likelihood estimates[2] of the parameters μ and σ^2 can be calculated using the iterative equations[3] (14.9.3) to (14.9.6) (Cohen [14]). The suffix n denotes the nth approximation to the maximum likelihood estimate and the symbols ϕ and Φ refer to the ordinate and cumulative area of the unit normal curve.

- $$\alpha_n = (A - \mu_n)/\sigma_n,$$ (14.9.3)

- $$\beta_n = (B - \mu_n)/\sigma_n,$$ (14.9.4)

- $$\mu_{n+1} = \bar{X} - \sigma_n(N_1/N_2)\,\phi(\alpha_n)/\Phi(\alpha_n)$$
  $$+ \sigma_n(N_3/N_2)\,\phi(\beta_n)/(1 - \Phi(\beta_n)),$$ (14.9.5)

- $$\sigma_{n+1}^2 = S^2 + (\bar{X} - \mu_{n+1})^2 - \sigma_n^2(N_1/N_2)\,\alpha_n\phi(\alpha_n)/\Phi(\alpha_n)$$
  $$+ \sigma_n^2(N_3/N_2)\,\beta_n\phi(\beta_n)/(1 - \Phi(\beta_n)).$$ (14.9.6)

The sample moments $\bar{X}$ and S^2 can be used as the initial approximations μ_0 and σ_0^2. In view of the effects of truncation, it is probably better to use a value of σ_0^2 somewhat larger than S^2, and an intelligent adjustment to $\bar{X}$ to obtain μ_0 might also be made.

The above formulae simplify a little with a single censorship point. If the single censorship point is above at B, N_1 will be zero, the terms involving α_n will disappear from (14.9.5) and (14.9.6), and (14.9.3) is not needed. If the

---

[2] Section 13.2.
[3] Iterative equations (14.9.5), (14.9.6) are simple rearrangements of the maximum likelihood equations (see section 3.5).

single censorship point is below at A, N_3 will be zero, the term involving β_n will disappear from (14.9.5) and (14.9.6) and (14.9.4) is not needed.

Other methods for dealing with censored normal populations are outlined by Johnson and Kotz [47] 77–87.

Example 14.9.1. A random sample of thirty observations is drawn from a normal population censored below at 50 and above at 150. The following results are recorded:

$$N_1 = 4, \qquad \bar{X} = 93.4,$$
$$N_2 = 25, \qquad S^2 = 543.68,$$
$$N_3 = 1, \qquad S = 23.316.$$

Obtain maximum likelihood estimates of the mean μ and variance σ^2.

Clearly $A = 50$ and $B = 150$. As initial estimates, let us try $\mu_0 = 93$; $\sigma_0 = 25$. We substitute these values into equations (14.9.3) to (14.9.6) and obtain

$$\alpha_0 = (50-93)/25 = -1.72,$$
$$\beta_0 = (150-93)/25 = 2.28,$$
$$\mu_1 = 93.4 - 25\,(4/25)\,(0.090\,89)/(0.042\,72)$$
$$\qquad + 25\,(1/25)\,(0.029\,65)/(0.011\,30) = 87.51,$$
$$\sigma_1^2 = 543.68 + (93.4-87.51)^2 - (25)^2\,(4/25)\,(-1.72)\,(0.090\,89)/(0.042\,72)$$
$$\qquad + (25)^2\,(1/25)\,(2.28)\,(0.029\,65)/(0.011\,30) = 1093.88.$$

We now repeat the calculations using μ_1 instead of μ_0 and σ_1^2 instead of σ_0^2. With successive iterations we obtain

$$\mu_2 = 87.77, \qquad \sigma_2 = 32.57;$$
$$\mu_3 = 87.75, \qquad \sigma_3 = 32.97;$$
$$\mu_4 = 87.74, \qquad \sigma_4 = 33.01;$$
$$\mu_5 = 87.75, \qquad \sigma_5 = 32.99.$$

We conclude that $\hat{\mu} = 87.75$ and $\hat{\sigma} = 33.00$.

Further reading: Cohen [14]; Cohen [15]; Johnson and Kotz [48] 77–87.

14.10 Estimating the parameters of a truncated normal distribution

A random sample of size N_2 is drawn from a truncated normal distribution, truncated below at A and above at B. The values A and B are known, but the normal parameters μ and σ^2 are unknown and need to be estimated.

We begin by calculating the sample moments $\bar{X}$ and S^2 using (14.9.1) and (14.9.2). Maximum likelihood estimates[4] of the parameters μ and σ^2 can be calculated using the iterative equations[5] (14.10.1) to (14.10.4) (Cohen [14]).

---

[4] Section 13.2.

[5] Iterative equations (14.10.3), (14.10.4) are simple rearrangements of the maximum likelihood equations (see section 3.5).

The suffix n denotes the nth approximation to the maximum likelihood estimate and the symbols ϕ and Φ refer to the ordinate and cumulative area of the unit normal curve.

■　　$\alpha_n = (A - \mu_n)/\sigma_n,$　　　　　　　　　　　　　　　　(14.10.1)

■　　$\beta_n = (B - \mu_n)/\sigma_n,$　　　　　　　　　　　　　　　　(14.10.2)

■　　$\mu_{n+1} = \overline{X} + \sigma_n(\phi(\beta_n) - \phi(\alpha_n))/(\Phi(\beta_n) - \Phi(\alpha_n)),$　　(14.10.3)

■　　$\sigma_{n+1}^2 = S^2 + (\overline{X} - \mu_{n+1})^2 + \sigma_n^2(\beta_n\phi(\beta_n) - \alpha_n\phi(\alpha_n))/(\Phi(\beta_n) - \Phi(\alpha_n)).$

　　　　　　　　　　　　　　　　　　　　　　　　　　(14.10.4)

Again, the sample moments $\overline{X}$ and S^2 can be used as the initial approximations μ_0 and σ_0^2. A value of σ_0^2 slightly greater than S^2 is probably better because of the effects of truncation. An intelligent adjustment to $\overline{X}$ to obtain μ_0 might also be made.

　　Formulae (14.10.3) and (14.10.4) simplify a little with a single truncation point. If the single truncation point is above at B, the terms $\phi(\alpha_n)$, $\alpha_n\phi(\alpha_n)$ and $\Phi(\alpha_n)$ disappear, and there is no need for (14.10.1). If the single truncation point is below at A, the terms $\phi(\beta_n)$ and $\beta_n\phi(\beta_n)$ disappear, $\Phi(\beta_n)$ becomes one, and there is no need for (14.10.2).

　　Occasionally the truncation points A and B are unknown. If this is so and the sample size is sufficiently large, choose a value of A slightly less than the smallest observed value and a value of B slightly bigger than the largest observed value.

　　Other methods for dealing with truncated normal populations are outlined by Johnson and Kotz [48] 77–87.

　　Example 14.10.1. A random sample of twenty-five observations is drawn from a truncated normal distribution, truncated below at 50 and above at 150. The following results are obtained: $\overline{X} = 93.4$; $S^2 = 543.68$. Obtain maximum likelihood estimates of the parameters μ and σ^2.

　　Again $A = 50$ and $B = 150$. As initial estimates, let us try $\mu_0 = 93$, $\sigma_0 = 25$. We substitute these values into equations (14.10. 1) to (14.10.4) and obtain

$$\alpha_0 = (50 - 93)/25 = -1.72,$$
$$\beta_0 = (150 - 93)/25 = 2.28,$$
$$\mu_1 = 93.4 + 25(0.02965 - 0.09089)/(0.98870 - 0.04272) = 91.78,$$
$$\sigma_1^2 = 543.68 + (93.4 - 91.78)^2$$
$$\qquad + 25^2\{(2.28)(0.02965) - (-1.72)(0.09089)\}$$
$$\qquad\qquad\qquad \div(0.98870 - 0.04272)$$

$$= 694.25.$$

We now repeat the calculation with μ_1 instead of μ_0 and σ_1^2 instead of σ_0^2.

With successive interations we obtain

$$\mu_2 = 91.10, \qquad \sigma_2 = 27.21;$$
$$\mu_3 = 90.76, \qquad \sigma_3 = 27.77;$$
$$\mu_4 = 90.51, \qquad \sigma_4 = 28.16;$$
$$\mu_5 = 90.33, \qquad \sigma_5 = 28.47;$$
$$\mu_6 = 90.19, \qquad \sigma_6 = 28.69.$$

Convergence is slow. The approximations $\{\mu_n\}$ are steadily decreasing and the $\{\sigma_n\}$ are increasing. We might save a few iterations if we try

$$\mu_6 = 89.00, \qquad \sigma_6 = 30.00.$$
Then $\quad \mu_7 = 89.29, \qquad \sigma_7 = 29.88;$
$$\mu_8 = 89.44, \qquad \sigma_8 = 29.77.$$

The $\{\mu_n\}$ are now increasing and the $\{\sigma_n\}$ decreasing. Let us try

$$\mu_8 = 89.70, \qquad \sigma_8 = 29.33.$$
Then $\quad \mu_9 = 89.74, \qquad \sigma_9 = 29.34;$
$$\mu_{10} = 89.74, \qquad \sigma_{10} = 29.35.$$

We conclude that $\hat{\mu} = 89.74$ and $\hat{\sigma} = 29.35$. The large amount of arithmetic involved should be noted, and the final answer should be compared with that of example 14.9.1 in which additional information was available.

Further reading: Cohen [14]; Cohen [15]; Ferguson *et al.* [22]; Johnson and Kotz [48] 77–87.

14.11 Estimating the parameter of a censored Poisson sample

Techniques for estimating the parameters of censored Poisson variates are outlined by Johnson and Kotz [47] 104–9. When the parameter λ is not too small, however, a modification of the method of section 14.9 can be used. Reasonably accurate results will be obtained provided λ is greater than about 8 or 9.

Let us imagine that a sample of size N is drawn from a Poisson population with mean λ which is unknown. Observations less than a or greater than b are counted but not recorded. Both a and b are integers, and $b > a$. Of the N observations, N_1 turn out to be less than a, N_2 fall in the recording range and N_3 turn out to be greater than b ($N = N_1 + N_2 + N_3$). The N_2 recorded observations are denoted by $x_1, ..., x_{N_2}$ and we calculate the sample mean $\bar{X}$ via formula (14.9.1).

We define $A = a - \frac{1}{2}$ and $B = b + \frac{1}{2}$. An approximate maximum likelihood estimate[6] for λ can be calculated using the iterative equations (14.11.1) to (14.11.3). The suffix n denotes the nth approximation to the maximum likelihood estimate and the symbols ϕ and Φ refer to the ordinate and cumulative area of the unit normal curve.

[6] Section 13.2.

- $\alpha_n = (A - \lambda_n)/\sqrt{\lambda_n}$, (14.11.1)

- $\beta_n = (B - \lambda_n)/\sqrt{\lambda_n}$, (14.11.2)

- $\lambda_{n+1} = \bar{X} - \lambda_n^{\frac{1}{2}}(N_1/N_2)\,\phi(\alpha_n)/\Phi(\alpha_n)$

$$+ \lambda_n^{\frac{1}{2}}(N_3/N_2)\,\phi(\beta_n)/(1 - \Phi(\beta_n)). \qquad (14.11.3)$$

Equation (14.11.3) makes use of the normal approximation to the Poisson (section 10.8). It is derived from (14.9.5) by writing λ_n instead of μ_n and σ_n^2. The sample mean $\bar{X}$ can be used as the initial approximation λ_0. When N_3 is much larger than N_1, choose λ_0 somewhat larger than $\bar{X}$. Convergence will be faster. Conversely, when N_1 is much larger than N_3, choose λ_0 less than $\bar{X}$.

The above formulae simplify a little with a single censorship point. If the single censorship point is above at b, N_1 will be zero, the term involving α_n will disappear from (14.11.3) and there is no need for (14.11.1). If the single censorship point is below at a, N_3 will be zero, the term involving β_n will disappear from (14.11.3) and there is no need for (14.11.2).

Example 14.11.1. A sample of size twenty-nine is drawn from a Poisson population with unknown mean λ. Five observations are greater than 12 and the mean of the twenty-four observations 12 or less is 8.125. Obtain the maximum likelihood estimate of the parameter λ.

In this example there is a single censorship point (above), and we use (14.11.3) with $N_1 = 0$. We note that $B = 12.5$. Let us use as our initial approximation $\lambda_0 = \bar{X} = 8.125$. We calculate

$$\beta_0 = (12.5 - 8.125)/(8.125)^{\frac{1}{2}} = 1.5349.$$

The approximation λ_1 is then obtained via (14.11.3):

$$\lambda_1 = 8.125 + (8.125)^{\frac{1}{2}}(5/24)(0.122\,84)/(0.062\,41) = 9.294.$$

We are now able to calculate β_1 and λ_2. With successive iterations we obtain

$$\lambda_2 = 9.120,$$
$$\lambda_3 = 9.145,$$
$$\lambda_4 = 9.141,$$
$$\lambda_5 = 9.142,$$
$$\lambda_6 = 9.142.$$

The approximate maximum likelihood estimate is therefore $\hat{\lambda} = 9.142$.

Further reading: Johnson and Kotz [47] 104-9.

14.12 Estimating the parameter of a truncated Poisson distribution

Techniques for estimating the parameters of truncated Poisson distributions are outlined by Johnson and Kotz [47] 104-9. When the parameter λ is not too small, however, a modification of the method of section 14.10 can be used. Reasonably accurate results will be obtained provided λ is greater than about 8 or 9.

Let us imagine that a random sample of size N_2 is drawn from the truncated Poisson distribution over the range a to b inclusive (a and b are known integers and $b > a$). The Poisson parameter λ is unknown and needs to be estimated. The sample values are $x_1, x_2, ..., x_{N_2}$.

We define $A = a - \frac{1}{2}$ and $B = b + \frac{1}{2}$, and calculate the sample mean $\bar{X}$ via formula (14.9.1). An approximate maximum likelihood estimate for λ can be calculated using the iterative equations (14.12.1) to (14.12.3). The suffix n denotes the nth approximation to the maximum likelihood estimate and the symbols ϕ and Φ refer to the ordinate and cumulative area of the unit normal curve.

■ $\alpha_n = (A - \lambda_n)/\sqrt{\lambda_n},$ (14.12.1)

■ $\beta_n = (B - \lambda_n)/\sqrt{\lambda_n},$ (14.12.2)

■ $\lambda_{n+1} = \bar{X} + \lambda_n^{\frac{1}{2}}(\phi(\beta_n) - \phi(\alpha_n))/(\Phi(\beta_n) - \Phi(\alpha_n)).$ (14.12.3)

Equation (14.12.3) makes use of the normal approximation to the Poisson (section 10.8). It is derived from (14.10.3) by writing λ_n instead of μ_n and σ_n^2. The sample mean $\bar{X}$ can be used as the initial approximation λ_0.

The above formulae simplify a little with a single truncation point. If the single truncation point is above at b, $\phi(\alpha_n)$ and $\Phi(\alpha_n)$ become zero, and (14.12.1) is not needed. If the single truncation point is below at a, $\phi(\beta_n)$ becomes zero, $\Phi(\beta_n)$ becomes one, and (14.12.2) is not needed.

Example 14.12.1. A sample of size twenty-four is drawn from a truncated Poisson distribution on the non-negative integers 0 to 12. The sample mean $\bar{X} = 8.125$. Obtain the maximum likelihood estimate of the parameter λ.

There is a single truncation point (above), and we use (14.12.3) with $\phi(\alpha_n)$ and $\Phi(\alpha_n)$ set equal to zero. B is again 12.5. Let us use as our initial approximation $\lambda_0 = \bar{X} = 8.125$. We calculate

$$\beta_0 = (12.5 - 8.125)/(8.125)^{\frac{1}{2}} = 1.5349,$$
$$\lambda_1 = 8.125 + (8.125)^{\frac{1}{2}}(0.12284)/(0.93759) = 8.498.$$

We can now calculate β_1 and repeat the process. Successive approximations are:

$\lambda_2 = 8.620,$
$\lambda_3 = 8.665,$
$\lambda_4 = 8.681,$
$\lambda_5 = 8.687,$
$\lambda_6 = 8.690,$
$\lambda_7 = 8.691,$
$\lambda_8 = 8.691.$

Table 14.13.1. *The fat thicknesses (mm) of 387 Australian pig carcases**

Fat thickness (mm)	Number	Fat thickness (mm)	Number
⩽ 6	46	15	13
7	25	16	14
8	32	17	7
9	50	18	5
10	37	19	8
11	50	20	0
12	44	21	2
13	33	⩾ 22	0
14	21		—
		Total	387

* *Source*: Personal communication from H. C. Kirton, New South Wales State Department of Agriculture, Sydney, Australia.

The approximate maximum likelihood estimate is therefore $\hat{\lambda} = 8.691$. This result should be compared with the answer in example 14.11.1, where additional information was available. It is perhaps worth noting that the common sample came from a Poisson population with mean 9. The actual observations appear in table 14.6.1.

Further reading: Johnson and Kotz [47] 104–9.

14.13   Exercises

1. Describe a procedure for generating random observations on the t_3-distribution.

2. A triangular distribution on the unit interval (0, 1) is defined by the probability-density function $f(x) = 2x$ or the cumulative distribution function $F(x) = x^2$. Describe a procedure for generating random observations on this distribution.

3. In a binomial experiment of eighteen trials, fourteen 'successes' are observed. Use the arcsine transformation (14.5.1) to obtain a 95 per cent confidence interval for p (the probability of success of a single trial). Compare your answer with that given in example 13.5.1.

4. The fat thicknesses of 387 pig carcases are given in table 14.13.1. Fit a normal curve to these data.

Part III
The method of least squares

15

Simple linear regression and the method of least squares

Summary   The numerical technique of least squares for fitting a straight line through a set of points in the xy-plane is described in sections 15.1 and 15.2 of this chapter. Then, in sections 15.4 and 15.5, the usual statistical model for simple linear regression and tests for the significance of the regression line are outlined. Estimation methods for the intercept parameter β_0, the slope parameter β_1 and the error variance σ^2 are described in sections 15.6–15.7, and tests for the values of these parameters are outlined in section 15.8. Estimation methods for the mean value of y, given x, are described in section 15.10. Lack-of-fit, unequal variances and weighted least squares are dealt with in sections 15.11 and 15.12. Matrix methods are not essential for simple linear regression; nevertheless, the matrix approach is outlined in sections 15.3 and 15.9 to prepare the reader for chapters 16 and 17.

15.1   Introduction – the method of least squares

Let us imagine that we wish to fit a straight line as closely as possible to the four points in fig. 15.1.1:

$$(x_1, y_1) = (1, 10),$$
$$(x_2, y_2) = (3, 20),$$
$$(x_3, y_3) = (4, 18),$$
$$(x_4, y_4) = (5, 20).$$

These points are denoted by crosses. One way to do this is to place a straight edge near the points and then draw a line. But what do we mean by 'close to the points' and how do we know that the line is as close as possible to them? Even in the simple situation depicted in fig. 15.1.1, the best location for the line is far from obvious.

The *method of least squares* makes use of an objective criterion and provides a unique solution to the problem. The method may be explained as follows. Consider a straight line passing near the data points like the one given in fig. 15.1.1. If we were to adopt this as our line, the fitted values at $x = 1, x = 3, x = 4$ and $x = 5$ would be those indicated by small circles. We will have obtained a good fit if all the distances between the observed values (indicated by crosses) and the corresponding fitted values (indicated by circles) are small. The least-squares line minimises the sum of the squares of these distances.

The method of least squares, as a numerical technique, can be traced back

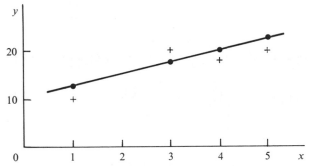

Fig. 15.1.1. A straight line to fit four points. The data points are indicated by crosses and the fitted points by circles.

to Legendre (1806) in his *Nouvelles méthodes pour la détermination des orbites des comètes*. The name can also be attributed to him. The first writer to connect the method with the mathematical theory of probability was Gauss (1809) [29]. Gauss also mentions that he had used the method since 1795. Laplace established the method in an entirely different manner in 1811 (Plackett [80]; Whittaker and Robinson [100] 209-28).

The term 'regression' was coined by Francis Galton in 1886. Galton observed that on the average the sons of tall fathers are not as tall as their fathers and the sons of short fathers are not as short as their fathers. There seemed to be a 'regression towards mediocrity'. The error in this argument is explained for example by Brownlee [7] 407. See also Hoel [45] 201.

> *Further reading*: Balaam [3] 185-7; Campbell [8] 31-2, 272-5; Draper and Smith [19] 1-6; Gauss [30], [31], [32], [33]; Hoel [45] 195-6; Johnson and Leone [50] 377-81; Mather [66] 116-20.

15.2 Fitting a straight line[1] by least squares

Let us denote the fitted value at x_i by Y_i, and assume that

$$Y_i = b_0 + b_1 x_i, \tag{15.2.1}$$

where b_0 and b_1 are two constants yet to be determined. Under the method of least squares, we wish to minimise

$$\sum_{i=1}^{4} (y_i - Y_i)^2. \tag{15.2.2}$$

That is, we wish to minimise[2]

$$\sum_{i=1}^{4} (y_i - b_0 - b_1 x_i)^2 \tag{15.2.3}$$

---

[1] If the straight line is required to pass through the origin or some other fixed point, the method of section 16.11 should be used.

[2] In the general situation, we have n points, and the sum in (15.2.3) is from 1 to n.

by a suitable choice of b_0 and b_1 (the x- and y-values are known). Criterion (15.2.3) may be written

$$(y_1-b_0-b_1x_1)^2+(y_2-b_0-b_1x_2)^2+(y_3-b_0-b_1x_3)^2$$
$$+(y_4-b_0-b_1x_4)^2.$$

At the optimal point, the partial derivative with respect to b_0 (treating b_1 as a constant) and the partial derivative with respect to b_1 (treating b_0 as a constant) must both be zero (section 1.6). Thus,

$$-2(y_1-b_0-b_1x_1)-2(y_2-b_0-b_1x_2)$$
$$-2(y_3-b_0-b_1x_3)-2(y_4-b_0-b_1x_4) = 0,$$

$$-2x_1(y_1-b_0-b_1x_1)-2x_2(y_2-b_0-b_1x_2)$$
$$-2x_3(y_3-b_0-b_1x_3)-2x_4(y_4-b_0-b_1x_4) = 0.$$

These equations may be rearranged as follows:

$$\left.\begin{array}{c}4b_0+\left(\displaystyle\sum_{i=1}^{4} x_i\right)b_1 = \displaystyle\sum_{i=1}^{4} y_i, \\[2mm] \left(\displaystyle\sum_{i=1}^{4} x_i\right)b_0+\left(\displaystyle\sum_{i=1}^{4} x_i^2\right)b_1 = \displaystyle\sum_{i=1}^{4} x_i y_i.\end{array}\right\} \tag{15.2.4}$$

We solve these two linear equations (the *normal equations*) to obtain b_0 and b_1. In the more general situation involving n points, the normal equations take the form:

■
$$\left.\begin{array}{c}nb_0+\left(\displaystyle\sum_{i=1}^{n} x_i\right)b_1 = \displaystyle\sum_{i=1}^{n} y_i, \\[2mm] \left(\displaystyle\sum_{i=1}^{n} x_i\right)b_0+\left(\displaystyle\sum_{i=1}^{n} x_i^2\right)b_1 = \displaystyle\sum_{i=1}^{n} x_i y_i.\end{array}\right\} \tag{15.2.5}$$
■

Simple explicit solutions to these simultaneous equations exist, namely

■
$$\left.\begin{array}{c}b_1=\left(\displaystyle\sum_{i=1}^{n} x_i y_i-n\bar{x}\bar{y}\right)\Big/\left(\displaystyle\sum_{i=1}^{n} x_i^2 -n\bar{x}^2\right), \\[2mm] b_0 = \bar{y}-b_1\bar{x}.\end{array}\right\} \tag{15.2.6}$$
■

It should be noted that the least-squares line passes through the mean point $(\bar{x}, \bar{y})$, and is sometimes written in the form

$$y = \bar{y}+r(s_y/s_x)(x-\bar{x}), \tag{15.2.7}$$

where s_x and s_y denote the sample standard deviations of the x- and y-values, and r denotes the correlation coefficient. Formula (15.2.7) should be compared with (9.6.2).

Example 15.2.1. For the four points in section 15.1,

$\Sigma x_i = 1+3+4+5 = 13,$
$\Sigma x_i^2 = 1+9+16+25 = 51,$
$\Sigma y_i = 10+20+18+20 = 68,$
$\Sigma x_i y_i = (1 \times 10)+(3 \times 20)+(4 \times 18)+(5 \times 20) = 242.$

The normal equations are

$4b_0+13b_1 = 68,$
$13b_0+51b_1 = 242.$

When we solve these equations, we find that $b_0 = 9.2$ and $b_1 = 2.4$, so that the fitted line is $Y = 9.2+2.4x$. The fitted values are

$Y_1 = 9.2+2.4 \times 1 = 11.6,$
$Y_2 = 9.2+2.4 \times 3 = 16.4,$
$Y_3 = 9.2+2.4 \times 4 = 18.8,$
$Y_4 = 9.2+2.4 \times 5 = 21.2.$

The more usual approach is to calculate b_0 and b_1 directly from (15.2.6):

$b_1 = \{242 - 4 \times(13/4))(68/4)\}/ \{51 - 4 \times (13/4)^2\} = 2.4,$
$b_0 = (68/4) - 2.4 \times (13/4) = 9.2.$

Further reading: Balaam [3] 185–7; Draper and Smith [19] 7–12; Fraser [26] 296–304; Johnson and Leone [50] 382–98; Mather [65] 109–28; Mather [66] 121–30; Remington and Schork [86] 254–6, 260–4; Sokolnikoff and Sokolnikoff [92] 536–43; Spiegel [93] 217–40; Wilks [102] 240–4.

15.3   Least squares – the use of matrix notation
Matrix notation is hardly necessary for the simple straight-line example 15.2.1, but it does allow us to generalise the results to curvilinear regression and multiple regression. In our simple four-point example, we define a matrix $\mathbf{X}$ and vectors $\mathbf{y}$, $\mathbf{Y}$ and $\mathbf{b}$ as follows:

$$\mathbf{X} = \begin{pmatrix} 1 & x_1 \\ 1 & x_2 \\ 1 & x_3 \\ 1 & x_4 \end{pmatrix}, \quad \mathbf{y} = \begin{pmatrix} y_1 \\ y_2 \\ y_3 \\ y_4 \end{pmatrix}, \quad \mathbf{Y} = \begin{pmatrix} Y_1 \\ Y_2 \\ Y_3 \\ Y_4 \end{pmatrix}, \quad \mathbf{b} = \begin{pmatrix} b_0 \\ b_1 \end{pmatrix}.$$

We note that

$$\mathbf{Xb} = \begin{pmatrix} b_0+b_1x_1 \\ b_0+b_1x_2 \\ b_0+b_1x_3 \\ b_0+b_1x_4 \end{pmatrix} \quad \text{and} \quad \mathbf{y} - \mathbf{Xb} = \begin{pmatrix} y_1-b_0-b_1x_1 \\ y_2-b_0-b_1x_2 \\ y_3-b_0-b_1x_3 \\ y_4-b_0-b_1x_4 \end{pmatrix},$$

so that the criterion we minimise (15.2.3) may be written

■ $(\mathbf{y} - \mathbf{Xb})' (\mathbf{y} - \mathbf{Xb})$. (15.3.1)

The fitted values given by (15.2.1) are listed in

■ $\mathbf{Y} = \mathbf{Xb}$. (15.3.2)

Let us now examine the matrix $\mathbf{S} = \mathbf{X'X}$ and the vector $\mathbf{X'y}$. We see that $\mathbf{S}$ is a 2×2 matrix and $\mathbf{X'y}$ is a column vector of length 2. In fact,

$$\mathbf{S} = \begin{pmatrix} 4 & \sum_{i=1}^{4} x_i \\ \sum_{i=1}^{4} x_i & \sum_{i=1}^{4} x_i^2 \end{pmatrix} \quad \text{and} \quad \mathbf{X}\,\mathbf{y} = \begin{pmatrix} \sum_{i=1}^{4} y_i \\ \sum_{i=1}^{4} x_i y_i \end{pmatrix}$$

It follows that the *normal equations* (15.2.4) may be summarised in matrix notation as follows:

■ $\mathbf{Sb} = \mathbf{X'y}$ (15.3.3)

with solution

$\mathbf{b} = \mathbf{S}^{-1}\mathbf{X'y}$. (15.3.4)

These matrices and vectors are readily determined in the general situation of n points. We shall see later that most of the common curvilinear and multiple regression models result in simultaneous linear equations which can be summarised in the form (15.3.3); the dimensions of $\mathbf{X}$ will then be greater than $n \times 2$.

Further reading: Draper and Smith [19] 44–52.

15.4 Simple linear regression – the statistical model

In sections 15.1, 15.2 and 15.3, the method of least squares was developed as a numerical technique. The method will normally be applied to experimental data. Experimental observations, however, are subject to statistical variation: different observations will be obtained when an experiment is repeated and the resulting least-squares lines will also differ.

For a statistical analysis we set up a mathematical model. We assume that

■ $y_i = \beta_0 + \beta_1 x_i + e_i$, (15.4.1)

where β_0 and β_1 are unknown constants, and the statistical error terms $\{e_i\}$ are independent normal random variables with zero means and equal unknown variances σ^2; x is referred to as the *independent variable*, assumed free of error, and y is called the *dependent variable*. The values b_0 and b_1 which we calculate are estimates of β_0 and β_1 based on a particular set of experimental results. If β_1 is zero, there is no linear relationship between x and y.

> *Further reading*: Brownlee [7] 334–7; Draper and Smith [19] 17; Remington and Schork [86] 257–9.

15.5   Testing the significance of a regression line

Even in the situation where x and y are unrelated, it is possible to draw the observed points on a graph and fit a least-squares line. The line will usually have a non-zero slope. The non-zero slope is brought about by chance, and does not imply a linear relationship between x and y. How then do we test whether regression is significant or not?

Let us begin by examining the total sum of squares of the $\{y_i\}$ values about the mean $\bar{y}$:

$$\sum_{i=1}^{n} (y_i - \bar{y})^2. \tag{15.5.1}$$

It may be shown that for the least squares fit

$$\sum_{i=1}^{n} (y_i - \bar{y})^2 = \sum_{i=1}^{n} (Y_i - \bar{y})^2 + \sum_{i=1}^{n} (y_i - Y_i)^2. \tag{15.5.2}$$

Thus, the total sum of squares (15.5.1) can be partitioned into two positive components:

(i) a sum of squares due to regression; and
(ii) a sum of squares about the regression line.

The regression sum of squares is the sum of the squared differences between the fitted values and the mean. The sum of squares about the regression line is merely the sum of the squared distances between the observed and fitted points (15.2.2). If the fitted line passes through all the original data points, the line fits perfectly and the sum of squares about the regression line is zero; all the variation in the $\{y_i\}$ values is then *explained* by regression. If, on the other hand, there is no real evidence of a linear trend in the data, the sum of squares due to regression will be small and most of the variability in the $\{y_i\}$ values will be explained as variation about the regression line. It would seem therefore that regression is significant when the regression sum of squares is large relative to the sum of squares about the regression line.

Formally, we need to test the null hypothesis $H_0: \beta_1 = 0$ against the alternative $H_1: \beta_1 \neq 0$. If we consider that β_1 is non-zero, the regression is significant. The test calculations are best carried out in a so-called *analysis of variance table*. In this table, the total sum of squares is partitioned into a regression sum of squares and a residual sum of squares (or sum of squares about the regression line) as shown in (15.5.2). The convenient formulae[3]

---

[3] Alternative, mathematically identical, formulae exist.

Table 15.5.1. *Simple linear regression analysis of variance table*

Source of variation (1)	Sum of squares (2)	Degrees of freedom (3)	Mean square (4) = (2)/(3)
Regression	$b_1^2 \left(\sum_{1}^{n} x_i^2 - n\bar{x}^2 \right)$	1	$\left(\begin{array}{c} \text{By} \\ \text{division} \end{array} \right)$
Residual	By difference	$n-2$	
Total	$\sum_{1}^{n} y_i^2 - n\bar{y}^2$	$n-1$	—

for calculation purposes are given in table 15.5.1.[4] The regression and residual mean squares in column (4) are obtained by dividing the sums of squares in column (2) by their respective degrees of freedom in column (3).

We then calculate

■ $$F_{1,n-2} = \frac{\text{regression mean square}}{\text{residual mean square}}. \qquad (15.5.3)$$

Under the null hypothesis $\beta_1 = 0$, this statistic has the F-distribution. If the calculated statistic is significantly large, we reject the null hypothesis and conclude that β_1 is non-zero; the regression is significant.

This F-test is equivalent to the commonly quoted two-tailed t-test statistic for simple linear regression

■ $$t_{n-2} = (n-2)^{\frac{1}{2}} r/(1-r^2)^{\frac{1}{2}}, \qquad (15.5.4)$$

where r is the observed correlation between the x- and y-values (section 8.8). In fact, the F-statistic is just the square of this t-statistic (section 9.4). Formula (15.5.4) is identical to the correlation test statistic (12.31.2).

Example 15.5.1. To test the significance of the regression line in the simple example 15.2.1, we calculate

$$\Sigma y_i^2 = 10^2 + 20^2 + 18^2 + 20^2 \quad = 1224,$$

regression s.s. $= (2.4)^2 \{51 - 4(13/4)^2\} = 50.4,$

total s.s. $= 1224 - 4(68/4)^2 \qquad = 68.0,$

residual s.s. $= 68.0 - 50.4 \qquad = 17.6.$

We then complete the analysis of variance table (15.5.2) and calculate

---

[4] In subsequent tables, 'sum of squares', 'degrees of freedom' and 'mean square' will be denoted by s.s., d.f. and m.s. respectively.

Table 15.5.2. *The analysis of variance for simple
linear regression example 15.5.1*

Source of variation	s.s.	d.f.	m.s.
Regression	50.4	1	50.4
Residual	17.6	2	8.8
Total	68.0	3	—

$F_{1,2} = 50.4/8.8 = 5.73$. This value is not significant at the 5 per cent level; so we have no evidence that β_1 is non-zero. *ie poor fit by line ; No linear relationship*

The proportion of the total sum of squares explained by regression is referred to as the *coefficient of determination*, and is usually denoted by R^2 because it also happens to be the square of the correlation between the fitted and observed y-values (the *coefficient of multiple correlation*[5]). From table 15.5.2, we see that

$$R^2 = 50.4/68.0 = 0.7412.$$

In this simple example we calculated the sums of squares by the usual formulae (table 15.5.1). The reader may wish to confirm that the same sums of squares are obtained using formula (15.5.2).

> *Further reading*: Balaam [3] 188–93; Draper and Smith [19] 24–6; Zar [105] 205–9.

15.6 Point estimators for β_0, β_1 and σ^2

The underlying parameters β_0, β_1 and σ^2 are unknown. Unbiased estimates are available, however. The regression coefficients b_0 and b_1 estimate β_0 and β_1, respectively, and the residual mean square in the analysis of variance table estimates σ^2. The variances and covariance of b_0 and b_1 are given in section 15.9.

15.7 Confidence intervals[6] for β_0, β_1 and σ^2

Let us denote the residual mean square in the analysis of variance table by s^2. Ninety-five per cent confidence intervals for β_0 and β_1 can be obtained by equating the following t-statistics to the upper and lower $2\frac{1}{2}$ per cent points of the t_{n-2}-distribution:

$$t_{n-2} = \frac{b_0 - \beta_0}{s\{(\Sigma x_i^2)/(n\Sigma x_i^2 - n^2\bar{x}^2)\}^{\frac{1}{2}}}, \qquad (15.7.1)$$

[5] See section 17.2. In the case of simple linear regression $R^2 = r^2$.
[6] Confidence intervals are defined in section 13.4.

Table 15.7.1. *Data for regression example 15.7.1*

x	y	x	y	x	y	x	y
1	17	6	28	11	44	16	60
2	13	7	26	12	47	17	58
3	22	8	28	13	45	18	61
4	20	9	34	14	54	19	64
5	20	10	46	15	55	20	70

Table 15.7.2. *Analysis of variance for example 15.7.1*

Source	s.s.	d.f.	m.s.
Regression	5877.487	1	5877.487
Residual	205.313	18	11.406 $= Var$
Total	6082. 800	19	–

■ $$t_{n-2} = \frac{b_1 - \beta_1}{s\,\{1/(\Sigma x_i^2 - n\bar{x}^2)\}^{\frac{1}{2}}}.$$ (15.7.2)

To obtain 95 per cent confidence limits for σ^2, we equate the following chi-square statistic to the upper and lower $2\frac{1}{2}$ per cent points of the χ^2_{n-2} distribution:

■ $$\chi^2_{n-2} = \frac{(n-2)s^2}{\sigma^2}.$$ (15.7.3)

Example 15.7.1. Obtain 95 per cent confidence limits for the intercept parameter β_0 and the error variance σ^2 using the data in table 15.7.1. We begin by calculating

$$\bar{x} = 10.5, \qquad SS\ \Sigma x_i y_i = 10\,503,$$
$$\bar{y} = 40.6, \qquad n = 20,$$
$$S2\ \Sigma x_i^2 = 2870, \qquad C\ b_0 = 9.384\,21,$$
$$S4\ \Sigma y_i^2 = 39\,050, \qquad M\ b_1 = 2.972\,93,$$

and forming the analysis of variance table 15.7.2.

To obtain 95 per cent confidence limits for β_0, we solve the equations

$$\frac{9.384\,21 - \beta_0}{[11.406 \times 2870/\{20 \times 2870 - 20^2 \times (10.5)^2\}]^{\frac{1}{2}}} = \pm 2.101,$$

which yield the limits 9.384 ± 3.296.

To obtain 95 per cent confidence limits for σ^2, we solve the equations

$$\frac{18 \times 11.406}{\sigma^2} = 31.53 \quad \text{and} \quad \frac{18 \times 11.406}{\sigma^2} = 8.23,$$

which yield the limits 6.51 and 24.95.

Further reading: Draper and Smith [19] 18–21; Kendall and Stuart [54] 362–3; Remington and Schork [86] 265–72; Zar [105] 209–10.

15.8 Tests for the values of the regression parameters β_0, β_1 and σ^2

The F- and t-methods for testing the null hypothesis $\beta_1 = 0$ were described in section 15.5. Let us now imagine that we wish to test the null hypothesis $\beta_1 = \delta$, where δ is some specified constant. The test can be carried out in several different ways, and the reader is advised to adopt whichever method is most convenient; they are all mathematically equivalent.

One approach is to set β_1 equal to δ in (15.7.2). We reject the null hypothesis $\beta_1 = \delta$ if the t-statistic falls in the upper or lower $2\frac{1}{2}$ per cent regions of the t_{n-2}-distribution. Alternatively, we can subtract δx_i from y_i to obtain a value z_i for each i, and then perform the t- or F-test of section 15.5 on the points $\{(x_i, z_i)\}$. The t-methods can also be used for testing one-tail hypotheses.

To test the null hypothesis $\beta_0 = \delta$, where δ is a specified constant, we set β_0 equal to δ in (15.7.1), and reject the null hypothesis if the test statistic falls in the upper or lower $2\frac{1}{2}$ per cent regions of the t_{n-2}-distribution.

To test the null hypothesis $\sigma^2 = \sigma_0^2$, where σ_0^2 is a specified constant, we set σ^2 equal to σ_0^2 in (15.7.3), and reject the null hypothesis if the test statistic falls in the upper or lower $2\frac{1}{2}$ per cent regions of the χ_{n-2}^2-distribution.

Example 15.8.1. Test the null hypothesis that the data in example 15.7.1 come from a population with slope parameter $\beta_1 = 3$.

The first approach is to use (15.7.2) and calculate the test statistic

$$t_{18} = (2.972\,93 - 3)/[11.406/\{2870 - 20 \times (10.5)^2\}]^{\frac{1}{2}} = -0.2067.$$

This value is not significant at the 5 per cent level; so we have no reason to reject the null hypothesis.

The alternative approach is to subtract $3x_i$ from each y_i in table 15.7.1 to obtain the $\{z_i\}$ values in table 15.8.1 and the analysis of variance table 15.8.2. The test statistic

$$F_{1,\,18} = 0.487/11.406 = 0.0427$$

is not significant at the 5 per cent level; so we have no reason to reject the null hypothesis.

The reader should note that the F-statistic is in fact the square of the

Table 15.8.1. *Numbers derived from table 15.7.1 by subtracting 3x from each y-value*

x	z	x	z	x	z	x	z
1	14	6	10	11	11	16	12
2	7	7	5	12	11	17	7
3	13	8	4	13	6	18	7
4	8	9	7	14	12	19	7
5	5	10	16	15	10	20	10

Table 15.8.2. *Analysis of variance for example 15.8.1*

Source	s.s.	d.f.	m.s.
Regression	0.487	1	0.487
Residual	205.313	18	11.406
Total	205.800	19	—

t-statistic, and the residual sums of squares in tables 15.7.2 and 15.8.2 are equal, so that the two tests are effectively equivalent. Furthermore the difference between the slope estimates for tables 15.7.1 and 15.8.1 is exactly 3.0.

> *Further reading*: Draper and Smith [19] 18–21; Kendall and Stuart [54] 362–3; Zar [105] 209–10.

15.9 The matrix method of regression analysis

In section 15.3 we showed how the numerical technique of least squares can be summarised using matrix notation. The matrix method can be carried over into the statistical analysis. Most of the vectors and matrices we require are defined in section 15.3, but we also need to define a two-dimensional column vector β with elements β_0 and β_1, and an n-dimensional column vector e with elements $\{e_i\}$ $(i = 1, 2, ..., n)$.

The statistical model (15.4.1) may be written

■ $$y = X\beta + e. \tag{15.9.1}$$

If we denote the inverse of the matrix S by S^{-1}, then it is not difficult to show by the method of cofactors (section 1.10) that

$$S^{-1} = \begin{pmatrix} S_{00}^{-1} & S_{01}^{-1} \\ S_{10}^{-1} & S_{11}^{-1} \end{pmatrix} = \begin{pmatrix} \Sigma x_i^2 & -\Sigma x_i \\ -\Sigma x_i & n \end{pmatrix} \Big/ (n\Sigma x_i^2 - n^2 \bar{x}^2). \tag{15.9.2}$$

The covariance matrix of b_0 and b_1 is $\sigma^2 \mathbf{S}^{-1}$ so that

- $$\text{var } b_0 = \sigma^2 S_{00}^{-1}, \tag{15.9.3}$$

- $$\text{var } b_1 = \sigma^2 S_{11}^{-1}, \tag{15.9.4}$$

- $$\text{cov } (b_0, b_1) = \sigma^2 S_{01}^{-1}. \tag{15.9.5}$$

It is easy to see that the t-statistics (15.7.1) and (15.7.2) for obtaining confidence intervals for β_0 and β_1 may be written in the form

- $$t_{n-2} = (b_0 - \beta_0)/(s^2 S_{00}^{-1})^{\frac{1}{2}}, \tag{15.9.6}$$

- $$t_{n-2} = (b_1 - \beta_1)/(s^2 S_{11}^{-1})^{\frac{1}{2}}. \tag{15.9.7}$$

The sums of squares in the analysis of variance table 15.5.1 can be represented in several useful forms, including the following:

- $$\text{regression s.s.} = \mathbf{b'X'y} - n\bar{y}^2, \tag{15.9.8}$$

- $$\text{regression s.s.} = \mathbf{b'Sb} - n\bar{y}^2, \tag{15.9.9}$$

- $$\text{total} \quad \text{s.s.} = \mathbf{y'y} - n\bar{y}^2. \tag{15.9.10}$$

The regression sum of squares in table 15.5.1 does not have the form (15.9.8) or (15.9.9), but it is in fact equal to them.

All these results generalise to curvilinear and multiple regression.

Further reading: Draper and Smith [19] 53–6; Kendall and Stuart [54] 354–5.

15.10 Confidence limits for the mean of the dependent variable for a given value of the independent variable

Suppose that, having fitted a linear regression model to n observations, we wish to estimate the expected value of y corresponding to the x-value x_0. The point estimator of the expected value of y is

- $$y_0 = b_0 + b_1 x_0. \tag{15.10.1}$$

If we define a row vector $\mathbf{x_0} = (1 \ x_0)$, similar to a row of $\mathbf{X}$, (15.10.1) may be written

- $$y_0 = \mathbf{x_0} \, \mathbf{b}. \tag{15.10.2}$$

The $\{b_j\}$ are of course only estimates of the $\{\beta_j\}$. To obtain a 95 per cent confidence interval for $\hat{y}$, the expected value of y at the point x_0, we equate (15.10.3) to the upper and lower $2\frac{1}{2}$ per cent points of the t_{n-2}-distribution.

- $$t_{n-2} = (\hat{y} - y_0)/\{s^2 (\mathbf{x_0} \, \mathbf{S}^{-1} \mathbf{x_0'})\}^{\frac{1}{2}}. \tag{15.10.3}$$

The reader should not need to be warned against the dangers of extrapolating from a fitted regression, however close, which has no theoretical basis.

These results generalise to curvilinear and multiple regression.

Example 15.10.1. Obtain 95 per cent confidence limits for the mean value of y at the point $x = 10$ for the data in table 15.7.1.

From example 15.7.1, we know that $b_0 = 9.384\,21$ and $b_1 = 2.972\,93$. It follows that the point estimate of the y-mean is

$$y_0 = 9.384\,21 + 10 \times 2.972\,93 = 39.1135.$$

The residual mean square s^2 is 11.406, and the inverse matrix

$$\mathbf{S}^{-1} = \begin{pmatrix} 20 & 210 \\ 210 & 2870 \end{pmatrix}^{-1} = \begin{pmatrix} 0.215\,789\,47 & -0.015\,789\,47 \\ -0.015\,789\,47 & 0.001\,503\,75 \end{pmatrix}.$$

Then

$$x_0 = (1 \quad 10); \quad \mathbf{S}^{-1}x_0' = \begin{pmatrix} 0.057\,894\,77 \\ -0.000\,751\,97 \end{pmatrix} \quad \text{and} \quad x_0 \mathbf{S}^{-1}x_0' = 0.050\,375.$$

To compute 95 per cent confidence limits for the mean $\hat{y}$, we solve the equations

$$(\hat{y} - 39.1135)/(11.406 \times 0.050\,375)^{\frac{1}{2}} = \pm 2.101,$$

which yield the limits 39.11 ± 1.59.

> *Further reading*: Draper and Smith [19] 21–4, 56; Kendall and Stuart [54] 363–5; Zar [105] 210–13.

15.11 Have we used the correct model?

Once a regression line has been fitted, the question naturally arises as to whether we have used a correct model. The techniques we now describe apply equally well to simple linear regression, curvilinear regression and multiple regression (and other models for that matter).

When we formulated our model in (15.4.1), we assumed that the errors $\{e_i\}$ were independent normal random variables with zero means and variances σ^2. The differences between the observed and fitted y-values or *residuals* should exhibit tendencies which are consistent with these assumptions. If they do not, we should have serious doubts about the appropriateness of our model.

The following checks should be carried out.

1. Do the residuals as a group seem to resemble a sample from a normal population with zero mean?

2. Plot the residuals one after the other in their natural sequence. Does the assumption of a constant variance seem reasonable or do the residuals at one end appear to have larger variances?

3. Does the residual plot suggest that the straight-line model is inappropriate? (The plot in fig. 15.11.1, for example, suggests that a quadratic curve should be fitted rather than a straight line.)

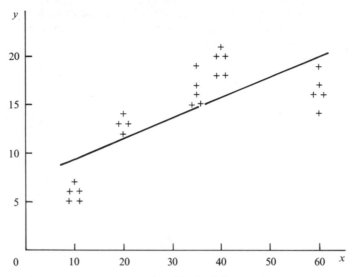

Fig. 15.11.1. An example of lack-of-fit.

4. Examine any outliers closely. These should only be rejected out of hand if they are the result of an obvious experimental blunder.

Lack-of-fit can be tested statistically when replicates of the observed $\{y_i\}$-values are available for each value of x. Let us imagine that there are n_i observations at x_i and these are denoted by $y_{i1}, y_{i2}, ..., y_{in_i}$. If the regression line provides an adequate fit, the mean of these n_i observations $\bar{y}_{i.}$ will be close to the fitted value Y_i and $(\bar{y}_{i.} - Y_i)^2$ will be small; if the fit is poor, this square will be large. A measure of *lack-of-fit* is therefore provided by
$$\sum_i n_i(\bar{y}_{i.} - Y_i)^2.$$

It so happens with replication that the residual sum of squares in table (15.5.1) may be further partitioned into two positive components: the above-mentioned lack-of-fit sum of squares and a *pure error* sum of squares. The latter is the sum of the squares of the $\{y_i\}$-values about their respective means $\{\bar{y}_{i.}\}$. The appropriate analysis of variance layout is given in table 15.11.1; p denotes the number of distinct x-values. Lack-of-fit is tested by calculating

$$\blacksquare \qquad F_{p-2,\,n-p} = \frac{\text{lack-of-fit mean square}}{\text{pure error mean square}}. \qquad (15.11.1)$$

The usual $F_{1,\,n-2}$ criterion is used to test the significance of the linear regression when the fit is adequate.

Computation is a little tedious. Let us denote the total of the $\{y_{ij}\}$-values at x_i by $T_{i.}$ and the grand total of the $\{y_{ij}\}$-values by $T_{..}$. The sums of squares are obtained in the following manner:

Table 15.11.1. *Simple linear regression: lack-of-fit analysis of variance. The mean squares (indicated by asterisks) are obtained by dividing the sums of squares in column (2) by the respective degrees of freedom in column (3)*

Source (1)	s.s. (2)		d.f. (3)	m.s. (4) = (2)/(3)
Regression	$b_1^2(\Sigma x^2 - n\bar{x}^2)$		1	**
Lack-of-fit	$\left(\begin{array}{c} \sum\limits_{i=1}^{p} (T_{i.}^2/n_i) - T_{..}^2/n \\ -\text{regress. s.s.} \end{array} \right.$	$\left. p-2 \right.$		$\left. ** \right.$
Residual	(By difference)		$n-2$	**
Pure error	(By difference)		$n-p$	**
Total	$\Sigma y^2 - n\bar{y}^2$		$n-1$	—

(1) Calculate the regression sum of squares (table 15.5.1).
(2) Calculate the total sum of squares (table 15.5.1).
(3) Calculate the residual sum of squares by subtraction: (2)−(1).
(4) Calculate[7] the sum of the regression and lack-of-fit sums of squares using (15.11.2).
(5) Calculate the lack-of-fit sum of squares by subtraction: (4)−(1).
(6) Calculate the pure error sum of squares by subtraction: (3)−(5).

∎ $(T_{1.}^2/n_1 + ... + T_{p.}^2/n_p) - T_{..}^2/n.$ (15.11.2)

Example 15.11.1. Use the method of least squares to fit a straight line to the data in table 15.11.2, and test for lack-of-fit.

We begin by calculating

$$\Sigma\Sigma y^2 = 5^2 + 6^2 + ... + 16^2 = 5452,$$
$$\Sigma x = (10 \times 5) + ... + (60 \times 5) = 805,$$
$$\Sigma x^2 = (10^2 \times 5) + ... + (60^2 \times 5) = 34\,225,$$
$$\Sigma\Sigma xy = (10 \times 5) + ... + (60 \times 16) = 13\,000.$$

Using formula (15.2.6)

$$b_1 = \{13\,000 - 24 \times (805/24)\,(342/24)\}/\{34\,225 - 24 \times (805/24)^2\}$$
$$= 0.211\,6222.$$

Using table 15.5.1,

$$\text{regression s.s.} = (0.2116222)^2 \,\{34\,225 - 24 \times (805/24)^2\}$$
$$= 323.52,$$

---

[7] The computation of the sum of the regression and lack-of-fit sums of squares is really a simple one-way analysis of variance calculation (section 12.18).

Table 15.11.2. *Data for regression example 15.11.1. The twenty-four observations are mutually independent*

x_i	y_{ij}					n_i	$T_{i.}$
10	5	6	5	6	7	5	29
20	12	13	14	13		4	52
35	17	19	16	15	15	5	82
40	18	20	21	18	20	5	97
60	17	19	16	14	16	5	82
Total						24	342

Table 15.11.3. *Numerical example – lack-of-fit analysis of variance*

Source of variation	s.s.		d.f.		m.s.	
Regression	323.52		1		323.52	
Residual $\begin{cases}\text{Lack-of-fit}\\\text{Pure error}\end{cases}$	254.98	$\begin{cases}218.58\\36.40\end{cases}$	22	$\begin{cases}3\\19\end{cases}$	11.59	$\begin{cases}72.86\\1.92\end{cases}$
Total	578.50		23		—	

$$
\begin{aligned}
\text{total s.s.} \quad &= 5452 - 24 \times (342/24)^2 \\
&= 578.50, \\
\text{residual s.s.} \quad &= 578.50 - 323.52 \\
&= 254.98.
\end{aligned}
$$

Using (15.11.2),

$$
\begin{aligned}
\text{regression s.s. plus lack-of-fit s.s.} \\
&= (29^2/5) + (52^2/4) + \ldots + (82^2/5) - (342^2/24) \\
&= 542.10.
\end{aligned}
$$

By subtraction,

$$
\begin{aligned}
\text{lack-of-fit s.s.} &= 542.10 - 323.52 \\
&= 218.58, \\
\text{pure error s.s.} &= 254.98 - 218.58 \\
&= 36.40.
\end{aligned}
$$

These values are summarised in table 15.11.3, where mean squares are calculated. To test the adequacy of the straight-line model we calculate $F_{3, 19} = 72.86/1.92 = 37.95$. This value is highly significant, indicating the poor fit provided by a straight line. The lack-of-fit is also evident when the data points and least-squares line are plotted (fig. 15.11.1).

Further reading: Brownlee [7] 366–75; Draper and Smith [19] 26–32, 86–100; Zar [105] 215–23.

$$
F_{1,22} = \frac{323.51}{11.59} = 27.91
$$

a regression significant but very high lack of fit.

15.12 Unequal variances – weighted least squares

Experimental results are not always equally reliable and the research worker often has some feeling for the relative reliability of his observations. Sometimes, the lack of homogeneity of variance may not become evident until the completion of a straight-line fit when the residuals are plotted. In either situation, the assumption that the variances of the error terms $\{e_i\}$ in (15.4.1) are equal is clearly invalid. Unless the lack of homogeneity of variance is marked (some of the variances being say ten times as great as others), the additional work of using weighted least squares is probably not worth while; the ordinary least-squares coefficients b_0 and b_1 are still unbiased estimators of β_0 and β_1, but they are no longer minimum-variance estimators (section 13.1). If there is marked inhomogeneity, weighted least squares should be used.

Let us consider the case of n points and assume that the variance at the ith point is σ^2/w_i. The $\{w_i\}$ are referred to as *weights*.

If b_0 and b_1 are all that we require and we do not aim to carry out statistical calculations, the weighted least-squares calculation is straight-forward: we pretend[8] that the observation y_i is observed an integral number of times Nw_i, where N is a suitable (possibly moderately large) integer. Standard least-squares methodology will then give us the correct values of b_0 and b_1.

The method of the previous paragraph *cannot* be used for statistical purposes and another approach must be used. This method is best described in terms of matrix theory and we shall assume for demonstration purposes that $n = 4$. We define

$$\mathbf{X}_w = \begin{pmatrix} \sqrt{w_1} & x_1\sqrt{w_1} \\ \sqrt{w_2} & x_2\sqrt{w_2} \\ \sqrt{w_3} & x_3\sqrt{w_3} \\ \sqrt{w_4} & x_4\sqrt{w_4} \end{pmatrix}, \quad \mathbf{y}_w = \begin{pmatrix} y_1\sqrt{w_1} \\ y_2\sqrt{w_2} \\ y_3\sqrt{w_3} \\ y_4\sqrt{w_4} \end{pmatrix}, \quad \mathbf{S}_w = \mathbf{X}_w'\mathbf{X}_w.$$

Writing $\mathbf{X}_w$ instead of $\mathbf{X}$, $\mathbf{y}_w$ instead of $\mathbf{y}$, $\mathbf{S}_w$ instead of $\mathbf{S}$; replacing the elements of $\mathbf{S}^{-1}$ by the corresponding elements of $\mathbf{S}_w^{-1}$ and $\bar{y}$ by

$$\bar{y}_w = (y_1w_1 + y_2w_2 + \dots + y_nw_n)/(w_1 + w_2 + \dots + w_n),$$

we can use the following standard regression formulae: (15.3.1), (15.3.3), (15.3.4), (15.5.3), (15.7.3), (15.9.1), (15.9.3), (15.9.4), (15.9.5), (15.9.6), (15.9.7), (15.10.2) and (15.10.3). If we replace n by Σw_i in (15.9.8), (15.9.9) and (15.9.10), we can also use them. These equations cover most requirements.

[8] This result seems plausible when we recall that the sample mean of Nw_i observations from a population with variance σ^2 has variance $\sigma^2/(Nw_i)$.

Example 15.12.1. The following points are obtained from an experiment:

$(x_1, y_1) = (1, 7),$
$(x_2, y_2) = (3, 18),$
$(x_3, y_3) = (4, 30),$
$(x_4, y_4) = (5, 35).$

It is known that the standard deviations of the observations at x_1 and x_4 are approximately equal, but the standard deviations of the observations at x_2 and x_3 are about three times as large. Fit a straight line using the weighted least-squares method.

In this example, the observations at x_1 and x_4 are about three times as accurate (in terms of standard deviations) as the observations at x_2 and x_3. We therefore use $w_1 = w_4 = 9$ and $w_2 = w_3 = 1$. To compute b_0 and b_1, we may pretend that the observations at x_1 and x_4 are each repeated nine times. We therefore use $n = 20$ and

$$\Sigma x_i = 9 \times 1 + 3 + 4 + 9 \times 5 = 61,$$
$$\Sigma x_i^2 = 9 \times 1^2 + 3^2 + 4^2 + 9 \times 5^2 = 259,$$
$$\Sigma y_i = 9 \times 7 + 18 + 30 + 9 \times 35 = 426,$$
$$\Sigma x_i y_i = 9 \times 1 \times 7 + 3 \times 18 + 4 \times 30 + 9 \times 5 \times 35 = 1812.$$

From (15.2.6) we obtain

$$b_1 = \{1812 - 20 \times (61/20)(426/20)\} / \{259 - 20 \times (61/20)^2\}$$
$$= 7.0281,$$
$$b_0 = (426/20) - 7.0281 \times (61/20) = -0.1357.$$

Using (15.2.1), the following fitted values are obtained: 6.892, 20.949, 27.977 and 35.005. Note how the fitted line adheres closely to the observations at x_1 and x_4.

Example 15.12.2. Fit a weighted least-squares line to the data of example 15.12.1 and obtain confidence intervals for β_0 and β_1.
We need to define

$$X_w = \begin{pmatrix} 3 & 3 \\ 1 & 3 \\ 1 & 4 \\ 3 & 15 \end{pmatrix}, \quad y_w = \begin{pmatrix} 21 \\ 18 \\ 30 \\ 105 \end{pmatrix}, \quad S_w = \begin{pmatrix} 20 & 61 \\ 61 & 259 \end{pmatrix}.$$

We also need $\bar{y}_w = (7 \times 9 + 18 \times 1 + 30 \times 1 + 35 \times 9)/20 = 21.3,$

$$S_w^{-1} = \begin{pmatrix} 0.177\,518\,848 & -0.041\,809\,458 \\ -0.041\,809\,458 & 0.013\,708\,019 \end{pmatrix} \quad \text{and} \quad X_w'y_w = \begin{pmatrix} 426 \\ 1812 \end{pmatrix}.$$

Table 15.12.1. *Weighted regression example – analysis of variance*

Source	s.s.	d.f.	m.s.
Regression	3603.31	1	3603.31
Residual	12.89	2	6.45
Total	3616.20	3	–

From (15.3.4) we have

$$\begin{pmatrix} b_0 \\ b_1 \end{pmatrix} = S_w^{-1} X_w' y_w = \begin{pmatrix} -0.1357 \\ 7.0281 \end{pmatrix}$$

confirming the earlier answer.

Using (15.9.8) and (15.9.10) with subscripts w and Σw_i instead of n, we complete the analysis of variance table 15.12.1 and then calculate

$$F_{1,2} = 3603.31/6.45 = 559.$$

This value is highly significant; so it is certainly meaningful to calculate confidence intervals for β_0 and β_1.

To obtain 95 per cent confidence intervals for β_0 and β_1 we equate (15.9.6) and (15.9.7) to the upper and lower $2\frac{1}{2}$ per cent points of the t_2-distribution:

$$(-0.1357 - \beta_0)/(6.45 \times 0.177\,518\,848)^{\frac{1}{2}} = \pm 4.303,$$
$$(7.0281 - \beta_1)/(6.45 \times 0.013\,708\,019)^{\frac{1}{2}} = \pm 4.303.$$

We obtain confidence intervals of -0.1357 ± 4.6044 and 7.0281 ± 1.2795. The bounds are wide because of the small sample size.

Further reading: Draper and Smith [19] 77–80.

15.13 The similarity of correlation analysis and simple linear regression

The bivariate normal distribution was introduced in section 9.6, and tests of correlation based on this distribution were given in sections 12.31–12.33. The reader will have noticed certain similarities between the bivariate normal correlation techniques and simple linear regression: for example, the least-squares regression line (15.2.7) and the conditional expectation equation (9.6.2); the correlation test statistic (12.31.2) and the regression statistic (15.5.4).

Both techniques can only be used when the underlying relationship between the two variables is linear; the regression technique can, however, be extended to deal with curves (chapter 16). In correlation analysis, both x and y are random variables, but we need to make the strong assumption of bivariate normality; we are usually interested in measuring the closeness

Table 15.14.1. *Thirty-six independent outcomes of an experiment*

x	Observations y			
7	123	119	125	128
14	150	142	173	158
21	191	182	175	179
28	206	229	214	218
35	251	245	241	252
42	258	266	276	274
49	307	302	296	301
56	321	328	326	330
63	353	342	346	351

of the linear relation rather than the actual parameter values. In regression analysis we distinguish between the independent variable x, assumed free of error, and the dependent variable y whose conditional distribution, given x, is taken to be normal; we usually aim to learn all we can about the regression parameters. It is worth noting that for bivariate normal variables x and y, the distribution of y, given x, is normal with constant variance independent of x; it follows that regression techniques can often be used to solve problems with bivariate normal data.

Both techniques can be generalised to three or more variables. The regression generalisation is given in chapter 17.

Further reading: Campbell [8] 281–2; Johnson and Leone [50] 435–6; Zar [105] 198–9.

15.14   Exercises

1. Use the method of least squares to fit a straight line to the data in table 15.14.1.

2. Determine the matrix $\mathbf{S}$ and vector $\mathbf{X'y}$ for these data.

3. Calculate the correlation coefficient r for these data and use (15.5.4) to test the significance of linear regression.

4. Form the regression analysis of variance table for the straight-line fit of the table 15.14.1 data. Test the significance of linear regression and confirm numerically the relationship between t and F.

5. Obtain 95 per cent confidence intervals for β_0, β_1 and σ^2.

6. Have we used the correct model to fit these data?

7. Use the unweighted least-squares method to fit a straight line to the example 15.12.1 data. Compare your results with those given in the text.

16

Curvilinear regression

Summary   This chapter deals with curvilinear regression – a generalisation of simple linear regression. The general least-squares procedure is outlined in sections 16.1 and 16.2. Sections 16.3 and 16.4 then describe the usual statistical model and test of significance for the regression curve. Estimation procedures and hypothesis-testing are covered in sections 16.5–16.8. Sections 16.9 and 16.10 deal with lack-of-fit and weighted least squares, and the problem of linear regression through the origin or some other fixed point is covered in section 16.11. The chapter ends with two sections on orthogonal polynomial regression.

16.1   Introduction and simple examples of curvilinear least squares

The techniques of chapter 15 carry over to curvilinear regression. While the theory can be described without it, matrix notation allows the use of general equations and it greatly simplifies the work. The reader is advised to make sure that he is familiar with the relevant sections of chapter 15 (15.3 and 15.9 in particular).

A straight line involves the zeroth and first powers of x, and in fitting a straight line by the method of least squares we made use of a matrix $\mathbf{X}$ with two columns, one comprising the zeroth and the other the first powers of the $\{x_i\}$. The second-degree polynomial

$$Y = b_0 + b_1 x + b_2 x^2 \qquad (16.1.1)$$

involves the zeroth, first and second powers of x, and to fit such a polynomial close to n points by the method of least squares, we use a matrix $\mathbf{X}$ with three columns. The first, second and third columns contain respectively the zeroth, first and second powers of the $\{x_i\}$. Thus,

$$\mathbf{X} = \begin{pmatrix} 1 & x_1 & x_1^2 \\ 1 & x_2 & x_2^2 \\ \cdot & \cdot & \cdot \\ \cdot & \cdot & \cdot \\ \cdot & \cdot & \cdot \\ 1 & x_n & x_n^2 \end{pmatrix}. \qquad (16.1.2)$$

We need to determine three constants, b_0, b_1 and b_2, and these are listed in a three-dimensional column vector $\mathbf{b}$. As before, the $\{y_i\}$ observations are listed in a column vector $\mathbf{y}$ and the fitted values $\{Y_i\}$ are listed in a column vector **Y**. Both these vectors contain n elements. It is easy to see that

$$S = X'X = \begin{pmatrix} n & \Sigma x_i & \Sigma x_i^2 \\ \Sigma x_i & \Sigma x_i^2 & \Sigma x_i^3 \\ \Sigma x_i^2 & \Sigma x_i^3 & \Sigma x_i^4 \end{pmatrix}, \quad \text{and} \quad X'y = \begin{pmatrix} \Sigma y_i \\ \Sigma x_i y_i \\ \Sigma x_i^2 y_i \end{pmatrix}.$$

(16.1.3)

The least-squares constants b_0, b_1 and b_2 are obtained by solving the matrix equation $Sb = X'y$ (section 15.3).

The exponential curve[1]

$$Y = b_0 + b_1 e^x$$

(16.1.4)

involves the zeroth power of x and e^x, and to fit such a curve near n points, we use a matrix **X** with two columns. The first column contains the zeroth powers of the $\{x_i\}$ and the second the values $\{\exp(x_i)\}$. Thus

$$X = \begin{pmatrix} 1 & \exp(x_1) \\ 1 & \exp(x_2) \\ \cdot & \cdot \\ \cdot & \cdot \\ \cdot & \cdot \\ 1 & \exp(x_n) \end{pmatrix}.$$

(16.1.5)

We need to determine two constants b_0 and b_1, and these are listed in a two-dimensional column vector **b**. The $\{y_i\}$ observations and fitted values $\{Y_i\}$ are listed in n-dimensional column vectors **y** and **Y** respectively. Again, it is easy to see that

$$S = X'X = \begin{pmatrix} n & \Sigma \exp(x_i) \\ \Sigma \exp(x_i) & \Sigma \{\exp(x_i)\}^2 \end{pmatrix} \quad \text{and} \quad X'y = \begin{pmatrix} \Sigma y_i \\ \Sigma \{\exp(x_i)\} y_i \end{pmatrix}$$

(16.1.6)

The least-squares constants b_0 and b_1 are obtained by solving the matrix equation $Sb = X'y$ (section 15.3).

Example 16.1.1. Use the method of least squares to fit a quadratic of the form (16.1.1) close to the four points

$(x_1, y_1) = (1, 5)$,
$(x_2, y_2) = (3, 40)$,
$(x_3, y_3) = (4, 109)$,
$(x_4, y_4) = (5, 297)$.

---

[1] Exponential curves of the form $Y = Ae^{Bx}$ are usually estimated by fitting a straight line to $\ln y$. See also section 18.1.

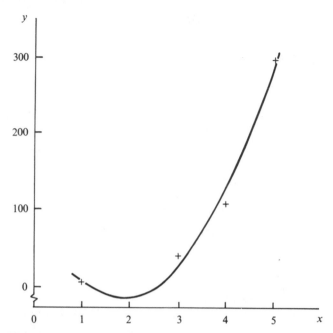

Fig. 16.1.1. A quadratic least-squares line. The data points are indicated by crosses.

From (16.1.3) we know that

$$\mathbf{S} = \begin{pmatrix} 4 & 13 & 51 \\ 13 & 51 & 217 \\ 51 & 217 & 963 \end{pmatrix} \quad \text{and} \quad \mathbf{X'y} = \begin{pmatrix} 451 \\ 2046 \\ 9534 \end{pmatrix}.$$

The normal equations (summarised by $\mathbf{Sb} = \mathbf{X'y}$) may be written

$$4b_0 + 13b_1 + 51b_2 = 451,$$
$$13b_0 + 51b_1 + 217b_2 = 2046,$$
$$51b_0 + 217b_1 + 963b_2 = 9534.$$

We find that $b_0 = 89.79988$, $b_1 = -113.12719$ and $b_2 = 30.636352$, so that the fitted values are 7.309, 26.145, 127.473 and 290.073. The fitted curve and the data points are shown in fig. 16.1.1.

Example 16.1.2. Use the method of least squares to fit an exponential curve of the form (16.1.4) close to the four points in example 16.1.1. According to (16.1.6),

$$\mathbf{S} = \begin{pmatrix} 4 & 225.8151278 \\ 225.8151278 & 25418.24162 \end{pmatrix} \quad \text{and} \quad \mathbf{X'y} = \begin{pmatrix} 451 \\ 50846.91948 \end{pmatrix}.$$

The normal equations (summarised by $\mathbf{Sb} = \mathbf{X'y}$) may be written

$$4b_0 \qquad\quad + \quad 225.815\,1278\,b_1 = \quad 451 \qquad ,$$
$$225.815\,1278\,b_0 + 25\,418.241\,62\,b_1 \quad = 50\,846.919\,48.$$

We find that $b_0 = -0.362\,598\,010$ and $b_1 = 2.003\,631\,893$, so that the fitted values are 5.084, 39.881, 109.032 and 297.003. The fit appears to be good.

Further reading: Brownlee [7] 447; Colquhoun [16] 243–71; Hoel [46] 177–80; Snedecor [91] 447–72; Wilks [102] 273–5.

16.2 General curvilinear least squares

The matrix methods outlined above in section 16.1 can be used to fit any functional form which is linear in the coefficients $\{b_i\}$. Thus, for example, the method can be used to fit a curve of the form

$$Y = b_0 + b_1 \sin(\pi e^x) + b_2 \ln x,$$

but it cannot be used to fit a curve of the form[2]

$$Y = b_0 + b_1 \exp(b_2 x).$$

In the general situation, we need to determine $k+1$ coefficients to fit a curve of the form

■ $$Y = b_0 g_0(x) + b_1 g_1(x) + \ldots + b_k g_k(x). \tag{16.2.1}$$

The functions $\{g_i(x)\}$ must not contain unknown parameters which need estimation.

To fit (16.2.1) close to n points by the method of least squares, we use a matrix with $k+1$ columns. The first column comprises the values $\{g_0(x_i)\}$, the second the values $\{g_1(x_i)\}$, ..., and the final column the values $\{g_k(x_i)\}$. It is not difficult to see that the matrix $\mathbf{S} = \mathbf{X'X}$ takes the form

■ $$\mathbf{S} = \begin{pmatrix} S_{00} & S_{01} & S_{02} & \cdots & S_{0k} \\ S_{10} & S_{11} & S_{12} & \cdots & S_{1k} \\ \vdots & \vdots & \vdots & & \vdots \\ S_{k0} & S_{k1} & S_{k2} & \cdots & S_{kk} \end{pmatrix}, \tag{16.2.2}$$

where

■ $$S_{rs} = S_{sr} = \sum_{i=1}^{n} g_r(x_i)\, g_s(x_i). \tag{16.2.3}$$

The unknown coefficients are listed in a $(k+1)$-dimensional column vector $\mathbf{b}$, the n observations $\{y_i\}$ in an n-dimensional vector $\mathbf{y}$, and the n fitted values $\{Y_i\}$ in an n-dimensional vector $\mathbf{Y}$. The vector $\mathbf{X'y}$ is a column vector of

[2] Non-linear methods are necessary (chapter 18). See also footnote 1.

dimension $k+1$ and it is easy to see that

$$\blacksquare \quad X'y = \begin{pmatrix} \sum_i y_i g_0(x_i) \\ \sum_i y_i g_1(x_i) \\ . \\ . \\ . \\ \sum_i y_i g_k(x_i) \end{pmatrix} \tag{16.2.4}$$

We solve the $k+1$ linear equations $Sb = X'y$ to obtain the $k+1$ coefficients $\{b_i\}$.

Example 16.2.1. For the quadratic curve (16.1.1), $g_0(x) = 1, g_1(x) = x$ and $g_2(x) = x^2$. The matrix S and the vector $X'y$ are given by (16.1.3).

Example 16.2.2. For the exponential curve (16.1.4), $g_0(x) = 1$ and $g_1(x) = e^x$. The matrix S and the vector $X'y$ are given by (16.1.6).

Example 16.2.3. Use the method of least squares to fit a curve of the form $Y = b_1 e^x$ close to the four points in example 16.1.1.

In this case we only need to fit one constant; so the matrix X has only one column comprising the values $\{\exp x_i\}$. The matrix $S = X'X$ will be a 1×1 matrix, and the single element is

$$\sum_{i=1}^{4} \{\exp x_i\}^2 = 25\,418.241\,62.$$

The column vector $X'y$ contains only one element, namely

$$\sum_{i=1}^{4} \{\exp x_i\} y_i = 50\,846.919\,48.$$

The normal equation $Sb = X'y$ is simply

$$25\,418.241\,62 b_1 = 50\,846.919\,48,$$

so that $b_1 = 2.000\,410\,580$. The fitted values are 5.438, 40.179, 109.219 and 296.887 respectively.

Further reading: Draper and Smith [19] 58–66; Johnson and Leone [50] 423–34; Snedecor [91] 447–72; Zar [105] 268–70.

16.3   Curvilinear regression – the statistical model

In sections 16.1 and 16.2, the method of least squares is described as a numerical technique. We now formulate the statistical model.

When we fit a quadratic like (16.1.1), we assume an underlying statistical model of the form

$$y_i = \beta_0 + \beta_1 x_i + \beta_2 x_i^2 + e_i, \tag{16.3.1}$$

where the $\{x_i\}$ are assumed free of error. The $\{\beta_i\}$ are unknown coefficients and the $\{e_i\}$ are independent normal random variables with zero means and equal variances σ^2. Estimates of β_0, β_1 and β_2 are provided by b_0, b_1 and b_2 respectively. For the exponential curve (16.1.4), we assume an underlying statistical model of the form

$$y_i = \beta_0 + \beta_1 \exp x_i + e_i. \tag{16.3.2}$$

The underlying statistical model for the general curvilinear fit (16.2.1) is

■ $$y_i = \beta_0 g_0(x_i) + \ldots + \beta_k g_k(x_i) + e_i. \tag{16.3.3}$$

In both (16.3.2) and (16.3.3), the $\{x_i\}$ are assumed free of error. All these models have the matrix form (15.9.1) when we list the unknown $\{\beta_i\}$ and errors $\{e_i\}$ in column vectors β and e respectively.

Further reading: Draper and Smith [19] 58–66.

16.4   Testing the significance of a regression curve

Even in the situation where x and y are unrelated, it is possible to draw the observed points on a graph and fit least-squares curves like (16.1.1) and (16.1.4). The calculated coefficients $\{b_i\}$ will usually be non-zero. We need to distinguish between non-zero values such as these and non-zero b-values which imply a genuine relationship between x and y.

The usual approach is to form an analysis of variance table and perform an F-test to see whether the underlying β-coefficients are zero or not. If the F-ratio calculated is significantly large, we reject the null hypothesis that the coefficients are all zero and conclude that the regression is significant.

Most regression models involve a coefficient β_0 (a mean) which is attached to a function not actually involving x (for example, the models (16.3.1) and (16.3.2)). A non-zero value for this coefficient does not imply a relationship between x and y; so we do not include it in our significance test. We use the analysis of variance table 16.4.1 to test the significance of the other regression coefficients. In this table, ν denotes the number of β-coefficients and we test whether $\nu - 1$ of these (all except β_0) are zero. The test statistic actually used is

■ $$F_{\nu-1,\, n-\nu} = \frac{\text{regression mean square}}{\text{residual mean square}}. \tag{16.4.1}$$

Occasionally, we need to test the null hypothesis that all the regression coefficients are zero (the model may not include a mean parameter β_0; or we might be interested in testing whether β_0 and the other β-parameters are all

Table 16.4.1. *Curvilinear regression - analysis of variance*

Source (1)	s.s. (2)	d.f. (3)	m.s. (4) = (2)/(3)
Regression	$\mathbf{b}'\mathbf{X}'\mathbf{y} - n\bar{y}^2$	$\nu - 1$	$\left(\begin{array}{c}\text{By}\\\text{division}\end{array}\right)$
Residual	By difference	$n - \nu$	
Total	$\mathbf{y}'\mathbf{y} - n\bar{y}^2$	$n - 1$	—

Table 16.4.2. *Curvilinear regression - uncorrected analysis of variance*

Source (1)	s.s. (2)	d.f. (3)	m.s. (4) = (2)/(3)
Regression	$\mathbf{b}'\mathbf{X}'\mathbf{y}$	ν	$\left(\begin{array}{c}\text{By}\\\text{division}\end{array}\right)$
Residual	By difference	$n - \nu$	
Total	$\mathbf{y}'\mathbf{y}$	n	—

zero). In this situation, we form the *uncorrected*[3] analysis of variance table 16.4.2 and use the test statistic

$$F_{\nu,\, n-\nu} = \frac{\text{regression mean square}}{\text{residual mean square}}. \qquad (16.4.2)$$

Example 16.4.1. Test the significance of the regression coefficients in the simple quadratic example 16.1.1.

In this case, $n = 4$ and $\nu = 3$. We compute the sums of squares for the analysis of variance using the formulae in table 16.4.1:

$$
\begin{aligned}
\text{total s.s.} &= \mathbf{y}'\mathbf{y} - n\bar{y}^2 \\
&= \Sigma y_i^2 - n\bar{y}^2 \\
&= 101\,715 - 50\,850.25 \\
&= 50\,864.75, \\
\text{regression s.s.} &= \mathbf{b}'\mathbf{X}'\mathbf{y} - n\bar{y}^2 \\
&= (89.799\,88 \quad -113.127\,19 + 30.636\,352) \begin{pmatrix} 451 \\ 2046 \\ 9534 \end{pmatrix} \\
&\quad - 50\,850.25 \\
&= 101\,128.48 - 50\,850.25 \\
&= 50\,278.23, \\
\text{residual s.s.} &= 50\,864.75 - 50\,278.23 \\
&= 586.52.
\end{aligned}
$$

[3] Not corrected for the mean $\bar{y}$.

Table 16.4.3. *Analysis of variance for example 16.4.1*

Source	s.s.	d.f.	m.s.
Regression	50 278.23	2	25 139.11
Residual	586.52	1	586.52
Total	50 864.75	3	—

The reader should verify that this last figure is the sum of the squared deviations between the observed and fitted values (section 15.5). The analysis of variance is given as table 16.4.3. To test the significance of regression, we compute

$$F_{2,1} = \frac{25\,139.11}{586.52} = 42.86.$$

This value is not significant at the 5 per cent level; if (16.3.1) is the correct model, we have no evidence that β_1 and β_2 are non-zero.

Example 16.4.2. Test the significance of the regression coefficients in the exponential example 16.1.2.

In this case $n = 4$ and $v = 2$. The data are the same as in the previous example; so we know that the total sum of squares is 50 864.75. We use the formula in table 16.4.1 to compute the regression sum of squares and obtain the residual sum of squares by difference. Thus,

$$\text{regression s.s.} = (-0.362\,598\,010 + 2.003\,631\,893)\begin{pmatrix} 451 \\ 50\,846.919\,48 \end{pmatrix}$$

$$-50\,850.25$$
$$= 101\,714.9778 - 50\,850.25$$
$$= 50\,864.7278,$$
$$\text{residual s.s.} = 50\,864.75 - 50\,864.7278$$
$$= 0.0222.$$

The reader should verify that this last figure is (apart from the effects of rounding) the sum of the squared deviations between the observed and fitted values (section 15.5). The analysis of variance is given as table 16.4.4. To test the significance of regression, we compute

$$F_{1,2} = \frac{50\,864.7278}{0.0111} = 458 \times 10^4.$$

This value is highly significant; so we have clear evidence that the coefficient β_1 in the model (16.3.2) is non-zero.

Table 16.4.4. *Analysis of variance for example 16.4.2*

Source	s.s.	d.f.	m.s.
Regression	50 864.7278	1	50 864.7278
Residual	0.0222	2	0.0111
Total	50 864.7500	3	—

Table 16.4.5. *Analysis of variance for example 16.4.3*

Source	s.s.	d.f.	m.s.
Regression	101 714.71	1	101 714.71
Residual	0.29	3	9.66×10^{-2}
Total	101 715.00	4	—

Example 16.4.3. Test the significance of the regression coefficient in example 16.2.3.

In this case $n = 4$ and $\nu = 1$. There is no regression coefficient β_0 to correct for, and we therefore use the formulae in table 16.4.2:

$$\text{total s.s.} = \mathbf{y'y} = \Sigma y_i^2 = 101\,715,$$
$$\text{regression s.s.} = \mathbf{b'X'y} = 2.000\,410\,580 \times 50\,846.919\,48 = 101\,714.71,$$
$$\text{residual s.s.} = 0.29.$$

The reader should verify that this last figure is (apart from the effects of rounding) the sum of the squared deviations between the observed and fitted values (section 15.5). The analysis of variance is given as table 16.4.5. To test the significance of regression, we compute

$$F_{1,\,3} = \frac{101\,714.71}{9.66 \times 10^{-2}} = 1.05 \times 10^6.$$

This value is highly significant; if we have used the correct model, there is clear evidence that β_1 is non-zero.

Further reading: Draper and Smith [19] 61–2.

16.5 Point estimators for σ^2 and the β-parameters

The underlying parameters $\{\beta_i\}$ and σ^2 are unknown; one can, however, obtain unbiased estimates of them. The regression coefficients $\{b_i\}$ estimate the $\{\beta_i\}$ and the residual mean square in the analysis of variance table (16.4.1 or 16.4.2) estimates σ^2.

The covariance matrix of the $\{b_i\}$ is $\sigma^2 \mathbf{S}^{-1}$. If we denote the elements[4] of the inverse matrix $\mathbf{S}^{-1}$ by $\{S_{ij}^{-1}\}$,

■ $$\text{var } b_i = \sigma^2 S_{ii}^{-1}, \tag{16.5.1}$$

■ $$\text{cov }(b_i, b_j) = \sigma^2 S_{ij}^{-1}. \tag{16.5.2}$$

Formulae (16.5.1) and (16.5.2) are generalisations of (15.9.3), (15.9.4) and (15.9.5).

Further reading: Draper and Smith [19] 61.

16.6   Confidence intervals for σ^2 and the β-parameters

Let us denote the residual mean square in the analysis of variance table (16.4.1 or 16.4.2) by s^2, and the elements[5] of the inverse matrix $\mathbf{S}^{-1}$ by $\{S_{ij}^{-1}\}$. To obtain 95 per cent confidence limits for β_i, we equate (16.6.1) to the upper and lower $2\frac{1}{2}$ per cent points of the $t_{n-\nu}$-distribution (ν is the number of estimated β-parameters).

■ $$t_{n-\nu} = (b_i - \beta_i)/(s^2 S_{ii}^{-1})^{\frac{1}{2}}. \tag{16.6.1}$$

To obtain 95 per cent confidence limits for σ^2, we equate (16.6.2) to the upper and lower $2\frac{1}{2}$ per cent points of the $\chi^2_{n-\nu}$-distribution.

■ $$\chi^2_{n-\nu} = \frac{(n-\nu)s^2}{\sigma^2}. \tag{16.6.2}$$

Formula (16.6.1) is a generalisation of (15.9.6) and (16.6.2) a generalisation of (15.7.3).

Example 16.6.1. Obtain 95 per cent confidence intervals for β_1 and σ^2 in example 16.1.2.

In this case $n = 4$ and $\nu = 2$; the residual mean square s^2 is 0.0111 (table 16.4.4), and

$$\mathbf{S}^{-1} = \begin{pmatrix} 0.501\,54 & -0.445\,57 \times 10^{-2} \\ -0.445\,57 \times 10^{-2} & 0.789\,25 \times 10^{-4} \end{pmatrix}.$$

To obtain 95 per cent confidence limits for β_1, we solve the equations

$$(2.003\,631\,893 - \beta_1)/(0.0111 \times 0.789\,25 \times 10^{-4})^{\frac{1}{2}} = \pm 4.303,$$

and obtain $\beta_1 = 2.003\,63 \pm 0.004\,03$.

[4] If the β-parameters are $\beta_0, \beta_1, ...$, the rows and columns of $\mathbf{S}$ and $\mathbf{S}^{-1}$ are numbered 0, 1, ...; if the β-parameters are $\beta_1, \beta_2, ...$, the rows and columns of these matrices are numbered 1, 2, ...

[5] See footnote 4.

To obtain 95 per cent confidence limits for σ^2, we solve the equations

$$\frac{2 \times 0.0111}{\sigma^2} = 7.38, \quad \frac{2 \times 0.0111}{\sigma^2} = 0.0506,$$

and obtain the limits 0.003 01 and 0.439.

Further reading: Draper and Smith [19] 64–5.

16.7   Hypothesis testing

The t-statistic (16.6.1) can be used to test hypotheses about the regression parameters. Let us imagine that we wish to test the null hypothesis $\beta_i = \delta$, where δ is some specified constant, given that the model is of the assumed form with ν parameters. We set β_i equal to δ in (16.6.1) and reject the null hypothesis if the t-statistic falls in the upper or lower $2\frac{1}{2}$ per cent regions of the $t_{n-\nu}$-distribution (example 15.8.1).

Very often we wish to test the null hypothesis $\beta_i = 0$. The t-statistic (16.6.1) can always be used, but an alternative equivalent F-test is available which can be more convenient. The calculations are as follows:

1. Fit the ν-parameter model and form the ν-parameter analysis of variance table.

2. Fit the $(\nu-1)$-parameter model involving all the terms except the one we are interested in (b_i), and compute the $(\nu-1)$-parameter analysis of variance table. The residual sum of squares will exceed the ν-parameter residual sum of squares, and the difference is the additional reduction in the residual sum of squares due to the inclusion of the b_i term.

3. Calculate the test statistic

■ $$F_{1,n-\nu} = \frac{\text{additional reduction due to } b_i}{\nu\text{-parameter residual mean square}}.$$ (16.7.1)

The null hypothesis is rejected when the test statistic falls in the upper critical region of the $F_{1,\,n-\nu}$-distribution.

This test procedure is often used sequentially in programs for finding the '*best*' *regression equation* (section 17.3). It should be noted that the additional reduction in sum of squares due to the b_i term will usually depend upon which terms have already been fitted. Only when the terms are orthogonal will the reduction be independent of the previously-fitted terms (section 16.13).

The computational procedure leading to (16.7.1) is a particular example of a more general technique for analysing the effect of introducing q new terms into a model already involving p terms. In the more general situation, we compute the p-parameter and $(p+q)$-parameter analysis of variance tables, and combine them into a single table like 16.7.1. The additional sum of squares due to the additional q terms is equal to the difference between the residual sums of squares in the p- and $(p+q)$-tables, and the residual sum of squares is

Table 16.7.1. *The effect of introducing q additional terms into a regression model with p terms – analysis of variance*

Source (1)	s.s. (2)	d.f. (3)	m.s. (4) = (2)/(3)
Regression on p terms	$\left(\begin{array}{c}\text{Regression s.s. in}\\p\text{-parameter model}\end{array}\right)$	p	
Additional reduction due to q new terms	(By difference)	q	
Residual	$\left(\begin{array}{c}\text{Residual s.s. in}\\(p+q)\text{-parameter model}\end{array}\right)$	$n-p-q$	$\left(\begin{array}{c}\text{By}\\\text{division}\end{array}\right)$
Total	$y'y$	n	

equal to the $(p+q)$-parameter residual sum of squares. The statistic for testing whether the additional terms produce a significantly better fit is

$$\blacksquare \quad F_{q,\,n-p-q} = \frac{\text{additional reduction mean square}}{\text{residual mean square}}. \tag{16.7.2}$$

Table 16.7.1 displays the uncorrected[6] analysis of variance based on table 16.4.2. The corrected table is constructed in a similar manner using table 16.4.1 (the degrees of freedom in column (3) become $p-1, q, n-p-q$ and $n-1$ respectively).

It is worth noting that the test statistic (16.7.1) can also be used to test the null hypothesis $\beta_i = \delta$ (where δ is some specified constant) provided δX_{ji} is first subtracted from each y_j value (X_{ji} is the jth element of the ith column of X). A simple linear regression example is given in section 15.8. A test for the general linear hypothesis is described, for example, by Draper and Smith [19] 72–6.

The reader's attention is drawn to section 12.2 which warns of the dangers of applying several tests to the one set of data.

> *Further reading*: Brownlee [7] 441–6; Draper and Smith [19] 67–8, 71–6, 163–94; Zar [105] 269–71.

16.8   The expected value of y for a given value of x

Suppose that, having fitted a $(k+1)$-parameter regression line of the form (16.2.1), we wish to estimate the expected value of y corresponding to the specific value x_0 of x. Formula (15.10.2) can be used to obtain a point estimate y_0, provided we define a $(k+1)$-dimensional row vector

$$x_0 = (g_0(x_0) \quad g_1(x_0) \quad \cdots \quad g_k(x_0)). \tag{16.8.1}$$

---

[6] Section 16.4.

(This vector has the same form as a row of **X**.) We can obtain 95 per cent confidence limits for the mean $\hat{y}$ by equating (16.8.2) to the upper and lower $2\frac{1}{2}$ per cent points of the t_{n-k-1}-distribution.

■ $\qquad t_{n-k-1} = (\hat{y} - y_0)/\{s^2 (\mathbf{x}_0 \mathbf{S}^{-1} \mathbf{x}_0')\}^{\frac{1}{2}}.$ $\qquad\qquad$ (16.8.2)

The warning in section 15.10 about extrapolation from a fitted regression should be noted.

> **Example 16.8.1.** Obtain 95 per cent confidence limits for the mean value of y at the point $x = 2$ using the data and model of example 16.2.3 and analysis of variance table 16.4.5.

In this case $n = 4$ and $k = 0$; the vectors **b**, $\mathbf{x}_0$ and matrix **S** are trivial:

$\mathbf{b} = 2.000\,410\,580, \qquad \mathbf{x}_0 = e^2 = 7.38905;$
$\mathbf{S} = 25\,418.241\,62, \qquad \mathbf{S}^{-1} = 1/25\,418.241\,62 = 3.93 \times 10^{-5}.$

The residual mean square s^2 is 0.0966 (table 16.4.5). From (15.10.2),

$y_0 = \mathbf{x}_0 \mathbf{b} = 7.389\,05 \times 2.000\,410\,580 = 14.781.$

This is the point estimate of the y-mean at $x = 2$.

To obtain 95 per cent confidence limits for the y-mean at $x = 2$, we solve the equations

$(\hat{y} - 14.781)/\{0.0966 \times (7.389\,05 \times 0.000\,0393 \times 7.389\,05)\}^{\frac{1}{2}}$
$= \pm 3.182,$

and obtain $\hat{y} = 14.781 \pm 0.046$.

> *Further reading*: Draper and Smith [19] 61.

16.9 Have we used the correct model?

Once a regression line has been fitted, the question naturally arises as to whether the correct model has been used. The reader is referred to section 15.11. The lack-of-fit test described in that section can also be applied to curvilinear regression (replication is of course necessary). The same computational procedure[7] is followed:
1. Calculate the regression sum of squares (table 16.4.1).
2. Calculate the total sum of squares (table 16.4.1).
3. Calculate the residual sum of squares by subtraction: (2)−(1).
4. Calculate the sum of the regression and lack-of-fit sums of squares by means of formula (15.11.2).
5. Calculate the lack-of-fit sum of squares by subtraction: (4)−(1).

[7] On rare occasions the uncorrected analysis of variance in table 16.4.2 will be required. The formulae in table 16.4.2 will then replace those from table 16.4.1 in this lack-of-fit calculation, and the term $T^2_{..}/n$ will be omitted from (15.11.2). The degrees of freedom in table 16.9.1 remain unchanged except those for regression and total which become ν and n respectively. See example 16.11.1.

Table 16.9.1. *Curvilinear regression: lack-of-fit analysis of variance. The mean squares (indicated by asterisks) are obtained by dividing the sums of squares in column (2) by the respective degrees of freedom in column (3)*

Source (1)	s.s. (2)	d.f. (3)	m.s. (4) = (2)/(3)
Regression	$b'X'y - n\bar{y}^2$	$v-1$	**
Residual ⎧ Lack-of-fit	$\left(\sum\limits_{i=1}^{p} (T_{i.}^2/n_i) - T_{..}^2/n\right.$	$p-v$	**
(By difference) ⎨	$\left. - \text{regression s.s.}\right.$	$n-v$	**
⎩ Pure error	(By difference)	$n-p$	**
Total	$y'y - n\bar{y}^2$	$n-1$	—

6. Calculate the pure error sum of squares by subtraction: (3) − (5).

The appropriate numbers of degrees of freedom are shown in table 16.9.1. In this table v denotes the number of regression coefficients and p denotes the number of distinct x-values. Lack-of-fit is tested by calculating

$$F_{p-v,\, n-p} = \frac{\text{lack-of-fit mean square}}{\text{pure error mean square}}. \tag{16.9.1}$$

Example 16.9.1. Fit a quadratic curve to the data in table 15.11.2, and test for lack-of-fit.

The forms of the matrix S and the vector $X'y$ are exhibited in (16.1.3). Summing the various powers of the $\{x_i\}$ and $\{y_i\}$ we obtain the normal equations:

$$
\begin{aligned}
24b_0 + && 805b_1 + && 34\,225b_2 = && 342, \\
805b_0 + && 34\,225b_1 + && 1\,651\,375b_2 = && 13\,000, \\
34\,225b_0 + && 1\,651\,375b_1 + && 85\,793\,125b_2 = && 574\,550,
\end{aligned}
$$

whence $b_0 = -2.262\,958\,508$,
$b_1 = 0.931\,520\,296$,
$b_2 = -0.010\,330\,543$,

and the fitted values at the five distinct x-values are 6.019, 12.235, 17.685, 18.469 and 16.438 respectively. The fit appears to be good.

We now calculate the various sums of squares:

$$
\begin{aligned}
\text{regression s.s.} &= b'X'y - n\bar{y}^2 \quad \text{(table 16.4.1)} \\
&= 342b_0 + 13\,000b_1 + 574\,550b_2 - 4873.5 \\
&= 526.92, \\
\text{total s.s.} &= 578.50 \quad \text{(example 15.11.1),}
\end{aligned}
$$

Table 16.9.2. *Analysis of variance for example 16.9.1*

Source	s.s.	d.f.	m.s.
Regression	526.92	2	263.5
Residual $\begin{cases}$ Lack-of-fit Pure error $\end{cases}$	$51.58 \begin{cases} 15.18 \\ 36.40 \end{cases}$	$21 \begin{cases} 2 \\ 19 \end{cases}$	$2.5 \begin{cases} 7.6 \\ 1.9 \end{cases}$
Total	578.50	23	—

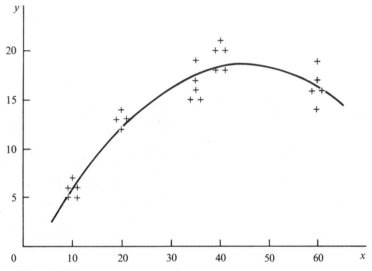

Fig. 16.9.1. A quadratic least-squares line.

$$\text{residual s.s.} = 578.50 - 526.92$$
$$= 51.58,$$
$$\text{regression s.s. plus lack-of-fit s.s.}$$
$$= 542.10 \quad (\text{example } 15.11.1)$$
$$\text{lack-of-fit s.s.} = 542.10 - 526.92$$
$$= 15.18,$$
$$\text{pure error s.s.} = 51.58 - 15.18,$$
$$= 36.40.$$

The analysis of variance is table 16.9.2. To test for lack-of-fit, we compute

$$F_{2, 19} = 7.6/1.9 = 4.0.$$

This value is significant at the 5 per cent level but not at the $2\frac{1}{2}$ per cent level. Some lack-of-fit is still evident in fig. 16.9.1.

Further reading: Draper and Smith [19] 26–32, 86–100.

$$F_{2,21} = \frac{263 \cdot 5}{2 \cdot 5} = 105 \cdot 40$$
ie Regression highly significant

16.10 Unequal variances – weighted least squares

This problem was discussed in section 15.12 in connection with simple linear regression, and the reader should refer to that section. The methods are also applicable to curvilinear models. Thus, if we only need to determine the coefficients $\{b_i\}$, and the variance of the observation at x_i is σ^2/w_i, we can pretend that the observation y_i is observed Nw_i times, where N is a suitable integer, and standard least-squares methods will give us the correct values of the $\{b_i\}$ (example 15.12.1).

This method *cannot* be used for statistical purposes; we must use the curvilinear generalisation of the second method in section 15.12. In the case of n observations, the usual X matrix has n rows. We define a matrix X_w which is obtained from X by multiplying all the elements in the first row by $\sqrt{w_1}$, all the elements in the second row by $\sqrt{w_2}$, ..., and all the elements in the nth row by $\sqrt{w_n}$. Matrix S_w, vector y_w and mean $\bar{y}_w$ are defined as in section 15.12. Then, writing X_w instead of X, y_w instead of y, S_w instead of S, replacing the elements of S^{-1} by the corresponding elements of S_w^{-1} and $\bar{y}$ by $\bar{y}_w$, the following standard regression formulae can be used: (15.3.1), (15.3.3), (15.3.4), (15.9.1), (15.10.2), table 16.4.2, (16.4.1), (16.4.2), (16.5.1), (16.5.2), (16.6.1), (16.6.2), table 16.7.1, (16.7.1), (16.7.2), (16.8.2). Table 16.4.1, (15.9.8), (15.9.9) and (15.9.10) can be used if we replace n in the sums of squares by Σw_i. This approach was demonstrated in example 15.12.2.

Further reading: Draper and Smith [19] 77–80.

16.11 Linear regression through the origin or some other fixed point

Sometimes we know *a priori* that a straight line in a regression should pass through the origin; that is, our statistical model takes the form

■ $\quad y_i = \beta_1 x_i + e_i,$ i.e. $\beta_0 = 0$ (16.11.1)

where the $\{x_i\}$ are assumed free of error. We fit a straight line of the form

■ $\quad Y_i = b_1 x_i,$ $b_0 = 0$ (16.11.2)

and to do this, we make use of a matrix X with only one column which lists the n x-values. The main regression vectors and matrices are all trivial (they are all scalars):

$$\beta = \beta_1, \qquad S = \Sigma x_i^2, \qquad X'y = \Sigma x_i y_i,$$

$$b = b_1, \qquad S^{-1} = 1/(\Sigma x_i^2), \qquad y'y = \Sigma y_i^2.$$

The matrix results of sections 16.2–16.10 are easy to apply. From section 16.4, for example, we know that b is obtained by solving the normal equation $Sb = X'y$, and we deduce that

■ $\quad b_1 = (\Sigma x_i y_i)/(\Sigma x_i^2).$ (16.11.3)

Equation (16.5.1) tells us that

■ $\mathrm{var}\, b_1 = \sigma^2/(\Sigma x_i^2).$ (16.11.4)

95 per cent confidence limits for β_1 are obtained using (16.6.1) with $\nu = 1$.

The reader should note that the appropriate analysis of variance is given by the uncorrected table 16.4.2 with $\nu = 1$. The coefficient of determination is obtained by dividing the regression sum of squares by the total sum of squares; the value obtained will not in general be equal to the square of the ordinary correlation coefficient of the observed and fitted values (as defined by (8.8.10)).

Sometimes we know *a priori* that a straight line in a regression should pass through a fixed point $(\bar{x}, \bar{y})$. The simplest approach is to subtract $\bar{x}$ from each x-value and $\bar{y}$ from each y-value and proceed as above. The fitted line takes the form

■ $(Y_i - \bar{y}) = b_i(x_i - \bar{x}).$ (16.11.5)

Example 16.11.1. Fit a straight line of the form (16.11.2) close to the twelve points in table 16.11.1 and test for lack-of-fit. We begin by calculating

$$\Sigma x^2 = 255, \quad \Sigma xy = 266, \quad \Sigma y^2 = 295.$$

It follows from (16.11.3) that $b_1 = 266/255 = 1.043\,137$.

The basic analysis of variance takes the uncorrected form in table 16.4.2. Thus,

$$
\begin{aligned}
\text{regression s.s.} \; &= \mathbf{b'X'y} \\
&= 1.043\,137 \times 266 \\
&= 277.475, \\
\text{total s.s.} &= \mathbf{y'y} \\
&= 295.000, \\
\text{residual s.s.} &= 295.000 - 277.475 \\
&= 17.525.
\end{aligned}
$$

According to footnote 7 and section 16.9, the sum of the regression and lack-of-fit sums of squares is

$$2^2/3 + 4^2/3 + 14^2/3 + 25^2/3 = 280.333.$$

$\dfrac{T_1^2}{n_1} + \dfrac{T_2^2}{n_2} + \dfrac{T_3^3}{n_3} + \dfrac{T^4}{n^4}$

We deduce that

$$
\begin{aligned}
\text{lack-of-fit s.s.} &= 280.333 - 277.475 \\
&= 2.858, \\
\text{pure error s.s.} &= 17.525 - 2.858 \\
&= 14.667.
\end{aligned}
$$

The analysis of variance is shown as table 16.11.2.

Table 16.11.1. *Data for regression example*
16.11.1. The twenty-four observations
are mutually independent

x_i	y_{ij}			n_i	$T_{i.}$
1	0	1	1	3	2
2	1	1	2	3	4
4	3	5	6	3	14
8	6	9	10	3	25
Total				12	45

Table 16.11.2. *Analysis of variance for example 16.11.1*

Source		s.s.		d.f.		m.s.	
Regression		277.475		1		277.475	
Residual $\begin{cases} \text{Lack-of-fit} \\ \text{Pure error} \end{cases}$		17.525 $\begin{cases} 2.858 \\ 14.667 \end{cases}$		11 $\begin{cases} 3 \\ 8 \end{cases}$		1.593 $\begin{cases} 0.953 \\ 1.833 \end{cases}$	
Total		295.000		12		—	

To test for lack-of-fit, we calculate

$$F_{3,8} = 0.953/1.833 = 0.520.$$

This value is not significant at the 5 per cent level; so we have no evidence of lack-of-fit. Regression is highly significant because

$$F_{1,11} = 277.475/1.593 = 174.2.$$

To obtain 95 per cent confidence limits for β_1, we solve the equations

$$(1.043\,137 - \beta_1)/\{1.593 \times (1/255)\}^{\frac{1}{2}} = \pm 2.201,$$

and obtain $\beta_1 = 1.043 \pm 0.174$.

Further reading: Brownlee [7] 358–61; Wilks [102] 273; Zar [105] 214–15.

16.12 Orthogonal polynomial least squares

Let us imagine that we wish to fit a polynomial curve to a set of data (x_i, y_i), where $i = 1, 2, ..., n$. We may begin by attempting to fit a straight line using the methods of chapter 15. If the resulting line is unsatisfactory, we may then attempt to fit a quadratic using the method of example 16.1.1; if this curve is still unsatisfactory, we may try a cubic, and so on. Each time, we need to recalculate *all* the regression coefficients, and the amount of work involved is considerable.

When the $\{x_i\}$ are equally spaced it is not necessary to re-calculate the regression coefficients and substantial savings in time and effort can be made if *orthogonal (Chebychev) polynomials* are used. It is convenient to standardise the x-values so that they are spaced symmetrically about the origin at unit intervals. When $n = 3$, for example, the x-values are assumed to be $-1, 0$ and 1; when $n = 4$, they are assumed to be $-\frac{3}{2}, -\frac{1}{2}, \frac{1}{2}$ and $\frac{3}{2}$; when $n = 5$, they are assumed to be $-2, -1, 0, 1$ and 2. We shall assume from here on that the x-values have been standardised in this manner.

Let us denote the rth-degree orthogonal polynomial by $\phi_r(x)$. These polynomials have the important property that for $r \neq s$ and standardised $\{x_i\}$,

$$\sum_{i=1}^{n} \phi_r(x_i) \phi_s(x_i) = 0. \tag{16.12.1}$$

This property uniquely determines all the orthogonal polynomials, apart from a constant factor. In fact, it may be shown[8] that

$$\phi_0(x) = 1, \tag{16.12.2}$$

$$\phi_1(x) = \lambda_1 x, \tag{16.12.3}$$

$$\phi_2(x) = \lambda_2 \{x^2 - \tfrac{1}{12}(n^2 - 1)\}, \tag{16.12.4}$$

$$\phi_3(x) = \lambda_3 \{x^3 - \tfrac{1}{20}(3n^2 - 7)x\}. \tag{16.12.5}$$

Higher-order polynomials are given, for example, in the *Biometrika Tables* [76] 212. The constants $\{\lambda_r\}$ are usually chosen so that the numerical values of the orthogonal polynomials at the standardised points are integers. Values (at the standardised points) of the orthogonal polynomials of degree four and less for the case $n = 13$ are given in table 16.12.1. These entries come from the *Biometrika Tables* [76] 212–21, and the reader should verify that (16.12.1) holds.

Any polynomial of degree k can be expressed as a linear combination of the orthogonal polynomials of degree k and lower; thus, to fit a polynomial of degree k, we may fit a curve of the form

$$Y = b_0\phi_0(x) + b_1\phi_1(x) + ... + b_k \phi_k(x). \tag{16.12.6}$$

Simplifications now become evident. The least-squares matrix $\mathbf{X}$ has as its jth column $(j = 0, 1, ..., k)$ the values at the standardised points of the jth-order orthogonal polynomial, and the elements of matrix $\mathbf{S} = \mathbf{X}'\mathbf{X}$ are given by (16.2.3) with ϕ replacing g. Equation (16.12.1) tells us that S_{rs} and S_{sr} are zero when $r \neq s$; so $\mathbf{S}$ is a diagonal matrix. The inverse $\mathbf{S}^{-1}$ is obtained by

---

[8] Exercise 7 of section 16.14.

Table 16.12.1. *Numerical values (at the standardised points) of the orthogonal polynomials of degree four and less for the case* n = 13*

	$\phi_0(x)$	$\phi_1(x)$	$\phi_2(x)$	$\phi_3(x)$	$\phi_4(x)$
	1	−6	22	−11	99
	1	−5	11	0	−66
	1	−4	2	6	−96
	1	−3	−5	8	−54
	1	−2	−10	7	11
	1	−1	−13	4	64
	1	0	−14	0	84
	1	1	−13	−4	64
	1	2	−10	−7	11
	1	3	−5	−8	−54
	1	4	2	−6	−96
	1	5	11	0	−66
	1	6	22	11	99
λ_r	1	1	1	$\frac{1}{6}$	$\frac{7}{12}$
$\sum\limits_{i=1}^{n} \{\phi_r(x_i)\}^2$	13	182	2002	572	68 068

replacing each diagonal element by its reciprocal (section 1.8). In the case $n = 13$, the diagonal elements are those values listed across the final row of table 16.12.1. To obtain the jth element ($j = 0, 1, ..., k$) of the least squares vector $\mathbf{X'y}$, we multiply each y-value by the corresponding value of the jth-order orthogonal polynomial and sum the products.

The use of orthogonal polynomials in regression analysis was extensively developed by R. A. Fisher [23] in 1921, and Fisher later introduced the proportionality constants $\{\lambda_i\}$ which ensure integer values at the standardised points.

Example 16.12.1. Use the method of orthogonal polynomials to fit curves of degree zero, one, two and three close to the thirteen points in table 16.12.2.

If the x-values had been standardised, we would calculate the elements of $\mathbf{X'y}$ as follows:

$$\Sigma y_i \phi_0(x_i) = \Sigma y_i = 30.3 + 61.2 + 92.7 + ... + 442.9 \quad = 2983.8,$$
$$\Sigma y_i \phi_1(x_i) = 30.3 \times (-6) + 61.2 \times (-5) + ... + 442.9 \times 6 = 6255.5,$$

Table 16.12.2. *Data for orthogonal polynomial example 16.12.1. The fitted values $\{Y_i\}$ are based on the least-squares cubic*

x_i	y_i	Y_i	x_i	y_i	Y_i
10.1	30.3	30.30	10.8	259.7	259.69
10.2	61.2	61.20	10.9	295.0	295.00
10.3	92.7	92.72	11.0	331.0	330.98
10.4	124.9	124.86	11.1	367.6	367.61
10.5	157.6	157.61	11.2	404.9	404.91
10.6	191.0	191.00	11.3	442.9	442.89
10.7	225.0	225.02			

$$\Sigma y_i \, \phi_2(x_i) = 30.3 \times 22 + 61.2 \times 11 + ... + 442.9 \times 22 \quad = \quad 643.5,$$
$$\Sigma y_i \, \phi_3(x_i) = 30.3 \times (-11) + 61.2 \times 0 + ... + 442.9 \times 11 \quad = \quad 3.8.$$

The diagonal elements of $\mathbf{S}^{-1}$ are the reciprocals of the entries along the final row of table 16.12.1, namely

$$1/13, \quad 1/182, \quad 1/2002, \quad 1/572,$$

and the off-diagonal elements are zero. Since $\mathbf{b} = \mathbf{S}^{-1}\mathbf{X'y}$,

$$
\begin{aligned}
b_0 &= 2983.8/13 &= 229.523\,0769, \\
b_1 &= 6255.5/182 &= 34.370\,879\,12, \\
b_2 &= 643.5/2002 = &0.321\,428\,571, \\
b_3 &= 3.8/572 = &0.006\,643\,356.
\end{aligned}
$$

We now recall that the x-values are not standardised; the average of the smallest and largest is 10.7 and the interval between successive x-values is 0.1. Standardised x-values are therefore obtained by subtracting 10.7 from the table 16.12.2 x-values and dividing by 0.1. We use this fact, together with formulae (16.12.2) to (16.12.5) and the λ-values in table 16.12.1 to deduce the least-squares lines of degree zero, one, two and three.

The least-squares line of degree zero is

$$Y = 229.523\,0769,$$

and the least-squares straight line is

$$Y = 229.523\,0769 + 34.370\,879\,12\left(\frac{x - 10.7}{0.1}\right).$$

The least-squares quadratic is

$$Y = 229.523\,0769 + 34.370\,879\,12\left(\frac{x - 10.7}{0.1}\right)$$
$$+ 0.321\,428\,571\left\{\left(\frac{x - 10.7}{0.1}\right)^2 - 14\right\},$$

$$\uparrow$$
$$\tfrac{1}{12}(n^2 - 1)$$
$$\tfrac{1}{12}(13^2 - 1)$$

while the least-squares cubic is

$$Y = 229.523\,0769 + 34.370\,879\,12\left(\frac{x-10.7}{0.1}\right)$$
$$+ 0.321\,428\,571\left\{\left(\frac{x-10.7}{0.1}\right)^2 - 14\right\}$$
$$+ \tfrac{1}{6}(0.006\,643\,356)\left\{\left(\frac{x-10.7}{0.1}\right)^3 - 25\left(\frac{x-10.7}{0.1}\right)\right\}.$$

The calculation of the fitted values at the standardised points is straightforward if we use the entries in table 16.12.1. The fitted value at the fourth point according to the least squares cubic is, for example, *Read arrows to 4th line of table*

$$229.523\,0769 + 34.370\,879\,12 \times (-3) + 0.321\,428\,571 \times (-5)$$
$$+ 0.006\,643\,356 \times 8 = 124.86.$$

Further reading: Chakravarti *et al.* [9] 173; Draper and Smith [19] 69–70, 150–5; Kendall and Stuart [54] 359–62; Mather [66] 131–4; Pearson and Hartley [76] 91–5, 212–21.

16.13 Orthogonal polynomial regression – the statistical analysis

We have seen in section 16.12 that the matrix $\mathbf{S}$ is diagonal when orthogonal polynomials are used. The inverse matrix $\mathbf{S}^{-1}$ can be written down immediately, and this leads to considerable savings in time and effort in the fitting of least-squares polynomials. The savings carry over to the statistical analysis, where we assume a model of the form

∎ $$y_i = \beta_0\phi_0(x_i) + \dots + \beta_k\phi_k(x_i) + e_i. \tag{16.13.1}$$

The $\{x_i\}$ are assumed free of error, and the $\{e_i\}$ are independent normal random variables with zero means and equal variances σ^2.

All the usual matrix formulae can be used, and most of them simplify considerably. $\mathbf{S}^{-1}$ is a diagonal matrix, and it follows from (16.5.2) that the regression coefficients are uncorrelated. The regression coefficients are normally distributed, and we deduce that they are independent (section 9.6).

The analysis of variance is usually drawn up in the uncorrected form (section 16.4). Each additional polynomial term results in a better fit and a reduction in the residual sum of squares; and, because the polynomials are orthogonal, the additional reduction in sum of squares due to the introduction of a particular polynomial does not depend on the polynomials previously fitted. The regression sum of squares $\mathbf{b}'\mathbf{X}'\mathbf{y}$ can therefore be partitioned uniquely into independent reductions for the various polynomials. These happen to be the individual products in $\mathbf{b}'\mathbf{X}'\mathbf{y}$. The reduction due to the jth-order polynomial is, for example,

$$b_j\left(\sum_{i=1}^{n} y_i\phi_j(x_i)\right).$$

The regression analysis of variance is shown in table 16.13.1.

Table 16.13.1. *Orthogonal polynomial regression – analysis of variance*

Source (1)	s.s. (2)	d.f. (3)	m.s. (4) = (2)/(3)
Mean (b_0)	$b_0 \sum_{i=1}^{n} y_i$	1	
Linear term (b_1)	$b_1 \sum_{i=1}^{n} y_i \phi_1(x_i)$	1	Same
			as
Quadratic term (b_2)	$b_2 \sum_{i=1}^{n} y_i \phi_2(x_i)$	1	column
...	...	...	(2)
kth degree term (b_k)	$b_k \sum_{i=1}^{n} y_i \phi_k(x_i)$	1	
Residual	(By difference)	$n-k-1$	(By division)
Total	$\sum_{i=1}^{n} y_i^2$	n	—

The significance of the regression coefficient b_j can be tested using the F-statistic

$$F_{1,n-k-1} = \frac{b_j \text{ mean square}}{\text{residual mean square}}.$$ (16.13.2)

Significance tests such as this are used to decide the degree of the polynomial to be fitted, and which (if any) of the lower-degree orthogonal polynomials to omit. Sequential procedures are often used. It must be remembered, however, that, even if a coefficient β_j is zero or very small, the coefficient β_{j+k} ($k = 1, 2, ...$) of a higher-order polynomial may still be large. If this is so, the residual sum of squares calculated by fitting a polynomial of degree j will not provide a correct residual mean square for the F-test on b_j. Care is required. If we are prepared to assume that the fitted function is of degree no greater than k, the analysis of variance table 16.13.1 can be formed and the F-statistic (16.13.2) used to test each b_j ($j \leqslant k$) for significance. The non-significant terms are then omitted from the regression equation and their sums of squares are added into the residual sum of squares for subsequent analysis (for example, estimation). The number of degrees of freedom for the residual is increased appropriately.

Example 16.13.1. Use the method of orthogonal polynomials to fit a low-order polynomial close to the thirteen points in table 16.12.2.

Let us fit a polynomial of degree four or less. We already know from example 16.12.1 that

$$\Sigma y_i \phi_0(x_i) = 2983.8, \quad b_0 = 229.523\,0769;$$
$$\Sigma y_i \phi_1(x_i) = 6255.5, \quad b_1 = 34.370\,879\,12;$$
$$\Sigma y_i \phi_2(x_i) = 643.5, \quad b_2 = 0.321\,428\,571;$$
$$\Sigma y_i \phi_3(x_i) = 3.8, \quad b_3 = 0.006\,643\,356;$$

and it is easy to compute

$$\Sigma y_i \phi_4(x_i) = 0.2, \quad b_4 = 0.000\,002\,938;$$
$$\Sigma y_i^2 = 900\,064.86.$$

The reductions in the residual sum of squares due to the various polynomials are calculated as follows:

$$b_0: 229.523\,0769 \times 2983.8 = 684\,850.9568,$$
$$b_1: 34.370\,879\,12 \times 6255.5 = 215\,007.0343,$$
$$b_2: 0.321\,428\,571 \times 643.5 = 206.8393,$$
$$b_3: 0.006\,643\,356 \times 3.8 = 0.0252,$$
$$b_4: 0.000\,002\,938 \times 0.2 = 5.88 \times 10^{-7}.$$

The analysis of variance and F-values are shown in table 16.13.2. The coefficient b_4 is not significant, but all the others (including b_3) are highly significant. We therefore fit a function involving $\phi_0(x)$, $\phi_1(x)$, $\phi_2(x)$ and $\phi_3(x)$, namely the least-squares cubic of example 16.12.1. The fitted values are shown in table 16.12.2.

Confidence intervals for $\beta_0, \beta_1, \beta_2, \beta_3$ and σ^2 can be obtained using (16.6.1) and (16.6.2), and for this purpose we note that

$$S_{00}^{-1} = \tfrac{1}{13}, \qquad S_{22}^{-1} = \tfrac{1}{2002},$$
$$S_{11}^{-1} = \tfrac{1}{182}, \qquad S_{33}^{-1} = \tfrac{1}{572}.$$

The quartic coefficient b_4 is not significant and the residual mean square s^2 should be obtained by adding the quartic reduction in sum of squares to the residual sum of squares and dividing by 9 rather than 8 degrees of freedom. We then use (16.6.1) and (16.6.2) with $n = 13$, $v = 4$.

Further reading: Draper and Smith [19] 150–5; Mather [66] 131–4; Pearson and Hartley [76] 91–5, 212–21.

16.14   Exercises

1. Use the method of section 15.2 and the four points (x_1, y_1), (x_2, y_2), (x_3, y_3), (x_4, y_4) to derive the normal equations for fitting the quadratic least-squares line (16.1.1). Confirm formulae (16.1.3).

Table 16.13.2. *Orthogonal polynomial example 16.13.1 – analysis of variance*

Source	s.s.	d.f.	m.s.	F	
Mean (b_0)	684 850.9568	1	6.85×10^5	1.25×10^9	
Linear term (b_1)	215 007.0343	1	2.15×10^5	3.91×10^8	
Quadratic term (b_2)	206.8393	1	2.07×10^2	3.76×10^5	
Cubic term (b_3)	0.0252	1	2.52×10^{-2}	4.58×10^1	
Quartic term (b_4)	0.0000*	1	5.88×10^{-7}	1.07×10^{-3}	
Residual	0.0044	8	5.50×10^{-4}	—	
Total	900 064.8600	13	—	—	

Handwritten annotations: $n = 13$; $K = 0 \to 4$; regression m.s. / residual m.s. $F_{1,n-k-1}$; Sig; $\leftarrow$ Fit this curve, lowest significant F value; NS.

* 5.88×10^{-7}.

2. Use the method of section 15.2 and the four points (x_1, y_1), (x_2, y_2), (x_3, y_3), (x_4, y_4) to derive the normal equations for fitting the exponential least-squares line (16.1.4). Confirm formulae (16.1.6).

3. Which of the following models are linear in the least-squares sense (section 16.2)?
   (a) $y_i = b_0 + b_1 \sqrt{x_i} + e_i$;
   (b) $y_i = b_1 (1 + x_i)^{\frac{3}{2}} + e_i$;
   (c) $y_i = b_1 \sin(b_2 x_i)$;
   (d) $y_i = b_0 + b_1/x_i + b_2/x_i^2 + e_i$.

4. Rearrange the following statistical models so that the standard linear least-squares approach can be used:
   (a) $y_i = a_1 (x_i - a_0) + e_i$;
   (b) $y_i = k \sin(\pi x_i + \alpha) + e_i$.

5. Use the method of least squares and matrix notation to fit a third-degree polynomial through the four points in example 16.1.1.

6. Fit a curve of the form $Y = b_0 + b_1 x^3$ to the data in table 16.12.2. Test the regression coefficient b_1 for significance.

7. The zeroth-order orthogonal polynomial on the five standardised points -2, $-1, 0, 1, 2$ is $\phi_0(x) = 1$. If we assume that $\phi_1(x) = Ax + B$ and invoke the orthogonality condition (16.12.1), we have

$$\sum_{x=-2}^{2} \phi_0(x) \phi_1(x) = \sum_{x=-2}^{2} (Ax + B) = 5B = 0.$$

So B is zero and $\phi_1(x)$ is proportional to x. We may write $\phi_1(x) = \lambda_1 x$. Assume that $\phi_2(x) = ax^2 + bx + c$, and use the fact that $\phi_2(x)$ must be orthogonal to both $\phi_0(x)$ and $\phi_1(x)$ to deduce the form of $\phi_2(x)$ when $n = 5$. Use the same method to find $\phi_3(x)$.

17
Multiple linear regression

Summary   In this chapter, we show how the matrix methods of chapters 15 and 16 can be used to study the regression of a dependent variable y on two or more independent variables. We also discuss briefly the problem of choosing the 'best' regression equation.

17.1   Introduction and simple examples of multiple regression

In chapters 15 and 16 we described the methods of simple linear regression and curvilinear regression of a dependent variable y on a single independent variable x. The same techniques can be used to study the regression of a dependent variable y on two or more independent variables $(x, z, \text{etc.})$ when the regression equation is linear[1] in the coefficients $\{b_i\}$. The analysis is then referred to as *multiple linear regression*. Matrix methods are essential; all the formulae and methods have already been given in the earlier chapters, and references to them are listed in table 17.1.1.

Examples 17.1.1–17.1.5 show how the regression vectors and matrices $\mathbf{y}, \mathbf{b}, \mathbf{X}$ and $\mathbf{S}$ are obtained. They also demonstrate the following techniques: the corrected and uncorrected analysis of variance for testing the significance of regression (examples 17.1.1 and 17.1.3), confidence limits for the regression parameters $\{\beta_i\}$ and σ^2 (example 17.1.1), hypothesis-testing (example 17.1.3), confidence limits for the expected value of the dependent variable given the values of the independent variables (example 17.1.2), testing for lack-of-fit (example 17.1.5), and weighted least squares (example 17.1.4).

> **Example 17.1.1.** Use the method of least squares to fit a curve of the form
>
> $$Y = b_0 + b_1 x + b_2 \mathrm{e}^z \tag{17.1.1}$$

close to the four (three-dimensional) points

$$(x_1, z_1, y_1) = (3, 1, 8.2),$$
$$(x_2, z_2, y_2) = (20, 3, 60.3),$$
$$(x_3, z_3, y_3) = (1, 0, 3.1),$$
$$(x_4, z_4, y_4) = (55, 4, 164.3).$$

[1] Section 16.2.

300

Table 17.1.1. *References for multiple linear regression techniques*

Technique	Reference
(1) Least squares –	
Function minimised	(15.3.1)
Formula for fitted values	(15.3.2)
Normal equations	(15.3.3)
(2) Statistical model	(15.9.1)
(3) Testing the significance of regression	section 16.4
Corrected analysis of variance	table 16.4.1, (16.4.1)
Uncorrected analysis of variance	table 16.4.2, (16.4.2)
(4) Point estimators of $\{\beta_i\}$ and σ^2	section 16.5
Covariance matrix of the $\{b_i\}$	(16.5.1), (16.5.2)
(5) Confidence intervals for $\{\beta_i\}$ and σ^2	section 16.6
(6) Hypothesis-testing	section 16.7, example 17.1.3
(7) Expected value of y given the values of the independent variables –	
Point estimator y_0	(15.10.2), example 17.1.2
Confidence limits for $\hat{y}$	(16.8.2), example 17.1.2
(8) Lack-of-fit	section 16.9, example 17.1.5
(9) Weighted least squares	section 16.10, example 17.1.4

The independent variables are x and z, and y is the dependent variable. Test the significance of the regression equation, and obtain confidence limits for the underlying parameters β_0 and σ^2.

A straight line involves the zeroth and first power of x, and in fitting a straight line by the method of least squares (section 15.3) we used a matrix X with two columns, one comprising the zeroth powers of the $\{x_i\}$ and the other the first powers of the $\{x_i\}$. When we fitted the exponential curve (16.1.4) involving a constant and a multiple of e^x, we used a matrix X with two columns, one composed entirely of ones and the other comprising the values $\{\exp x_i\}$. To fit a curve of the form (17.1.1) we use a matrix X with three columns. The first, second and third columns contain respectively ones, the values $\{x_i\}$ and the values $\{\exp z_i\}$. As usual, the $\{y_i\}$ are listed in a column vector y and the regression coefficients are listed in a three-dimensional column vector b. Thus,

$$X = \begin{pmatrix} 1 & 3 & 2.718\,281\,828 \\ 1 & 20 & 20.085\,536\,92 \\ 1 & 1 & 1.000\,000\,000 \\ 1 & 55 & 54.598\,150\,03 \end{pmatrix} \quad \text{and} \quad y = \begin{pmatrix} 8.2 \\ 60.3 \\ 3.1 \\ 164.3 \end{pmatrix},$$

so that

$$S = X'X = \begin{pmatrix} 4 & 79 & 78.401\,968\,78 \\ 79 & 3435 & 3413.763\,836 \\ 78.401\,968\,78 & 3413.763\,836 & 3392.775\,836 \end{pmatrix}$$

and $X'y = \begin{pmatrix} 235.9 \\ 10\,270.2 \\ 10\,207.023\,84 \end{pmatrix}$.

The normal equations are given by (15.3.3), namely:

$$
\begin{aligned}
4b_0 && +79b_1 && +78.401\,968\,78b_2 &= 235.9, \\
79b_0 && +3435b_1 && +3413.763\,836b_2 &= 10\,270.2, \\
78.401\,968\,78b_0 &+ 3413.763\,836b_1 &+ 3392.775\,836b_2 &= 10\,207.023\,84.
\end{aligned}
$$

We find that $b_0 = -0.001\,022\,462$, $b_1 = 0.293\,743\,778$ and $b_2 = 2.712\,920\,796$, so that the fitted values are 8.255, 60.364, 3.006 and 164.275.

The statistical model is given by (15.9.1) or

$$y_i = \beta_0 + \beta_1 x_i + \beta_2 \exp z_i + e_i, \tag{17.1.2}$$

where the $\{x_i\}$ and $\{z_i\}$ are assumed free of error. The $\{e_i\}$ are independent normal random variables with zero expectations and common variance σ^2. To test the significance of the regression coefficients b_1 and b_2 we use the analysis of variance table 16.4.1 with $\nu = 3$. Thus,

$$
\begin{aligned}
\text{total s.s.} &= y'y - n\bar{y}^2 \\
&= \Sigma y_i^2 - 4(235.9/4)^2 \\
&= 30\,707.43 - 13\,912.2025 \\
&= 16\,795.2275;
\end{aligned}
$$

$$
\begin{aligned}
\text{regression s.s.} &= b'X'y - n\bar{y}^2 \\
&= (b_0 \quad b_1 \quad b_2) \begin{pmatrix} 235.9 \\ 10\,270.2 \\ 10\,207.023\,84 \end{pmatrix} - 13\,912.2025 \\
&= 16\,795.2109;
\end{aligned}
$$

and the analysis of variance[2] is given in table 17.1.2. The test statistic is

$$F_{2,1} = 8397.6054/0.0166 = 5.06 \times 10^5$$

which is highly significant: if (17.1.2) is the appropriate model, we have ample evidence that β_1 and β_2 are not both zero.

---

[2] The reader should confirm that, apart from the effects of rounding, the residual sum of squares is equal to the sum of the squared differences between the observed and fitted values.

Table 17.1.2. *Multiple regression example 17.1.1*
– analysis of variance

Source	s.s.	d.f.	m.s.
Regression	16 795.2109	2	8397.6054
Residual	0.0166	1	0.0166
Total	16 795.2275	3	—

Confidence limits for β_0 can be obtained using (16.6.1) with $n = 4$ and $v = 3$. We note that

$$S^{-1} = \begin{pmatrix} 0.480\,70 & -0.459\,17 & 0.450\,90 \\ -0.459\,17 & 8.889\,50 & -8.933\,88 \\ 0.450\,90 & -8.933\,88 & 8.979\,02 \end{pmatrix}$$

and the residual mean square $s^2 = 0.0166$. When we solve the equations

$$(-0.001\,022\,462 - \beta_0)/(0.0166 \times 0.480\,70)^{\frac{1}{2}} = \pm 12.706,$$

we obtain the 95 per cent limits $-0.001\,02 \pm 1.135$.

Confidence limits for σ^2 are obtained using (16.6.2). When we solve the equations

$$\frac{(4-3)(0.0166)}{\sigma^2} = 5.02 \quad \text{and} \quad \frac{(4-3)(0.0166)}{\sigma^2} = 0.000\,982,$$

we obtain the 95 per cent limits 0.003 31 and 16.9. The interval is wide because the sample size is so small.

Example 17.1.2. Use the method of least squares to fit a curve of the following form close to the four data points in example 17.1.1:

$$Y = b_0 + b_1(xe^z)^{\frac{1}{2}}. \tag{17.1.3}$$

Test the significance of the regression equation and obtain 95 per cent confidence limits for the expected value of y when $x = 10$ and $z = 2$.
In this case we use a matrix X with two columns, the first composed entirely of ones, and the other comprising the values $\{(x_i \exp z_i)^{\frac{1}{2}}\}$. The matrix $S = X'X$ is therefore

$$\begin{pmatrix} n & \Sigma(x_i \exp z_i)^{\frac{1}{2}} \\ \Sigma(x_i \exp z_i)^{\frac{1}{2}} & \Sigma x_i \exp z_i \end{pmatrix} = \begin{pmatrix} 4 & 78.697\,098\,47 \\ 78.697\,098\,47 & 3413.763\,833 \end{pmatrix},$$

and $\quad X'y = \begin{pmatrix} \Sigma y_i \\ \Sigma y_i (x_i \exp z_i)^{\frac{1}{2}} \end{pmatrix} = \begin{pmatrix} 235.9 \\ 10\,238.520\,17 \end{pmatrix}.$

The normal equations (15.3.3) are

$$4b_0 \qquad\qquad + 78.697\,098\,47 b_1 = \qquad 235.9,$$
$$78.697\,098\,47 b_0 + 3413.763\,833 b_1 = 10\,238.520\,17,$$

so that $b_0 = -0.058\,288\,883$ and $b_1 = 3.000\,531\,915$, and the fitted values are 8.510, 60.081, 2.942 and 164.367 respectively.

To test the significance of the regression coefficient b_1, we use the analysis of variance[3] table 16.4.1 with $n = 4$ and $\nu = 2$. The total sum of squares is already known to be 16\,795.2275 and the

$$\text{regression s.s.} = \mathbf{b'X'y} - n\bar{y}^2$$
$$= (b_0 \quad b_1) \begin{pmatrix} 235.9 \\ 10\,238.520\,17 \end{pmatrix} - 13\,912.2025$$
$$= 16\,795.0536.$$

We form the analysis of variance table 17.1.3 and compute the test statistic

$$F_{1,2} = \frac{16\,795.0536}{0.0870} = 1.93 \times 10^5$$

which is highly significant. If the appropriate model is

$$y_i = \beta_0 + \beta_1 (x_i \exp z_i)^{\frac{1}{2}} + e_i, \tag{17.1.4}$$

we have ample evidence that β_1 is non-zero.

To compute the point estimate of the expected value of y at the point $x = 10$, $z = 2$, we form the row vector

$$\mathbf{x_0} = (1 \quad \sqrt{(10e^2)}) = (1 \quad 8.595\,9619).$$

The reader should note that this vector is formed in the same manner as a row of $\mathbf{X}$. The point estimate of the expected value of y when $x = 10$ and $z = 2$ is given by (15.10.2):

$$y_0 = \mathbf{x_0 b} = (1 \quad 8.595\,9619) \begin{pmatrix} -0.058\,288\,883 \\ 3.000\,531\,915 \end{pmatrix} = 25.7342.$$

To obtain 95 per cent confidence limits for the mean value of y when $x = 10$ and $z = 2$, we first note that

$$\mathbf{S}^{-1} = \begin{pmatrix} 0.457\,497\,354 & -0.010\,546\,633 \\ -0.010\,546\,633 & 0.000\,536\,062 \end{pmatrix},$$

and compute $\mathbf{x_0 S^{-1} x_0'} = 0.3158$. The residual mean square in the analysis of variance table 17.1.3 is 0.0870. The regression equation involves two b-values; so we make use of (16.8.2) with $k + 1 = 2$, and solve the equations

$$(\hat{y} - 25.7342)/(0.0870 \times 0.3158)^{\frac{1}{2}} = \pm 4.303.$$

The limits are 25.73 ± 0.71.

[3] See footnote 2.

Table 17.1.3. *Multiple regression example 17.1.2 – analysis of variance*

Source	s.s.	d.f.	m.s.
Regression	16 795.0536	1	16 795.0536
Residual	0.1739	2	0.0870
Total	16 795.2275	3	—

Table 17.1.4. *Multiple regression example 17.1.3 – analysis of variance*

Source	s.s.	d.f.	m.s.
Regression	30 707.2488	1	30 707.2488
Residual	0.1812	3	0.0604
Total	30 707.4300	4	—

Example 17.1.3. Use the method of least squares to fit a curve of the following form close to the four data points in example 17.1.1:

$$Y = b_1(x \exp z)^{\frac{1}{2}} \qquad (17.1.5)$$

Test the null hypothesis that the underlying regression coefficient β_1 is 2.7183.

In this case matrix X has only one column comprising the values $\{(x_i \exp z_i)^{\frac{1}{2}}\}$, and the single element in matrix S is ˙

$$S = \Sigma x_i \exp z_i = 3413.763\,833.$$

The single element in $X'y$ is $\Sigma y_i (x_i \exp z_i)^{\frac{1}{2}} = 10\,238.520\,17$, and the normal equation (15.3.3) is

$$3413.763\,833 b_1 = 10\,238.520\,17,$$

with solution $b_1 = 2.999\,188\,189$. The fitted values are 8.565, 60.112, 2.999 and 164.352 respectively. Equation (17.1.5) contains no overall mean term b_0; so the analysis of variance table 17.1.4 is drawn up using the uncorrected formulae in table 16.4.2.

The single element of the inverse matrix S^{-1} is 0.000 292 93. To test the null hypothesis $\beta_1 = 2.7183$, we can use (16.6.1) and compute

$$t_3 = (2.999\,188\,189 - 2.7183)/(0.0604 \times 0.000\,292\,93)^{\frac{1}{2}} = 66.8.$$

This value is highly significant; so we reject the null hypothesis and conclude that $\beta_1 \neq 2.7183$.

11

Table 17.1.5. *Values of the dependent variable y corres- ponding to different values of the independent variables x and z (example 17.1.4)*

	z		
x	1	8	32
1	23	52	112
2	31	63	102
5	28	73	125
10	50	67	144
50	112	167	197
100	213	261	308

Example 17.1.4. Use the method of weighted least squares to fit an equation of the form

$$Y = b_1 x + b_2 \sqrt{z} \qquad (17.1.6)$$

close to the eighteen points in table 17.1.5. It is known that the variance of the dependent variable y at a given point is approximately proportional to its expectation. Test the null hypothesis $\beta_1 = 2$.

The variance at the point (x_i, z_i, y_i) is approximately proportional to the expected value of y_i, and an approximation to the expected value of y_i is provided by y_i itself. Let us therefore use $1/y_i$ as the weight w_i at the point (x_i, z_i, y_i). The regression vectors and matrices take the following forms (section 16.10):

$$\mathbf{X}_w = \begin{pmatrix} x_1/\sqrt{y_1} & \sqrt{z_1}/\sqrt{y_1} \\ x_2/\sqrt{y_2} & \sqrt{z_2}/\sqrt{y_2} \\ \cdot & \cdot \\ x_{18}/\sqrt{y_{18}} & \sqrt{z_{18}}/\sqrt{y_{18}} \end{pmatrix}; \qquad \mathbf{y}_w = \begin{pmatrix} y_1/\sqrt{y_1} \\ y_2/\sqrt{y_2} \\ \cdot \\ y_{18}/\sqrt{y_{18}} \end{pmatrix};$$

$$\mathbf{S}_w = \mathbf{X}'_w \mathbf{X}_w = \begin{pmatrix} \Sigma x_i^2/y_i & \Sigma (x_i/y_i)\sqrt{z_i} \\ \Sigma (x_i/y_i)\sqrt{z_i} & \Sigma z_i/y_i \end{pmatrix}$$

$$= \begin{pmatrix} 173.637\,59 & 8.145\,9947 \\ 8.145\,9947 & 2.077\,4465 \end{pmatrix};$$

$$\mathbf{X}'_w \mathbf{y}_w = \begin{pmatrix} \Sigma x_i \\ \Sigma \sqrt{z_i} \end{pmatrix} = \begin{pmatrix} 504 \\ 56.911\,688\,25 \end{pmatrix};$$

$$\mathbf{S}_w^{-1} = \begin{pmatrix} 0.007\,057\,37 & -0.027\,673\,05 \\ -0.027\,673\,05 & 0.589\,870\,57 \end{pmatrix}.$$

The normal equations (15.3.3) have solution $b_1 = 1.981\,9937$, $b_2 = 19.623\,311$.

Table 17.1.6. *Multiple regression example 17.1.4 – analysis of variance*

Source	s.s.	d.f.	m.s.
Regression	2115.72	2	1057.860
Residual	12.28	16	0.768
Total	2128.00	18	—

Table 17.1.7. *Data for lack-of-fit example 17.1.5*

Independent variables x_i	z_i	Observations on the dependent variable $y_{i,j}$				Number of observations at (x_i, z_i) n_i	Total of observations at (x_i, z_i) $T_i.$
1	2	2	5	3	2	4	12
3	4	15	14	12	15	4	56
5	4	34	30	36		3	100
9	6	70	66	64	67	4	267
Totals						15	435

Equation (17.1.6) does not contain an overall mean term b_0. To form the regression analysis of variance, we therefore use the uncorrected formulae in table 16.4.2:

$$\text{total s.s.} = y'_w y_w$$
$$= y_1 + y_2 + \dots + y_{18}$$
$$= 2128;$$
$$\text{regression s.s.} = b'(X'_w y_w)$$
$$= 1.981\,9937 \times 504 + 19.623\,311 \times 56.911\,688\,25$$
$$= 2115.72.$$

The analysis of variance is given in table 17.1.6. To test the null hypothesis $\beta_1 = 2$, we can use (16.6.1) with $n = 18$, $\nu = 2$ and compute

$$t_{16} = (1.981\,9937 - 2)/(0.007\,057\,37 \times 0.768)^{\frac{1}{2}} = -0.245.$$

This value is not significant at the 5 per cent level; so we accept the null hypothesis.

Example 17.1.5. Fit an equation of the following form to the data in table 17.1.7.

$$Y = b_1 x + b_2 z. \qquad (17.1.7)$$

Test for lack-of-fit.

Table 17.1.8. *Regression example 17.1.5 – analysis of variance*

Source	s.s.		d.f.		m.s.	
Regression	21 969.219		2		10 985	
Residual {Lack-of-fit / Pure error	55.781	{ 6.364 / 49.417	13	{ 2 / 11	4.291	{ 3.182 / 4.492
Total	22 025.000		15			

The normal equations are easily shown to be

$$\begin{pmatrix} 439 & 332 \\ 332 & 272 \end{pmatrix} \begin{pmatrix} b_1 \\ b_2 \end{pmatrix} = \begin{pmatrix} 3083 \\ 2250 \end{pmatrix},$$

and we find that $b_1 = 9.971\,2544$, $b_2 = -3.898\,7369$. Equation (17.1.7) has no overall mean term b_0; so we perform an uncorrected lack-of-fit analysis of variance (table 16.4.2 and footnote 7 of section 16.9). The calculations are as follows:

total s.s.
$= 2^2 + 5^2 + 3^2 + \ldots + (67)^2$
$= 22\,025;$
regression s.s.
$= 9.971\,2544 \times 3083 - 3.898\,7369 \times 2250$
$= 21\,969.219;$
residual s.s.
$= 22\,025 - 21\,969.219$
$= 55.781;$
regression s.s. plus lack-of-fit s.s.
$= (12)^2/4 + (56)^2/4 + (100)^2/3 + (267)^2/4$
$= 21\,975.583;$
lack-of-fit s.s.
$= 21\,975.583 - 21\,969.219$
$= 6.364;$
pure error s.s.
$= 55.781 - 6.364$
$= 49.417.$

The analysis of variance is table 17.1.8. To test for lack-of-fit, we compute

$$F_{2,11} = 3.182/4.492 = 0.71.$$

This value is not significant: we have no evidence of lack-of-fit.

Further reading: Brownlee [7] 419–65; Draper and Smith [19] 104–27; Hoel [45] 207–8; Hoel [46] 174–6; Johnson and Leone [50] 413–22; Mather [66] 135; Zar [105] 252–80.

17.2 The coefficient of determination

Most computer regression packages print the *coefficient of determination.* This coefficient is merely the proportion of the total sum of squares explained by regression, or the ratio of the regression sum of squares to the total sum of squares in the analysis of variance. It must lie between zero and one. When the corrected analysis of variance table 16.4.1 is used, the coefficient of determination happens to be the square of the correlation between the observed and fitted values, and it is called the *coefficient of multiple correlation.* It is usually denoted by R^2.

The coefficient of determination is sometimes used as a criterion for selecting the 'best' regression equation (section 17.3).

> **Example 17.2.1.** The coefficient of determination in example 17.1.4 is 2115.72/2128.00 or 0.9942.
>
> *Further reading*: Draper and Smith [19] 26; Zar [105] 260.

17.3 Which model shall we use?

Three different regression equations (17.1.1), (17.1.3) and (17.1.5) have been fitted to the data of example 17.1.1, and regression is highly significant in each case. Which one of the underlying regression models shall we use?

The following points need to be taken into consideration when we make our choice:

1. there may be good theoretical reasons which lead us to believe that the regression equation has a particular mathematical form;
2. the regression equation should provide fitted values which are as close as possible to the data (the equation can then be used with confidence for predictive purposes);
3. the regression equation should be as simple as possible.

Requirements 2 and 3 usually conflict and a compromise needs to be reached. The final choice of '*best*' *regression equation* tends therefore to be subjective.

Equation (17.1.5) is a special case of (17.1.3); so we really only have to choose between two basic forms: (17.1.1) and (17.1.3). We are given no information of a theoretical nature to indicate which mathematical form is preferable, and there is little to choose between the two equations on the grounds of simplicity. Equation (17.1.1) provides a marginally better fit (a smaller residual sum of squares), but that is all. Let us choose it.

Equation (17.1.1) involves three constants b_0, b_1 and b_2. It may be that one (or more) of the underlying parameters β_0, β_1 and β_2 is very small or zero. If we omit that term from the model, the residual sum of squares will increase slightly, but we will have a simpler model. The regression and residual sums of squares for all the reduced equations of (17.1.1) are shown in table 17.3.1. Uncorrected[4] sums of squares have been used because we wish to investigate the possibility of removing b_0. We see that the two-constant equation (4)

[4] Table 16.4.2.

Table 17.3.1. *Alternative regression equations for the data in example 17.1.1*

Regression equation	Uncorrected regression sum of squares	Residual sum of squares	Residual mean square
(1) $b_0 + b_1 x + b_2 e^z$	30 707.4134	0.0166	0.0166
(2) $b_0 + b_1 x$	30 706.5936	0.8364	0.4182
(3) $b_0 + b_2 e^z$	30 707.4037	0.0263	0.0132
(4) $b_1 x + b_2 e^z$	30 707.4133	0.0167	0.0084
(5) b_0	13 912.2025	16 795.2275	5598.4091
(6) $b_1 x$	30 706.5526	0.8774	0.2925
(7) $b_2 e^z$	30 707.4032	0.0268	0.0089

provides practically the same fit as the three-constant equation (1), and the one-constant equation (7) is almost as good.

A number of objective methods have been suggested for selecting the 'best' regression equation. They do not all necessarily lead to the same conclusion. The residual mean square estimates the variance σ^2, and one objective method is to choose the equation with the smallest residual mean square (equation (4) in table 17.3.1). Such an approach requires the computation of residual mean squares for all possible reduced equations ($2^\nu - 1$ regressions when the initial model involves ν parameters). The amount of computation involved is prohibitive even for moderate ν (1023 regressions when $\nu = 10$).

Various sequential procedures have been developed to select the 'best' regression equation without looking at all possible regressions. We shall confine our attention to the popular *backward elimination procedure*.

Imagine that a ν-parameter model has been fitted to some data. The F-statistic (16.7.1) (or the equivalent t-statistic (16.6.1)) can be used to test the null hypothesis $\beta_i = 0$ for each i. A simpler model involving $\nu - 1$ terms can then be obtained by removing the term with the smallest F-value, provided that F-value is not significant. The process is then repeated with the $(\nu - 1)$-parameter model to obtain a model with $\nu - 2$ terms, and so on. The procedure terminates as soon as there are no more non-significant terms to eliminate. If we start with ν terms and the procedure terminates leaving p terms, $\frac{1}{2}\{\nu(\nu + 1) - p(p - 1)\}$ regressions are required (52 when $\nu = 10$ and $p = 3$). Other short-cuts are also sometimes used.

When we apply the backward elimination procedure to the three-parameter model in table 17.3.1, we obtain equation (7) as our 'best' regression equation. The calculations are summarised in table 17.3.2.

The backward elimination procedure usually gives reasonable results, but it is not perfect. Draper and Smith [19] 163–77 describe some alternative

Table 17.3.2. *The backward elimination procedure*

Regression equation	Residual mean square	Term removed	Increase in residual sum of squares	F
$b_0 + b_1 x + b_2 e^z$	0.0166	$b_2 e^z$	0.8198	$F_{1,1} = 49.4$
		$b_1 x$	0.0097	$F_{1,1} = 0.6$
		b_0	0.0001	$F_{1,1} = 0.006$
$b_1 x + b_2 e^z$	0.0084	$b_2 e^z$	0.8607	$F_{1,2} = 102.5$
		$b_1 x$	0.0101	$F_{1,2} = 1.2$

procedures and make comparisons between them (see also Hocking [44]). The backward elimination method appears to be the best of the simpler procedures. The reader should note the warning in section 12.2 about applying several tests to the one set of data.

Uncorrected[5] sums of squares have been used in this analysis because we want to investigate the possibility of omitting b_0. If it is intended that an overall mean should be automatically included in the regression equation, corrected[6] sums of squares should be used.

In our numerical example e^x and z are highly correlated. It is not surprising therefore to find that a reasonably good fit can be achieved using only one of them in the regression equation. The high correlation causes some problems with the normal equations for regressions (1) and (4) in table 17.3.1, because the **S** matrices are almost singular.[7]

Further reading: Draper and Smith [19] 163–95, 234–42; Hocking [44]; Zar [105] 263–6.

17.4   Exercises

1. Calculate the sum of the squared differences between the observed and fitted values in examples 17.1.1, 17.1.2, and confirm that the sums are in fact the residual sums of squares.

2. Find the matrices **X** and **S** and the vector **X′y** which will be used to fit the following functions by the method of least squares:
   (a) $Y = b_0 + b_1 x + b_2 x^2 + b_3 z$;
   (b) $Y = b_0 + b_1 x + b_2 z + b_3 w$;
   (c) $Y = b_0 + b_1 \sin \pi x + b_2 \cos \pi z$.

3. Fit a function of the form $Y = b_0 + b_1 x + b_2 z$ to the data in table 17.4.1 and form the analysis of variance table.

4. The navy provides a course in deep-sea diving, and the marks $\{y_i\}$ of ten candidates in the final examination are given in table 17.4.2 together with the

[5] Table 16.4.2.   [6] Table 16.4.1.   [7] Section 1.8.

Table 17.4.1. *Data for exercise 3 of section 17.4*

x_i	z_i	y_i	x_i	z_i	y_i
1	1	2.1	6	6	11.9
2	8	9.9	7	2	8.9
3	4	6.9	8	7	15.1
4	9	13.1	9	3	11.9
5	5	10.1	10	10	20.1

Table 17.4.2. *Some naval data*

Lung size in litres x_i	Packets of cigarettes per week z_i	Final mark y_i
5.0	4	483
4.5	10	351
4.6	6	390
4.8	4	451
4.7	2	439
5.5	0	556
5.4	4	510
5.2	2	509
4.6	3	440
5.0	3	480

lung sizes $\{x_i\}$ and smoking intensities $\{z_i\}$ of the candidates. Fit a regression line of the form $Y = b_0 + b_1 x + b_2 z$ to these data and form the analysis of variance table. Do we need both the independent variables x and z in the regression equation?

18

Non-linear regression

Summary   The matrix methods of chapters 15, 16 and 17 fail as soon as a non-linear regression is encountered. In this chapter we show that the principle of least squares is still applicable. Any statistical analysis must be regarded as approximate.

18.1   Introduction and examples of non-linear regression

The matrix methods of chapters 15, 16 and 17 are applicable whenever the regression equation is linear in the coefficients $\{b_i\}$. They fail as soon as the model ceases to be 'linear'. Thus, the methods can be used to fit a function of the form

$$Y = b_0 + b_4 \sin(\pi e^x) + b_2 \ln z,$$

but they cannot be used to fit the relatively simple curve[1]

$$Y = b_0 + b_1 \exp(b_2 x). \tag{18.1.1}$$

Although the usual matrix methods fail with non-linear models, the numerical principle of least squares is still applicable. The normal equations turn out to be non-linear,[2] and iterative methods are usually needed to solve them (for example, the multi-dimensional Newton–Raphson method of section 3.6). Sometimes the least-squares solution is found by the method of steepest descent (section 7.5). It is clear that a computer is almost essential for non-linear regression.

The statistical methods of linear regression can be adapted to non-linear regression, but the analysis must be regarded as approximate. The total sum of squares in the analysis of variance table will be calculated in the usual manner and the computer can calculate the residual sum of squares by actually squaring and adding the differences between the observed and fitted

---

[1] Many workers would linearise (18.1.1) before proceeding to obtain regression results. The equation can be rewritten in the form

$$\ln(Y - b_0) = \ln b_1 + b_2 x$$

which gives a linear regression of $\ln(Y - b_0)$ on x. An appropriate value (often zero) is chosen for b_0; $\ln b_1$ and b_2 are then estimated by simple linear regression. Further details are given in example 18.1.2.

[2] In the usual mathematical sense.

313

values; the regression sum of squares is then found by difference. By analogy with linear regression, one degree of freedom is attached to each fitted constant.

> **Example 18.1.1.** Use the method of least squares to fit a curve of the form (18.1.1) close to the four points
>
> $(x_1, y_1) = (1, 5.4),$
> $(x_2, y_2) = (3, 15.3),$
> $(x_3, y_3) = (4, 26.2),$
> $(x_4, y_4) = (5, 47.4).$

We need to minimise

$$\sum_{i=1}^{4} \{y_i - b_0 - b_1 \exp(b_2 x_i)\}^2,$$

and at the minimum point the partial derivatives with respect to b_0, b_1 and b_2 will be zero (section 1.6). The resulting non-linear normal equations are

$$4b_0 + b_1 \sum \exp(b_2 x_i) = \sum y_i,$$
$$b_0 \sum \exp(b_2 x_i) + b_1 \sum \exp(2b_2 x_i) = \sum y_i \exp(b_2 x_i),$$
$$b_0 \sum x_i \exp(b_2 x_i) + b_1 \sum x_i \exp(2b_2 x_i) = \sum y_i x_i \exp(b_2 x_i),$$

which are somewhat tedious to solve by hand, but straightforward on a computer.

Another approach is possible in this case. Let us imagine for the moment that b_2 is known and we need to find b_0 and b_1; the least-squares line (18.1.1) is now linear in the unknown coefficients b_0 and b_1, and the usual matrix approach can be used to find them. In practice, of course, b_2 is not known, but various trial values can be used and the resulting least-squares curves fitted. The b_2-value and consequent b_0- and b_1-values producing the minimum residual sum of squares give the least-squares curve nearest the data points. We now demonstrate this approach.

An initial trial value of b_2 is required. If $Y(x)$ denotes the fitted value at the point x, it is easy to see from (18.1.1) that

$$\ln\left(\frac{Y(5) - Y(4)}{Y(4) - Y(3)}\right) = b_2 = \tfrac{1}{2} \ln\left(\frac{Y(5) - Y(3)}{Y(3) - Y(1)}\right).$$

Two different approximations to b_2 can therefore be obtained by substituting the observed y-values instead of the fitted values in this formula. We obtain 0.588 16 and 0.665 24. An initial trial value of 0.63 seems reasonable.

We now use the matrix methods of linear regression to compute the values of b_0 and b_1 corresponding to $b_2 = 0.63$. The first column of regression matrix **X** is a column of ones and the second column comprises the values

Table 18.1.1. *Successive approximations to a least-squares curve*

b_2	b_0	b_1	Regression s.s.
0.63	2.026 941 082	1.947 334 866	3196.217 331
0.62	2.056 273 539	1.767 252 438	3196.294 192
0.60	1.223 215 224	2.293 638 769	3196.304 055
0.58	0.643 464 106	2.559 808 774	3196.114 497
0.61	1.499 478 421	2.171 574 427	3196.323 481
0.608 987 683	1.471 904 985	2.183 616 581	3196.323 750
0.607 975 366	1.444 244 145	2.195 728 454	3196.323 518
0.609 025 063	1.472 924 743	2.183 170 672	3196.323 746
0.608 950 303	1.470 885 177	2.184 062 571	3196.323 751
0.608 956 533	1.471 055 142	2.183 988 233	3196.323 751

$\{\exp(0.63x_i)\}$. The y-values are listed in the column vector y in the usual way. We deduce the following normal equations for b_0 and b_1:

$$4b_0 + 44.261\,640\,46b_1 = 94.3,$$
$$44.261\,640\,46b_0 + 746.383\,3871b_1 = 1543.174\,130.$$

These equations have solution $b_0 = 2.026\,941\,082$, $b_1 = 1.947\,334\,866$.

Because the fitted line (18.1.1) has been made linear by assuming a known value of b_2, the uncorrected[3] regression sum of squares can be obtained in the usual manner by means of the formula

$$\mathbf{b'X'y} = (b_0 \quad b_1)\begin{pmatrix} 94.3 \\ 1543.174\,130 \end{pmatrix} = 3196.217\,331.$$

We need to find the value of b_2 which minimises the residual sum of squares (and maximises the regression sum of squares) and a trial-and-error method (varying the values of b_2) may be used. Successive approximations by the above method are summarised in table 18.1.1. The solution is

$$b_0 = 1.471\,055,$$
$$b_1 = 2.183\,988,$$
$$b_2 = 0.608\,957.$$

(The quadratic minimum formula (7.5.2) was used to obtain some of the later approximations to b_2 in table 18.1.1.)

Although the regression model is non-linear, it is possible to draw up an analysis of variance table for an approximate statistical analysis. The statistical model takes the form

$$y_i = \beta_0 + \beta_1 \exp(\beta_2 x_i) + e_i. \tag{18.1.2}$$

[3] Without the adjustment $n\bar{y}^2$ (section 16.4).

Table 18.1.2. *A non-linear least-squares example – analysis of variance*

Source	s.s.	d.f.	m.s.
Regression	973.201 251	2	486.6006
Residual	0.126 249	1	0.1262
Total	973.327 500	3	–

The $\{x_i\}$ are assumed free of error, and the $\{e_i\}$ are independent normal random variables with zero means and common variance σ^2. Depending on the objects of the investigation, the analysis of variance may be based on the corrected table 16.4.1 or the uncorrected table 16.4.2 with $n = 4$ and $\nu = 3$. The residual sum of squares is of course the same in both cases.

The analysis of variance table 18.1.2 has been drawn up in the corrected form in order to test the significance of the term $b_1 \exp(b_2 x)$. The total sum of squares was calculated using the formula in table 16.4.1, and the regression sum of squares was obtained by subtracting $n\bar{y}^2$ from the uncorrected regression sum of squares at the bottom of table 18.1.1. The ratio of the mean squares

$$486.6006/0.1262 = 3856$$

is much larger than the 5 per cent point of the $F_{2,1}$-distribution. The exponential term does seem to explain a significant proportion of the y-variability.

Example 18.1.2. The linearisation method used by many workers to fit a curve of the form (18.1.1) was described in footnote 1. Let us now apply the technique to the data of example 18.1.1.

The amount of effort put into determining b_0 depends upon the accuracy we require and the purposes to which the equation will be put. The constants b_1 and b_2 tend to compensate for an inaccurate value of b_0, and a rough estimate is often quite adequate. In table 18.1.1, for example, it is notable how little difference there is in the regression sum of squares when b_2 is varied, and yet what large proportional changes there are in b_0 and b_1. When greater precision is required, values of $\ln(y_i - b_0)$ are plotted against x_i for different values of b_0, and the value of b_0 producing the most linear scatter is chosen.

The choice $b_0 = 0$ does not seem unreasonable in our four-point example. When we apply the simple linear regression methods of chapter 15 to the regression of $\ln y$ on x, we obtain

$$\ln Y = 1.129\,455 + 0.540\,0626x,$$

Table 18.1.3. *The effects of temperature on the relative motion of the acyl fatty acids in chloroplast membranes**

Tempera-ture °C	$10^4\,K^{-1}$	Relative motion τ	Tempera-ture °C	$10^4\,K^{-1}$	Relative motion τ
0.1	3661	63.2	14.3	3480	24.3
1.6	3641	54.1	15.9	3461	23.6
3.1	3621	49.7	17.4	3443	21.6
4.7	3601	42.5	19.0	3424	19.6
6.1	3582	40.0	20.7	3404	19.1
7.6	3563	37.5	22.3	3386	17.5
9.0	3546	33.1	23.6	3371	16.7
10.0	3533	31.0	25.1	3354	16.6
11.1	3519	30.3	26.8	3335	15.8
11.9	3510	28.0	28.3	3318	14.5
12.7	3500	26.2	29.8	3302	13.7
13.5	3490	25.3			

* *Source*: communication from J. K. Raison of the Commonwealth Scientific and Industrial Research Organisation Division of Food Research, North Ryde, Australia.

or

$$Y = 3.093\,97 \; e^{0.540\,0626x}.$$

The fitted values are 5.3, 15.6, 26.8 and 46.1 respectively. It will be noted that the fitted values at $x = 1$ and $x = 5$ lie below the observed values, and the fitted values in-between exceed the observed values. A better fit will be obtained by choosing b_0 somewhat greater than zero; the regression curve will then rise more steeply (b_2 will be larger).

The statistical model underlying the linearisation method is

$$\ln (y_i - \beta_0) = \ln \beta_1 + \beta_2 x_i + e_i.$$

The $\{x_i\}$ are assumed free of error, and the $\{e_i\}$ are independent normal random variables with zero expectations and common variance. The model is *not* the same as (18.1.2). In fact, the two approaches always produce different *b*-coefficients, because they give different weights to the various observations.

Example 18.1.3. The data in table 18.1.3 represent the effects of temperature on the relative motion of the acyl fatty acids in chloroplast membranes. According to theory, $\log_{10} \tau$ (the logarithm of the relative motion τ) and the reciprocal of the absolute temperature are linearly related. The theory also says that the acyl fatty acids will melt at a certain

Table 18.1.4. *Fitting a line pair by the method of least squares.*
A record of the residual sum of squares for various partitions of
the 23 points. The symbol n denotes the number of points used
to fit the first line of the pair

n	(Residual s.s.) $\times 10^6$	n	(Residual s.s.) $\times 10^6$
3	6711	11	2250
4	6197	12	2243*
5	5300	13	2285
6	3879	14	2915
7	3426	15	3410
8	3166	16	3659
9	2507	17	5220
10	2245	18	6061

* The least-squares line pair.

temperature and a change in the slope of the graph will occur at the melting point.

Let us use the method of least squares to fit a pair of straight lines near the data and so detect the melting point of the fatty acids. The following approach can be used.

Fit a straight line near the first n points by the method of section 15.2 and compute the residual sum of squares for those n points using the formulae in table 15.5.1. Fit another straight line near the remaining $23 - n$ points and compute the residual sum of squares for these points. Compute the residual sum of squares for the line pair by adding together the residual sums of squares for the two lines. Perform this calculation for each value of n from 2 to 21 inclusive. Choose as the least-squares line pair, the pair having the smallest residual sum of squares.

A record of the residual sums of squares for various partitions of the 23 points is given in table 18.1.4. We see that the least squares line pair is obtained by fitting a line to the first twelve observations and another line to the remaining eleven values. If we write Y for the fitted value of $\log_{10} \tau$ and x instead of 10^4 times the reciprocal of the absolute temperature, the two lines are respectively

$$Y = 0.002\,242x - 6.424\,576 \quad \text{and} \quad Y = 0.001\,380x - 3.417\,718.$$

The fit is shown in fig. 18.1.1. The change in the slope at the fitted melting point does not appear to be substantial and it might seem that an adequate fit could be obtained using a single straight line. An F-method for testing whether the inclusion of certain additional terms in a linear regression results in a significant reduction in residual sum of squares was described in section

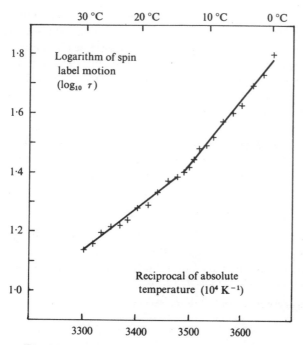

Fig. 18.1.1. The effect of temperature on the relative motion of acyl fatty acids in chloroplast membranes: a pair of straight lines.

Table 18.1.5. *Non-linear regression example 18.1.3 – analysis of variance*

Source	s.s.	d.f.	m.s.
Single straight line	0.776 602	1	—
Additional reduction due to line pair	0.013 384	3	0.004 4613
Residual	0.002 243	18	0.000 1246
Total	0.792 229	22	—

16.7. The uncorrected analysis of variance is table 16.7.1. Let us draw up a similar, but corrected, table for our non-linear model.

The regression sum of squares for a single line through the data points is easily shown to be 0.776 602, and the total sum of squares is 0.792 229 (the calculation formulae are given in table 15.5.1). The residual sum of squares when a line-pair is fitted is 0.002 243 (table 18.1.5). The additional reduction in sum of squares due to the introduction of a melting point and line pair is found by difference. These sums of squares are listed the analysis of variance table 18.1.5. The degrees of freedom are obtained by noting that a straight line imposes two constraints and the line pair approximately five.

The ratio of the mean squares is large (35.8). This suggests a significant reduction in sum of squares due to the melting point and line pair, but we cannot use an F-test to obtain conclusions at any stated level.

It is perhaps worth noting that the above calculations were performed on a programmable calculator having 51 storage registers and allowing 512 program steps. Fig. 18.1.1 was drawn automatically. The limited storage capacity of the calculator was overcome by storing two numbers per register (section 7.6).

Further reading: Draper and Smith [19] 263-301.

18.2   Exercises

1. Fit a low-order polynomial to the data in table 18.1.3 and compare the fit with the line-pair results in example 18.1.3.

2. Fit a curve of the form (18.1.1) to the data in table 4.2.1.

Appendix

Table A.1. *Normal ordinates and areas**

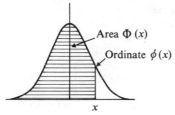

Area Φ (x)

Ordinate φ (x)

x

Note: $\phi(-x) = \phi(x)$, $\Phi(-x) = 1 - \Phi(x)$.

x	φ(x)	Φ(x)	x	φ(x)	Φ(x)	x	φ(x)	Φ(x)
0.00	.398 94	.500 00	0.15	.394 48	.559 62	0.30	.381 39	.617 91
0.01	.398 92	.503 99	0.16	.393 87	.563 56	0.31	.380 23	.621 72
0.02	.398 86	.507 98	0.17	.393 22	.567 49	0.32	.379 03	.625 52
0.03	.398 76	.511 97	0.18	.392 53	.571 42	0.33	.377 80	.629 30
0.04	.398 62	.515 95	0.19	.391 81	.575 35	0.34	.376 54	.633 07
0.05	.398 44	.519 94	0.20	.391 04	.579 26	0.35	.375 24	.636 83
0.06	.398 22	.523 92	0.21	.390 24	.583 17	0.36	.373 91	.640 58
0.07	.397 97	.527 90	0.22	.389 40	.587 06	0.37	.372 55	.644 31
0.08	.397 67	.531 88	0.23	.388 53	.590 95	0.38	.371 15	.648 03
0.09	.397 33	.535 86	0.24	.387 62	.594 83	0.39	.369 73	.651 73
0.10	.396 95	.539 83	0.25	.386 67	.598 71	0.40	.368 27	.655 42
0.11	.396 54	.543 80	0.26	.385 68	.602 57	0.41	.366 78	.659 10
0.12	.396 08	.547 76	0.27	.384 66	.606 42	0.42	.365 26	.662 76
0.13	.395 59	.551 72	0.28	.383 61	.610 26	0.43	.363 71	.666 40
0.14	.395 05	.555 67	0.29	.382 51	.614 09	0.44	.362 13	.670 03

* *Source*: Kenney, *Mathematics of Statistics,* Part One, 225-7, Van Nostrand, New York. Reprinted with permission of the publisher.

Table A.1 (*cont.*)

x	$\phi(x)$	$\Phi(x)$	x	$\phi(x)$	$\Phi(x)$	x	$\phi(x)$	$\Phi(x)$
0.45	.360 53	.673 64	0.85	.277 98	.802 34	1.25	.182 65	.894 35
0.46	.358 89	.677 24	0.86	.275 62	.805 11	1.26	.180 37	.896 17
0.47	.357 23	.680 82	0.87	.273 24	.807 85	1.27	.178 10	.897 96
0.48	.355 53	.684 39	0.88	.270 86	.810 57	1.28	.175 85	.899 73
0.49	.353 81	.687 93	0.89	.268 48	.813 27	1.29	.173 60	.901 47
0.50	.352 07	.691 46	0.90	.266 09	.815 94	1.30	.171 37	.903 20
0.51	.350 29	.694 97	0.91	.263 69	.818 59	1.31	.169 15	.904 90
0.52	.348 49	.698 47	0.92	.261 29	.821 21	1.32	.166 94	.906 58
0.53	.346 67	.701 94	0.93	.258 88	.823 81	1.33	.164 74	.908 24
0.54	.344 82	.705 40	0.94	.256 47	.826 39	1.34	.162 56	.909 88
0.55	.342 94	.708 84	0.95	.254 06	.828 94	1.35	.160 38	.911 49
0.56	.341 05	.712 26	0.96	.251 64	.831 47	1.36	.158 22	.913 09
0.57	.339 12	.715 66	0.97	.249 23	.833 98	1.37	.156 08	.914 66
0.58	.337 18	.719 04	0.98	.246 81	.836 46	1.38	.153 95	.916 21
0.59	.335 21	.722 40	0.99	.244 39	.838 91	1.39	.151 83	.917 74
0.60	.333 22	.725 75	1.00	.241 97	.841 34	1.40	.149 73	.919 24
0.61	.331 21	.729 07	1.01	.239 55	.843 75	1.41	.147 64	.920 73
0.62	.329 18	.732 37	1.02	.237 13	.846 14	1.42	.145 56	.922 20
0.63	.327 13	.735 65	1.03	.234 71	.848 50	1.43	.143 50	.923 64
0.64	.325 06	.738 91	1.04	.232 30	.850 83	1.44	.141 46	.925 07
0.65	.322 97	.742 15	1.05	.229 88	.853 14	1.45	.139 43	.926 47
0.66	.320 86	.745 37	1.06	.227 47	.855 43	1.46	.137 42	.927 86
0.67	.318 74	.748 57	1.07	.225 06	.857 69	1.47	.135 42	.929 22
0.68	.316 59	.751 75	1.08	.222 65	.859 93	1.48	.133 44	.930 56
0.69	.314 43	.754 90	1.09	.220 25	.862 14	1.49	.131 47	.931 89
0.70	.312 25	.758 04	1.10	.217 85	.864 33	1.50	.129 52	.933 19
0.71	.310 06	.761 15	1.11	.215 46	.866 50	1.51	.127 58	.934 48
0.72	.307 85	.764 24	1.12	.213 07	.868 64	1.52	.125 66	.935 74
0.73	.305 63	.767 30	1.13	.210 69	.870 76	1.53	.123 76	.936 99
0.74	.303 39	.770 35	1.14	.208 31	.872 86	1.54	.121 88	.938 22
0.75	.301 14	.773 37	1.15	.205 94	.874 93	1.55	.120 01	.939 43
0.76	.298 87	.776 37	1.16	.203 57	.876 98	1.56	.118 16	.940 62
0.77	.296 59	.779 35	1.17	.201 21	.879 00	1.57	.116 32	.941 79
0.78	.294 31	.782 30	1.18	.198 86	.881 00	1.58	.114 50	.942 95
0.79	.292 00	.785 24	1.19	.196 52	.882 98	1.59	.112 70	.944 08
0.80	.289 69	.788 14	1.20	.194 19	.884 93	1.60	.110 92	.945 20
0.81	.287 37	.791 03	1.21	.191 86	.886 86	1.61	.109 15	.946 30
0.82	.285 04	.793 89	1.22	.189 54	.888 77	1.62	.107 41	.947 38
0.83	.282 69	.796 73	1.23	.187 24	.890 65	1.63	.105 67	.948 45
0.84	.280 34	.799 55	1.24	.184 94	.892 51	1.64	.103 96	.949 50

Table A.1 (*cont.*)

x	$\phi(x)$	$\Phi(x)$	x	$\phi(x)$	$\Phi(x)$	x	$\phi(x)$	$\Phi(x)$
1.65	.102 26	.950 53	2.05	.048 79	.979 82	2.45	.019 84	.992 86
1.66	.100 59	.951 54	2.06	.047 80	.980 30	2.46	.019 36	.993 05
1.67	.098 93	.952 54	2.07	.046 82	.980 77	2.47	.018 89	.993 24
1.68	.097 28	.953 52	2.08	.045 86	.981 24	2.48	.018 42	.993 43
1.69	.095 66	.954 49	2.09	.044 91	.981 69	2.49	.017 97	.993 61
1.70	.094 05	.955 43	2.10	.043 98	.982 14	2.50	.017 53	.993 79
1.71	.092 46	.956 37	2.11	.043 07	.982 57	2.51	.017 09	.993 96
1.72	.090 89	.957 28	2.12	.042 17	.983 00	2.52	.016 67	.994 13
1.73	.089 33	.958 18	2.13	.041 28	.983 41	2.53	.016 25	.994 30
1.74	.087 80	.959 07	2.14	.040 41	.983 82	2.54	.015 85	.994 46
1.75	.086 28	.959 94	2.15	.039 55	.984 22	2.55	.015 45	.994 61
1.76	.084 78	.960 80	2.16	.038 71	.984 61	2.56	.015 06	.994 77
1.77	.083 29	.961 64	2.17	.037 88	.985 00	2.57	.014 68	.994 92
1.78	.081 83	.962 46	2.18	.037 06	.985 37	2.58	.014 31	.995 06
1.79	.080 38	.963 27	2.19	.036 26	.985 74	2.59	.013 94	.995 20
1.80	.078 95	.964 07	2.20	.035 47	.986 10	2.60	.013 58	.995 34
1.81	.077 54	.964 85	2.21	.034 70	.986 45	2.61	.013 23	.995 47
1.82	.076 14	.965 62	2.22	.033 94	.986 79	2.62	.012 89	.995 60
1.83	.074 77	.966 38	2.23	.033 19	.987 13	2.63	.012 56	.995 73
1.84	.073 41	.967 12	2.24	.032 46	.987 45	2.64	.012 23	.995 85
1.85	.072 06	.967 84	2.25	.031 74	.987 78	2.65	.011 91	.995 98
1.86	.070 74	.968 56	2.26	.031 03	.988 09	2.66	.011 60	.996 09
1.87	.069 43	.969 26	2.27	.030 34	.988 40	2.67	.011 30	.996 21
1.88	.068 14	.969 95	2.28	.029 65	.988 70	2.68	.011 00	.996 32
1.89	.066 87	.970 62	2.29	.028 98	.988 99	2.69	.010 71	.996 43
1.90	.065 62	.971 28	2.30	.028 33	.989 28	2.70	.010 42	.996 53
1.91	.064 39	.971 93	2.31	.027 68	.989 56	2.71	.010 14	.996 64
1.92	.063 16	.972 57	2.32	.027 05	.989 83	2.72	.009 87	.996 74
1.93	.061 95	.973 20	2.33	.026 43	.990 10	2.73	.009 61	.996 83
1.94	.060 77	.973 81	2.34	.025 82	.990 36	2.74	.009 35	.996 93
1.95	.059 59	.974 41	2.35	.025 22	.990 61	2.75	.009 09	.997 02
1.96	.058 44	.975 00	2.36	.024 63	.990 86	2.76	.008 85	.997 11
1.97	.057 30	.975 58	2.37	.024 06	.991 11	2.77	.008 61	.997 20
1.98	.056 18	.976 15	2.38	.023 49	.991 34	2.78	.008 37	.997 28
1.99	.055 08	.976 70	2.39	.022 94	.991 58	2.79	.008 14	.997 36
2.00	.053 99	.977 25	2.40	.022 39	.991 80	2.80	.007 92	.997 44
2.01	.052 92	.977 78	2.41	.021 86	.992 02	2.81	.007 70	.997 52
2.02	.051 86	.978 31	2.42	.021 34	.992 24	2.82	.007 48	.997 60
2.03	.050 82	.978 82	2.43	.020 83	.992 45	2.83	.007 27	.997 67
2.04	.049 80	.979 32	2.44	.020 33	.992 66	2.84	.007 07	.997 74

Table A.1 (*cont.*)

x	$\phi(x)$	$\Phi(x)$	x	$\phi(x)$	$\Phi(x)$	x	$\phi(x)$	$\Phi(x)$
2.85	.006 87	.997 81	3.25	.002 03	.999 42	3.65	.000 51	.999 87
2.86	.006 68	.997 88	3.26	.001 96	.999 44	3.66	.000 49	.999 87
2.87	.006 49	.997 95	3.27	.001 90	.999 46	3.67	.000 47	.999 88
2.88	.006 31	.998 01	3.28	.001 84	.999 48	3.68	.000 46	.999 88
2.89	.006 13	.998 07	3.29	.001 78	.999 50	3.69	.000 44	.999 89
2.90	.005 95	.998 13	3.30	.001 72	.999 52	3.70	.000 42	.999 89
2.91	.005 78	.998 19	3.31	.001 67	.999 53	3.71	.000 41	.999 90
2.92	.005 62	.998 25	3.32	.001 61	.999 55	3.72	.000 39	.999 90
2.93	.005 45	.998 31	3.33	.001 56	.999 57	3.73	.000 38	.999 90
2.94	.005 30	.998 36	3.34	.001 51	.999 58	3.74	.000 37	.999 91
2.95	.005 14	.998 41	3.35	.001 46	.999 60	3.75	.000 35	.999 91
2.96	.004 99	.998 46	3.36	.001 41	.999 61	3.76	.000 34	.999 92
2.97	.004 85	.998 51	3.37	.001 36	.999 62	3.77	.000 33	.999 92
2.98	.004 71	.998 56	3.38	.001 32	.999 64	3.78	.000 31	.999 92
2.99	.004 57	.998 61	3.39	.001 27	.999 65	3.79	.000 30	.999 92
3.00	.004 43	.998 65	3.40	.001 23	.999 66	3.80	.000 29	.999 93
3.01	.004 30	.998 69	3.41	.001 19	.999 68	3.81	.000 28	.999 93
3.02	.004 17	.998 74	3.42	.001 15	.999 69	3.82	.000 27	.999 93
3.03	.004 05	.998 78	3.43	.001 11	.999 70	3.83	.000 26	.999 94
3.04	.003 93	.998 82	3.44	.001 07	.999 71	3.84	.000 25	.999 94
3.05	.003 81	.998 86	3.45	.001 04	.999 72	3.85	.000 24	.999 94
3.06	.003 70	.998 89	3.46	.001 00	.999 73	3.86	.000 23	.999 94
3.07	.003 58	.998 93	3.47	.000 97	.999 74	3.87	.000 22	.999 95
3.08	.003 48	.998 97	3.48	.000 94	.999 75	3.88	.000 21	.999 95
3.09	.003 37	.999 00	3.49	.000 90	.999 76	3.89	.000 21	.999 95
3.10	.003 27	.999 03	3.50	.000 87	.999 77	3.90	.000 20	.999 95
3.11	.003 17	.999 06	3.51	.000 84	.999 78	3.91	.000 19	.999 95
3.12	.003 07	.999 10	3.52	.000 81	.999 78	3.92	.000 18	.999 96
3.13	.002 98	.999 13	3.53	.000 79	.999 79	3.93	.000 18	.999 96
3.14	.002 88	.999 16	3.54	.000 76	.999 80	3.94	.000 17	.999 96
3.15	.002 79	.999 18	3.55	.000 73	.999 81	3.95	.000 16	.999 96
3.16	.002 71	.999 21	3.56	.000 71	.999 81	3.96	.000 16	.999 96
3.17	.002 62	.999 24	3.57	.000 68	.999 82	3.97	.000 15	.999 96
3.18	.002 54	.999 26	3.58	.000 66	.999 83	3.98	.000 14	.999 97
3.19	.002 46	.999 29	3.59	.000 63	.999 83	3.99	.000 14	.999 97
3.20	.002 38	.999 31	3.60	.000 61	.999 84			
3.21	.002 31	.999 34	3.61	.000 59	.999 85			
3.22	.002 24	.999 36	3.62	.000 57	.999 85			
3.23	.002 16	.999 38	3.63	.000 55	.999 86			
3.24	.002 10	.999 40	3.64	.000 53	.999 86			

Table A.2. *The upper† 100α per cent points* of the chi-square distribution*

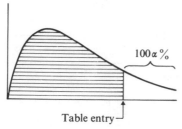

$100\alpha\%$

Table entry

The table entry is x, where $P(\chi_\nu^2 > x) = \alpha$.

	α									
ν	0.995	0.99	0.975	0.95	0.9	0.1	0.05	0.025	0.01	0.005
1	0.0^4393	0.0^3157	0.0^3982	0.0^2393	0.0158	2.71	3.84	5.02	6.63	7.88
2	0.0100	0.0201	0.0506	0.103	0.211	4.61	5.99	7.38	9.21	10.60
3	0.072	0.115	0.216	0.352	0.584	6.25	7.81	9.35	11.34	12.84
4	0.207	0.297	0.484	0.711	1.064	7.78	9.49	11.14	13.28	14.86
5	0.412	0.554	0.831	1.145	1.61	9.24	11.07	12.83	15.09	16.75
6	0.676	0.872	1.24	1.64	2.20	10.64	12.59	14.45	16.81	18.55
7	0.989	1.24	1.69	2.17	2.83	12.02	14.07	16.01	18.48	20.28
8	1.34	1.65	2.18	2.73	3.49	13.36	15.51	17.53	20.09	21.96
9	1.73	2.09	2.70	3.33	4.17	14.68	16.92	19.02	21.67	23.59
10	2.16	2.56	3.25	3.94	4.87	15.99	18.31	20.48	23.21	25.19
11	2.60	3.05	3.82	4.57	5.58	17.28	19.68	21.92	24.73	26.76
12	3.07	3.57	4.40	5.23	6.30	18.55	21.03	23.34	26.22	28.30
13	3.57	4.11	5.01	5.89	7.04	19.81	22.36	24.74	27.69	29.82
14	4.07	4.66	5.63	6.57	7.79	21.06	23.68	26.12	29.14	31.32
15	4.60	5.23	6.26	7.26	8.55	22.31	25.00	27.49	30.58	32.80
16	5.14	5.81	6.91	7.96	9.31	23.54	26.30	28.85	32.00	34.27
17	5.70	6.41	7.56	8.67	10.09	24.77	27.59	30.19	33.41	35.72
18	6.26	7.01	8.23	9.39	10.86	25.99	28.87	31.53	34.81	37.16
19	6.84	7.63	8.91	10.12	11.65	27.20	30.14	32.85	36.19	38.58
20	7.43	8.26	9.59	10.85	12.44	28.41	31.41	34.17	37.57	40.00
21	8.03	8.90	10.28	11.59	13.24	29.62	32.67	35.48	38.93	41.40
22	8.64	9.54	10.98	12.34	14.04	30.81	33.92	36.78	40.29	42.80
23	9.26	10.20	11.69	13.09	14.85	32.01	35.17	38.08	41.64	44.18
24	9.89	10.86	12.40	13.85	15.66	33.20	36.42	39.36	42.98	45.56
25	10.52	11.52	13.12	14.61	16.47	34.38	37.65	40.65	44.31	46.93
26	11.16	12.20	13.84	15.38	17.29	35.56	38.89	41.92	45.64	48.29
27	11.81	12.88	14.57	16.15	18.11	36.74	40.11	43.19	46.96	49.64
28	12.46	13.56	15.31	16.93	18.94	37.92	41.34	44.46	48.28	50.99
29	13.12	14.26	16.05	17.71	19.77	39.09	42.56	45.72	49.59	52.34

* *Source*: E. S. Pearson and C. M. Thompson, 'Tables of the Percentage Points of the Incomplete Beta Function and of the Chi-square Distribution', *Biometrika,* Vol. 32 (1941). Reprinted with permission of the publisher.

† The lower 100α per cent point is equal to the upper $100(1-\alpha)$ per cent point.

Table 2 (*cont.*)

ν	α									
	0.995	0.99	0.975	0.95	0.9	0.1	0.05	0.025	0.01	0.005
30	13.79	14.95	16.79	18.49	20.60	40.26	43.77	46.98	50.89	53.67
40	20.71	22.16	24.43	26.51	29.05	51.81	55.76	59.34	63.69	66.77
50	27.99	29.71	32.36	34.76	37.69	63.17	67.50	71.42	76.15	79.49
60	35.53	37.48	40.48	43.19	46.46	74.40	79.08	83.30	88.38	91.95
70	43.28	45.44	48.76	51.74	55.33	85.53	90.53	95.02	100.4	104.2
80	51.17	53.54	57.15	60.39	64.28	96.58	101.9	106.6	112.3	116.3
90	59.20	61.75	65.65	69.13	73.29	107.6	113.1	118.1	124.1	128.3
100	67.33	70.06	74.22	77.93	82.36	118.5	124.3	129.6	135.8	140.2

Note. For large ν use the approximation on p. 93.

Table A.3. *The upper† 100α per cent points* of Student's t-distribution*

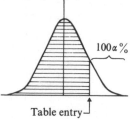

100α %

The table entry is x, where $P(t_\nu > x) = \alpha$.   Table entry⌐

ν	α					
	0.005	0.01	0.025	0.05	0.1	0.15
1	63.657	31.821	12.706	6.314	3.078	1.963
2	9.925	6.965	4.303	2.920	1.886	1.386
3	5.841	4.541	3.182	2.353	1.638	1.250
4	4.604	3.747	2.776	2.132	1.533	1.190
5	4.032	3.365	2.571	2.015	1.476	1.156
6	3.707	3.143	2.447	1.943	1.440	1.134
7	3.499	2.998	2.365	1.895	1.415	1.119
8	3.355	2.896	2.306	1.860	1.397	1.108
9	3.250	2.821	2.262	1.833	1.383	1.100
10	3.169	2.764	2.228	1.812	1.372	1.093
11	3.106	2.718	2.201	1.796	1.363	1.088
12	3.055	2.681	2.179	1.782	1.356	1.083
13	3.012	2.650	2.160	1.771	1.350	1.079
14	2.977	2.624	2.145	1.761	1.345	1.076
15	2.947	2.602	2.131	1.753	1.341	1.074
16	2.921	2.583	2.120	1.746	1.337	1.071
17	2.898	2.567	2.110	1.740	1.333	1.069
18	2.878	2.552	2.101	1.734	1.330	1.067
19	2.861	2.539	2.093	1.729	1.328	1.066
20	2.845	2.528	2.086	1.725	1.325	1.064
21	2.831	2.518	2.080	1.721	1.323	1.063
22	2.819	2.508	2.074	1.717	1.321	1.061
23	2.807	2.500	2.069	1.714	1.319	1.060
24	2.797	2.492	2.064	1.711	1.318	1.059
25	2.787	2.485	2.060	1.708	1.316	1.058
26	2.779	2.479	2.056	1.706	1.315	1.058
27	2.771	2.473	2.052	1.703	1.314	1.057
28	2.763	2.467	2.048	1.701	1.313	1.056
29	2.756	2.462	2.045	1.699	1.311	1.055
30	2.750	2.457	2.042	1.697	1.310	1.055
∞	2.576	2.326	1.960	1.645	1.282	1.036

* *Source: Statistical Methods for Research Workers,* with the generous permission of the author, Professor R. A. Fisher, and the publishers, Messrs. Oliver and Boyd.
† To obtain the lower 100α per cent point, change the sign of the upper 100α per cent point.

Table A.4(a). *The upper†* 5 *per cent points* of the F-distribution*

The table entry is x, where $P(F_{m, n} > x) = 0.05$.

n	m 1	2	3	4	5	6	7	8	9	10
1	161.4	199.5	215.7	224.6	230.2	234.0	236.8	238.9	240.5	241.9
2	18.51	19.00	19.16	19.25	19.30	19.33	19.35	19.37	19.38	19.40
3	10.13	9.55	9.28	9.12	9.01	8.94	8.89	8.85	8.81	8.79
4	7.71	6.94	6.59	6.39	6.26	6.16	6.09	6.04	6.00	5.96
5	6.61	5.79	5.41	5.19	5.05	4.95	4.88	4.82	4.77	4.74
6	5.99	5.14	4.76	4.53	4.39	4.28	4.21	4.15	4.10	4.06
7	5.59	4.74	4.35	4.12	3.97	3.87	3.79	3.73	3.68	3.64
8	5.32	4.46	4.07	3.84	3.69	3.58	3.50	3.44	3.39	3.35
9	5.12	4.26	3.86	3.63	3.48	3.37	3.29	3.23	3.18	3.14
10	4.96	4.10	3.71	3.48	3.33	3.22	3.14	3.07	3.02	2.98
11	4.84	3.98	3.59	3.36	3.20	3.09	3.01	2.95	2.90	2.85
12	4.75	3.89	3.49	3.26	3.11	3.00	2.91	2.85	2.80	2.75
13	4.67	3.81	3.41	3.18	3.03	2.92	2.83	2.77	2.71	2.67
14	4.60	3.74	3.34	3.11	2.96	2.85	2.76	2.70	2.65	2.60
15	4.54	3.68	3.29	3.06	2.90	2.79	2.71	2.64	2.59	2.54
16	4.49	3.63	3.24	3.01	2.85	2.74	2.66	2.59	2.54	2.49
17	4.45	3.59	3.20	2.96	2.81	2.70	2.61	2.55	2.49	2.45
18	4.41	3.55	3.16	2.93	2.77	2.66	2.58	2.51	2.46	2.41
19	4.38	3.52	3.13	2.90	2.74	2.63	2.54	2.48	2.42	2.38
20	4.35	3.49	3.10	2.87	2.71	2.60	2.51	2.45	2.39	2.35
21	4.32	3.47	3.07	2.84	2.68	2.57	2.49	2.42	2.37	2.32
22	4.30	3.44	3.05	2.82	2.66	2.55	2.46	2.40	2.34	2.30
23	4.28	3.42	3.03	2.80	2.64	2.53	2.44	2.37	2.32	2.27
24	4.26	3.40	3.01	2.78	2.62	2.51	2.42	2.36	2.30	2.25
25	4.24	3.39	2.99	2.76	2.60	2.49	2.40	2.34	2.28	2.24
26	4.23	3.37	2.98	2.74	2.59	2.47	2.39	2.32	2.27	2.22
27	4.21	3.35	2.96	2.73	2.57	2.46	2.37	2.31	2.25	2.20
28	4.20	3.34	2.95	2.71	2.56	2.45	2.36	2.29	2.24	2.19
29	4.18	3.33	2.93	2.70	2.55	2.43	2.35	2.28	2.22	2.18
30	4.17	3.32	2.92	2.69	2.53	2.42	2.33	2.27	2.21	2.16
40	4.08	3.23	2.84	2.61	2.45	2.34	2.25	2.18	2.12	2.08
60	4.00	3.15	2.76	2.53	2.37	2.25	2.17	2.10	2.04	1.99
120	3.92	3.07	2.68	2.45	2.29	2.17	2.09	2.02	1.96	1.19
∞	3.84	3.00	2.60	2.37	2.21	2.10	2.01	1.94	1.88	1.83

† To obtain the lower 5 per cent point of $F_{m, n}$, take the reciprocal of the upper 5 per cent point of $F_{n, m}$.

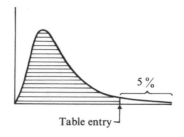

Table entry

								m	
12	15	20	24	30	40	60	120	∞	*n*
243.9	245.9	248.0	249.1	250.1	251.1	252.2	253.3	254.3	1
19.41	19.43	19.45	19.45	19.46	19.47	19.48	19.49	19.50	2
8.74	8.70	8.66	8.64	8.62	8.59	8.57	8.55	8.53	3
5.91	5.86	5.80	5.77	5.75	5.72	5.69	5.66	5.63	4
4.68	4.62	4.56	4.53	4.50	4.46	4.43	4.40	4.36	5
4.00	3.94	3.87	3.84	3.81	3.77	3.74	3.70	3.67	6
3.57	3.51	3.44	3.41	3.38	3.34	3.30	3.27	3.23	7
3.28	3.22	3.15	3.12	3.08	3.04	3.01	2.97	2.93	8
3.07	3.01	2.94	2.90	2.86	2.83	2.79	2.75	2.71	9
2.91	2.85	2.77	2.74	2.70	2.66	2.62	2.58	2.54	10
2.79	2.72	2.65	2.61	2.57	2.53	2.49	2.45	2.40	11
2.69	2.62	2.54	2.51	2.47	2.43	2.38	2.34	2.30	12
2.60	2.53	2.46	2.42	2.38	2.34	2.30	2.25	2.21	13
2.53	2.46	2.39	2.35	2.31	2.27	2.22	2.18	2.13	14
2.48	2.40	2.33	2.29	2.25	2.20	2.16	2.11	2.07	15
2.42	2.35	2.28	2.24	2.19	2.15	2.11	2.06	2.01	16
2.38	2.31	2.23	2.19	2.15	2.10	2.06	2.01	1.96	17
2.34	2.27	2.19	2.15	2.11	2.06	2.02	1.97	1.92	18
2.31	2.23	2.16	2.11	2.07	2.03	1.98	1.93	1.88	19
2.28	2.20	2.12	2.08	2.04	1.99	1.95	1.90	1.84	20
2.25	2.18	2.10	2.05	2.01	1.96	1.92	1.87	1.81	21
2.23	2.15	2.07	2.03	1.98	1.94	1.89	1.84	1.78	22
2.20	2.13	2.05	2.01	1.96	1.91	1.86	1.81	1.76	23
2.18	2.11	2.03	1.98	1.94	1.89	1.84	1.79	1.73	24
2.16	2.09	2.01	1.96	1.92	1.87	1.82	1.77	1.71	25
2.15	2.07	1.99	1.95	1.90	1.85	1.80	1.75	1.69	26
2.13	2.06	1.97	1.93	1.88	1.84	1.79	1.73	1.67	27
2.12	2.04	1.96	1.91	1.87	1.82	1.77	1.71	1.65	28
2.10	2.03	1.94	1.90	1.85	1.81	1.75	1.70	1.64	29
2.09	2.01	1.93	1.89	1.84	1.79	1.74	1.68	1.62	30
2.00	1.92	1.84	1.79	1.74	1.69	1.64	1.58	1.51	40
1.92	1.84	1.75	1.70	1.65	1.59	1.53	1.47	1.39	60
1.83	1.75	1.66	1.61	1.55	1.50	1.43	1.35	1.25	120
1.75	1.67	1.57	1.52	1.46	1.39	1.32	1.22	1.00	∞

Table A.4(b). *The upper† $2\frac{1}{2}$ per cent points* of the F-distribution*

The table entry is x where $P(F_{m,n} > x) = 0.025$.

n	m 1	2	3	4	5	6	7	8	9	10
1	647.8	799.5	864.2	899.6	921.8	937.1	948.2	956.7	963.3	968.6
2	38.51	39.00	39.17	39.25	39.30	39.33	39.36	39.37	39.39	39.40
3	17.44	16.04	15.44	15.10	14.88	14.73	14.62	14.54	14.47	14.42
4	12.22	10.65	9.98	9.60	9.36	9.20	9.07	8.98	8.90	8.84
5	10.01	8.43	7.76	7.39	7.15	6.98	6.85	6.76	6.68	6.62
6	8.81	7.26	6.60	6.23	5.99	5.82	5.70	5.60	5.52	5.46
7	8.07	6.54	5.89	5.52	5.29	5.12	4.99	4.90	4.82	4.76
8	7.57	6.06	5.42	5.05	4.82	4.65	4.53	4.43	4.36	4.30
9	7.21	5.71	5.08	4.72	4.48	4.32	4.20	4.10	4.03	3.96
10	6.94	5.46	4.83	4.47	4.24	4.07	3.95	3.85	3.78	3.72
11	6.72	5.26	4.63	4.28	4.04	3.88	3.76	3.66	3.59	3.53
12	6.55	5.10	4.47	4.12	3.89	3.73	3.61	3.51	3.44	3.37
13	6.41	4.97	4.35	4.00	3.77	3.60	3.48	3.39	3.31	3.25
14	6.30	4.86	4.24	3.89	3.66	3.50	3.38	3.29	3.21	3.15
15	6.20	4.77	4.15	3.80	3.58	3.41	3.29	3.20	3.12	3.06
16	6.12	4.69	4.08	3.73	3.50	3.34	3.22	3.12	3.05	2.99
17	6.04	4.62	4.01	3.66	3.44	3.28	3.16	3.06	2.98	2.92
18	5.98	4.56	3.95	3.61	3.38	3.22	3.10	3.01	2.93	2.87
19	5.92	4.51	3.90	3.56	3.33	3.17	3.05	2.96	2.88	2.82
20	5.87	4.46	3.86	3.51	3.29	3.13	3.01	2.91	2.84	2.77
21	5.83	4.42	3.82	3.48	3.25	3.09	2.97	2.87	2.80	2.73
22	5.79	4.38	3.78	3.44	3.22	3.05	2.93	2.84	2.76	2.70
23	5.75	4.35	3.75	3.41	3.18	3.02	2.90	2.81	2.73	2.67
24	5.72	4.32	3.72	3.38	3.15	2.99	2.87	2.78	2.70	2.64
25	5.69	4.29	3.69	3.35	3.13	2.97	2.85	2.75	2.68	2.61
26	5.66	4.27	3.67	3.33	3.10	2.94	2.82	2.73	2.65	2.59
27	5.63	4.24	3.65	3.31	3.08	2.92	2.80	2.71	2.63	2.57
28	5.61	4.22	3.63	3.29	3.06	2.90	2.78	2.69	2.61	2.55
29	5.59	4.20	3.61	3.27	3.04	2.88	2.76	2.67	2.59	2.53
30	5.57	4.18	3.59	3.25	3.03	2.87	2.75	2.65	2.57	2.51
40	5.42	4.05	3.46	3.13	2.90	2.74	2.62	2.53	2.45	2.39
60	5.29	3.93	3.34	3.01	2.79	2.63	2.51	2.41	2.33	2.27
120	5.15	3.80	3.23	2.89	2.67	2.52	2.39	2.30	2.22	2.16
∞	5.02	3.69	3.12	2.79	2.57	2.41	2.29	2.19	2.11	2.05

† To obtain the lower $2\frac{1}{2}$ per cent point of $F_{m,n}$, take the reciprocal of the upper $2\frac{1}{2}$ per cent point of $F_{n,m}$.

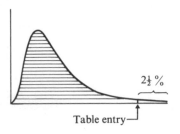

Table entry

$2\frac{1}{2}\%$

12	15	20	24	30	40	60	120	∞	n
976.7	984.9	993.1	997.2	1001	1006	1010	1014	1018	1
39.41	39.43	39.45	39.46	39.46	39.47	39.48	39.49	39.50	2
14.34	14.25	14.17	14.12	14.08	14.04	13.99	13.95	13.90	3
8.75	8.66	8.56	8.51	8.46	8.41	8.36	8.31	8.26	4
6.52	6.43	6.33	6.28	6.23	6.18	6.12	6.07	6.02	5
5.37	5.27	5.17	5.12	5.07	5.01	4.96	4.90	4.85	6
4.67	4.57	4.47	4.42	4.36	4.31	4.25	4.20	4.14	7
4.20	4.10	4.00	3.95	3.89	3.84	3.78	3.73	3.67	8
3.87	3.77	3.67	3.61	3.56	3.51	3.45	3.39	3.33	9
3.62	3.52	3.42	3.37	3.31	3.26	3.20	3.14	3.08	10
3.43	3.33	3.23	3.17	3.12	3.06	3.00	2.94	2.88	11
3.28	3.18	3.07	3.02	2.96	2.91	2.85	2.79	2.72	12
3.15	3.05	2.95	2.89	2.84	2.78	2.72	2.66	2.60	13
3.05	2.95	2.84	2.79	2.73	2.67	2.61	2.55	2.49	14
2.96	2.86	2.76	2.70	2.64	2.59	2.52	2.46	2.40	15
2.89	2.79	2.68	2.63	2.57	2.51	2.45	2.38	2.32	16
2.82	2.72	2.62	2.56	2.50	2.44	2.38	2.32	2.25	17
2.77	2.67	2.56	2.50	2.44	2.38	2.32	2.26	2.19	18
2.72	2.62	2.51	2.45	2.39	2.33	2.27	2.20	2.13	19
2.68	2.57	2.46	2.41	2.35	2.29	2.22	2.16	2.09	20
2.64	2.53	2.42	2.37	2.31	2.25	2.18	2.11	2.04	21
2.60	2.50	2.39	2.33	2.27	2.21	2.14	2.08	2.00	22
2.57	2.47	2.36	2.30	2.24	2.18	2.11	2.04	1.97	23
2.54	2.44	2.33	2.27	2.21	2.15	2.08	2.01	1.94	24
2.51	2.41	2.30	2.24	2.18	2.12	2.05	1.98	1.91	25
2.49	2.39	2.28	2.22	2.16	2.09	2.03	1.95	1.88	26
2.47	2.36	2.25	2.19	2.13	2.07	2.00	1.93	1.85	27
2.45	2.34	2.23	2.17	2.11	2.05	1.98	1.91	1.83	28
2.43	2.32	2.21	2.15	2.09	2.03	1.96	1.89	1.81	29
2.41	2.31	2.20	2.14	2.07	2.01	1.94	1.87	1.79	30
2.29	2.18	2.07	2.01	1.94	1.88	1.80	1.72	1.64	40
2.17	2.06	1.94	1.88	1.82	1.74	1.67	1.58	1.48	60
2.05	1.94	1.82	1.76	1.69	1.61	1.53	1.43	1.31	120
1.94	1.83	1.71	1.64	1.57	1.48	1.39	1.27	1.00	∞

References

[1]   Anscombe, F. J. (1948). The transformation of Poisson, binomial and negative binomial data. *Biometrika*, **35**, 246–54.

[2]   Bailey, N. T. J. (1959). *Statistical Methods in Biology*. English Universities Press.

[3]   Balaam, L. N. (1972). *Fundamentals of Biometry*. George Allen and Unwin Ltd.

[4]   Bortkiewicz, L. von (1898). *Das Gesetz der Kleinen Zahlen*. Leipzig: Teubner.

[5]   Box, G. E. P. (1953). Non-normality and tests on variances. *Biometrika*, **40**, 318–35.

[6]   Box, G. E. P. and Müller, M. E. (1958). A note on the generation of random normal deviates. *Annals of Mathematical Statistics*, **29**, 610–11.

[7]   Brownlee, K. A. (1965). *Statistical Theory and Methodology in Science and Engineering* (second edition). Wiley.

[8]   Campbell, R. C. (1974). *Statistics for Biologists*. Cambridge University Press.

[9]   Chakravarti, I. M., Laha, R. G. and Roy, J. (1967). *Handbook of Methods of Applied Statistics*, Vol. 1. Wiley.

[10]   Chakravarti, I. M., Laha, R. G. and Roy, J. (1967). *Handbook of Methods of Applied Statistics*, Vol. 2. Wiley.

[11]   Chemical Rubber Publishing Company (1963). *Standard Mathematical Tables* (C. D. Hodgman, Editor).

[12]   Clarke, R. D. (1946). An application of the Poisson distribution. *J. Institute of Actuaries*, **72**, 481.

[13]   Cochran, W. G. (1954). Some methods for strengthening the common χ^2 tests. *Biometrics*, **10**, 417–51.

[14]   Cohen, A. C. (1950). Estimating the mean and variance of normal populations from singly truncated and doubly truncated samples. *Annals of Mathematical Statistics*, **21**, 557–69.

[15]   Cohen, A. C. (1957). On the solution of estimating equations for truncated and censored samples from normal populations. *Biometrika*, **44**, 225–36.

[16]   Colquhoun, D. (1971). *Lectures on Biostatistics*. Clarendon Press.

[17]   Conte, S. D. (1965). *Elementary Numerical Analysis*. McGraw–Hill.

[18]   Craig, C. C. (1953). On the utilisation of marked specimens in estimating populations of flying insects. *Biometrika*, **40**, 170–6.

[19]   Draper, N. R. and Smith H. (1966). *Applied Regression Analysis*. Wiley.

[20]   Elderton, W. P. and Johnson, N. L. (1969). *Systems of Frequency Curves*. Cambridge University Press.

[21] Feller, W. (1950). *An Introduction to Probability Theory and Its Applications,* Vol. 1. Wiley.

[22] Ferguson, R. A., Fryer, J. G. and McWhinney, I. A. (1975). On the estimation of a truncated normal distribution. Contributed paper to the International Statistical Institute meeting in Warsaw.

[23] Fisher, R. A. (1921). Studies in crop variation. I. An examination of the yield of dressed grain from Broadbalk. *J. Agricultural Science,* 11, 107–35.

[24] Fisher, R. A., Corbet, A. S. and Williams, C. B. (1943). The relation between the number of species and the number of individuals in a random sample of an animal population. *J. Animal Ecology,* 12, 42–57.

[25] Fisher, R. A. and Yates, F. (1963). *Statistical Tables for Use in Biological, Agricultural and Medical Research* (sixth edition). Oliver and Boyd.

[26] Fraser, D. A. S. (1967). *Statistics – An Introduction.* Wiley.

[27] Freeman, H. (1967). *Finite Differences for Actuarial Students.* Cambridge University Press.

[28] Fröberg, C. (1965). *Introduction to Numerical Analysis.* Addison–Wesley.

[29] Gauss, K. F. (1809). *Theoria motus corporum coelestium in sectionibus conicus solem ambientium auctore C.F.G.*

[30] Gauss, K. F. (1823). *Theoria combinationis observationum erroribus minimis obnoxiae.* Göttingen.

[31] Gauss, K. F. (1828). *Supplementum theoriae combinationis observationum erroribus minimis obnoxiae.* Göttingen.

[32] Gauss, K. F. (1855). *Méthode des Moindres Carrés* (French translation by J. Bertrand). Paris.

[33] Gauss, K. F. (1857). *Theory of the Motion of the Heavenly Bodies, moving about the Sun in Conic Sections*: a Translation of Gauss' 'Theoria Motus' with an Appendix by Charles Henry Davis. Boston.

[34] Glasser, G. J. and Winter, R. F. (1961). Critical values of rank correlation for testing the hypothesis of independence. *Biometrika,* 48, 444–8.

[35] Greville, T. N. E. (1947) Actuarial note: adjusted average graduation formulas of maximum smoothness. *The Record, American Institute of Actuaries,* 36, 249–64.

[36] Greville, T. N. E. (1948). Actuarial note: tables of coefficients in adjusted average graduation formulas of maximum smoothness. *The Record, American Institute of Actuaries,* 37, 11–30.

[37] Greville, T. N. E. (1969). *Theory and Application of Spline Functions.* Academic Press.

[38] Grossman, S. I. and Turner, J. E. (1974). *Mathematics for the Biological Sciences.* Macmillan.

[39] Guenther, W. C. (1964). *Analysis of Variance.* Prentice-Hall.

[40] Hammersley, J. M. and Handscomb, D. C. (1967). *Monte Carlo Methods.* Methuen.

[41] Hartree, D. R. (1955). *Numerical Analysis.* Oxford University Press.

[42] Henrici, P. (1964). *Elements of Numerical Analysis.* Wiley.

[43] Hildebrand, F. B. (1956). *Introduction to Numerical Analysis.* McGraw-Hill.

[44] Hocking, R. R. (1976). The analysis and selection of variables in linear regression. *Biometrics,* 32, 1–50.

[45] Hoel, P. G. (1971). *Elementary Statistics.* Wiley.

[46] Hoel, P. G. (1971). *Introduction to Mathematical Statistics.* Wiley.

[47] Johnson, N. L. and Kotz, S. (1969). *Distributions in Statistics (Vol. 1) Discrete Distributions.* Houghton Mifflin, Boston.

[48] Johnson, N. L. and Kotz, S. (1970). *Distributions in Statistics (Vol. 2) Continuous Univariate Distributions 1.* Houghton Mifflin, Boston.

[49] Johnson, N. L. and Kotz, S. (1970). *Distributions in Statistics (Vol. 3) Continuous Univariate Distributions 2.* Houghton Mifflin, Boston.

[50] Johnson, N. L. and Leone, F. C. (1964). *Statistics and Experimental Design in Engineering and the Physical Sciences,* Vol. 1. Wiley.

[51] Johnson, N. L. and Leone, F. C. (1964). *Statistics and Experimental Design in Engineering and the Physical Sciences,* Vol. 2. Wiley.

[52] Kendall, M. G. (1955). *Rank Correlation Methods.* Griffin.

[53] Kendall, M. G. and Stuart, A. (1963). *The Advanced Theory of Statistics (Vol. 1) Distribution Theory* (second edition). Griffin.

[54] Kendall, M. G. and Stuart, A. (1967). *The Advanced Theory of Statistics (Vol. 2) Inference and Relationship* (second edition). Griffin.

[55] Kendall, M. G. and Stuart, A. (1968). *The Advanced Theory of Statistics (Vol. 3) Design and Analysis, and Time Series* (second edition). Griffin.

[56] Kruskal, W. H. and Wallis, W. A. (1952). Use of ranks in one-criterion variance analysis. *J. American Statistical Association,* **47,** 583–621.

[57] Lancaster, H. O. (1969). *The Chi-squared Distribution.* Wiley.

[58] Legendre, A. M. (1806). *Nouvelles méthodes pour la détermination des orbites des comètes; avec un supplément.* Paris.

[59] Liebermann, G. J. and Owen, D. B. (1961). *Tables of the Hypergeometric Probability Distribution.* Stanford University Press.

[60] Lotka, A. J. (1931). Population analysis – the extinction of families – I. *J. Washington Academy of Sciences,* **21,** 377–80.

[61] Lotka, A. J. (1931). Population analysis – the extinction of families – II. *J. Washington Academy of Sciences,* **21,** 453–9.

[62] Lyon, A. J. (1970). *Dealing with Data.* Pergamon.

[63] McCornack, R. L. (1965). Extended tables of the Wilcoxon matched pair signed rank statistic. *J. American Statistical Association,* **60,** 864–71.

[64] Massey, F. J. (1951). The Kolmogorov–Smirnov test of goodness of fit. *J. American Statistical Association,* **46,** 68–78.

[65] Mather, K. (1965). *Statistical Analysis in Biology.* Chapman and Hall.

[66] Mather, K. (1971). *The Elements of Biometry.* Chapman and Hall.

[67] Miller, M. D. (1946). *Elements of Graduation.* Actuarial Society of America.

[68] Milton, R. C. (1964). Extended table of critical values for the Mann–Whitney (Wilcoxon) two-sample statistic. *J. American Statistical Association,* **59,** 925–34.

[69] Molina, E. C. (1942). *Poisson's Exponential Limit.* Van Nostrand.

[70] Mood, A. M. and Graybill, F. A. (1963). *Introduction to the Theory of Statistics.* McGraw-Hill.

[71] National Bureau of Standards (1950). *Tables of the Binomial Probability Distribution.* US Government Printing Office.

[72] Neave, H. R. (1973). On using the Box–Müller transformation with multiplicative congruential pseudo-random number generators. *Applied Statistics,* **22,** 92–7.

[73] Nielsen, K. L. (1956). *Methods in Numerical Analysis.* Macmillan.

[74] Olds, E. G. (1938). Distribution of sums of squares of rank differences for small numbers of individuals. *Annals of Mathematical Statistics,* **9,** 133–48.

[75] Parzen, E. (1962). *Stochastic Processes.* Holden-Day.

[76] Pearson, E. S. and Hartley, H. O. (1958). *Biometrika Tables for Statisticians,* Vol. I. Cambridge University Press.

[77] Pearson, E. S. and Hartley, H. O. (1972). *Biometrika Tables for Statisticians,* Vol. II. Cambridge University Press.

[78] Peters, J. A. (Ed.) (1959). *Classic Papers in Genetics.* Prentice-Hall.

[79] Pielou, E. C. (1969). *An Introduction to Mathematical Ecology.* Wiley-Inter-science.

[80] Plackett, R. L. (1949). A historical note on the methods of least squares. *Biometrika,* 36, 458–60.

[81] Pollard, A. H. (1972). *Introductory Statistics – A Service Course.* Pergamon.

[82] Pollard, J. H. (1971). On distance estimators of density in randomly-distributed forests. *Biometrics,* 27, 991–1002.

[83] Pollard, J. H. (1973). *Mathematical Models for the Growth of Human Populations.* Cambridge University Press.

[84] Ralston, A. (1965). *A first Course in Numerical Analysis.* McGraw-Hill.

[85] Rand Corporation (1955). *A Million Random Digits and 100,000 Normal Deviates.* The Free Press, Glencoe, Illinois.

[86] Remington, R. D. and Schork, M. A. (1970). *Statistics with Applications to the Biological and Health Sciences.* Prentice-Hall.

[87] Samiuddin, M. and Atiqullah, M. (1976). A test for equality of variances. *Biometrika,* 63, 206–8.

[88] Scheffé, H. (1959). *Analysis of Variance.* Wiley.

[89] Scheid, F. (1968). *Numerical Analysis.* Schaum's Outline Series. McGraw-Hill.

[90] Skellam, J. G. (1952). Studies in statistical ecology. *Biometrika,* 39, 346–62.

[91] Snedecor, G. W. (1961). *Statistical Methods* (fifth edition). Iowa State University Press.

[92] Sokolnikoff, I. S. and Sokolnikoff, E. S. (1941). *Higher Mathematics for Engineers and Physicists.* McGraw-Hill.

[93] Spiegel, M. R. (1972). *Theory and Problems of Statistics.* Schaum's Outline Series. McGraw-Hill.

[94] 'Student' (1908 a). On the probable error of a mean. *Biometrika,* 6, 1–25.

[95] 'Student' (1908 b). On the probable error of a correlation ceofficient. *Biometrika,* 6, 302–10.

[96] Tetley, H. (1966). *Actuarial Statistics. Vol. 1. Statistics and Graduation.* Cambridge University Press.

[97] Thöni, H. (1967). *Transformation of Variables Used in the Analysis of Experimental and Observational Data. A Review.* Technical Report Number 7, Statistical Laboratory, Iowa State University, Ames, Iowa.

[98] U. S. Army Material Command (1972). *Engineering Design Handbook. Tables of the Cumulative Binomial Probabilities.*

[99] Weintraub, S. (1963). *Tables of the Cumulative Binomial Probability Distribution for Small Values of p.* The Free Press of Glencoe.

[100] Whittaker, E. and Robinson, G. (1954). *The Calculus of Observations.* Blackie.

[101] Wilks, S. S. (1962). *Mathematical Statistics.* Wiley.

[102] Wilks, S. S. (1966). *Elementary Statistical Analysis.* Princeton University Press.

[103] Williams, C. B. (1964). *Patterns in the Balance of Nature.* Academic Press.

[104] Williamson, E. and Bretherton, M. H. (1963). *Tables of the Negative Binomial Probability Distribution.* Wiley.

[105] Zar, J. H. (1974). *Biostatistical Analysis.* Prentice-Hall.

Author Index

Subject Index